Synchronized Phasor Measurements for Smart Grids, 2nd Edition

Synchronized Phasor Measurements for Smart Grids, 2nd Edition

You may also like

- PBPO190 | Kishor | Synchrophasor Technology: Real-time operation of power networks | June 2023
- PBPO097 | Mohanta | Synchronized Phasor Measurements for Smart Grids | Feb 2017
- PBPO077 | Messina *et al.* | Wide Area Monitoring of Interconnected Power Systems | May 2015
- PBPO0118 | Monti | Modelling and Simulation of Complex Power Systems | Aug 2022

We also publish a wide range of books on the following topics:
Computing and Networks
Control, Robotics and Sensors
Electrical Regulations
Electromagnetics and Radar
Energy Engineering
Healthcare Technologies
History and Management of Technology
IET Codes and Guidance
Materials, Circuits and Devices
Model Forms
Nanomaterials and Nanotechnologies
Optics, Photonics and Lasers
Production, Design and Manufacturing
Security
Telecommunications
Transportation

All books are available in print via https://shop.theiet.org or as eBooks via our Digital Library https://digital-library.theiet.org.

IET ENERGY ENGINEERING SERIES 272

Synchronized Phasor Measurements for Smart Grids, 2nd Edition

Edited by
Maddikara Jaya Bharata Reddy and
Dusmanta Kumar Mohanta

The Institution of Engineering and Technology

About the IET

This book is published by the Institution of Engineering and Technology (The IET).

We inspire, inform and influence the global engineering community to engineer a better world. As a diverse home across engineering and technology, we share knowledge that helps make better sense of the world, to accelerate innovation and solve the global challenges that matter.

The IET is a not-for-profit organisation. The surplus we make from our books is used to support activities and products for the engineering community and promote the positive role of science, engineering and technology in the world. This includes education resources and outreach, scholarships and awards, events and courses, publications, professional development and mentoring, and advocacy to governments.

To discover more about the IET please visit https://www.theiet.org/

About IET books

The IET publishes books across many engineering and technology disciplines. Our authors and editors offer fresh perspectives from universities and industry. Within our subject areas, we have several book series steered by editorial boards made up of leading subject experts.

We peer review each book at the proposal stage to ensure the quality and relevance of our publications.

Get involved

If you are interested in becoming an author, editor, series advisor, or peer reviewer please visit https://www.theiet.org/publishing/publishing-with-iet-books/ or contact author_support@theiet.org.

Discovering our electronic content

All of our books are available online via the IET's Digital Library. Our Digital Library is the home of technical documents, eBooks, conference publications, real-life case studies and journal articles. To find out more, please visit https://digital-library.theiet.org.

In collaboration with the United Nations and the International Publishers Association, the IET is a Signatory member of the SDG Publishers Compact. The Compact aims to accelerate progress to achieve the Sustainable Development Goals (SDGs) by 2030. Signatories aspire to develop sustainable practices and act as champions of the SDGs during the Decade of Action (2020–30), publishing books and journals that will help inform, develop, and inspire action in that direction.

In line with our sustainable goals, our UK printing partner has FSC accreditation, which is reducing our environmental impact to the planet. We use a print-on-demand model to further reduce our carbon footprint.

Published by The Institution of Engineering and Technology, London, United Kingdom

The Institution of Engineering and Technology (the "**Publisher**") is registered as a Charity in England & Wales (no. 211014) and Scotland (no. SC038698).

First published 2018

Second edition 2025

The Institution of Engineering and Technology
Futures Place,
Kings Way, Stevenage,
Herts, SG1 2UA
United Kingdom

www.theiet.org

British Library Cataloguing in Publication Data
A catalogue record for this product is available from the British Library

ISBN 978-1-83724-115-6 (hardback)
ISBN 978-1-83724-116-3 (PDF)

Typeset in India by MPS Limited

Contents

Foreword to the first edition

I and my colleagues at Virginia Tech have been involved in the development of phasor measurement units (PMUs) and wide-area measurement systems (WAMS) utilizing PMU-based measurements made on electric power grids. Following some of the major power blackouts in recent years, the efficacy of PMUs and WAMS has been well recognized to help identify causes and countermeasures for dealing with widespread disturbances on power grids. These measurements have provided a very accurate situational awareness of the current state of the grids, as well as their vulnerabilities. The promising applications arising out of this early work are improved monitoring, protection, and control of the power grids. Within a span of about 20 years after the invention of PMUs, many research teams around the world have been developing applications of this technology, and their work will surely lead to improved performance of electric power grids in the coming years.

This book, *Synchronized Phasor Measurements for Smart Grids*, presents a comprehensive view of synchronized phasor measurement technology and its applications. It combines academic rigor with pragmatic considerations in dealing with the emerging discipline of smart grid technologies. I am pleased to see the work presented in this book, and I am sure it will be a valuable reference for students, researchers, and practicing power system engineers of the future.

Prof. Arun G. Phadke
University Distinguished Research Professor
Department of Electrical and Computer Engineering
Virginia Tech
Blacksburg, VA, USA

Preface to the first edition

The modern electric power system is an amazingly complex system, having many synchronized generators swinging together almost at the same frequency, catering to consumers over a vast geographical area. Though phasor analysis is an effective theoretical tool for representation of AC power system, practically it was a daunting task to have real-time visualization of such widespread network due to lack of synchronized measurements of voltages and currents. The advent of global positioning system (GPS) paved the way for synchronized phasor measurements over large geographical sprawl and ushered the era for wide-area monitoring system (WAMS). Prof. A. G. Phadke and his team developed the phasor measurement unit (PMU) in Virginia Tech., USA during early 1980s. Subsequently, synchronized phasor measurements using PMUs have emerged as the edifice of WAMS for monitoring and control of "Smart Grids."

The envisioned smart grids necessitate effective real-time monitoring, communication, and control at generation, transmission, and distribution levels so as to ensure quality and reliability of power. Synchronized phasor measurements have emerged as indispensable real-time sensors which form the basis of decision-making for different control algorithms of smart grids. A PMU installed at a bus can, not only measure voltage and current phasors at that bus, as well as at branches incident to that bus. Thus, PMUs located at optimal locations make the power system completely observable.

The synchronized phasor measurements have the ability to capture the dynamic behavior of the power system through effective real-time state estimation (SE) to provide with enhanced situational awareness (SA) for initiating proactive remedial measures against catastrophic failures such as major blackouts. The real and reactive power imbalances manifest in terms of change of system frequency and voltage, respectively. The wide-area frequency and voltage monitoring have paved the way for improvised control mechanism.

Though there is a plethora of research contributions depicting the efficacy of synchronized measurements for smart grid, this book aims at providing with a synthetic and synoptic view.

Chapter 1 begins with an interesting overview of synchrophasor technology and its pragmatic applications for enhancing real-time monitoring and control of the power grid. Many case studies supported by field data corroborate real-time event analysis and low-frequency oscillation monitoring for identifying critical incidences for initiating commensurate remedial measures.

Chapter 2 discusses the optimal redundancy criterion for placement of PMUs, keeping in view of the fact that the upgradation monitoring of system for utilities is an expensive issue. Other challenges that arise during contingency situations such as failure of a PMU or a communication line, the observability of system is lost and thus situation awareness of the operator is decreased in terms of taking appropriate decisions. Thus, the concept of redundancy in measurements has been incorporated while formulating optimal placement of PMUs.

Chapter 3 deals with the efficacy of using WAMS data to prevent maloperation of distance relays during events such as power swings, load encroachment, and voltage instability. Synchronized measurement-based adaptive distance relay setting of series compensated and three-terminal lines to validate the methodology to discriminate faults from other events to facilitate Zone 3 operation of relays.

Chapter 4 explores the possibility of creating another visualization layer independent of supervisory control and data acquisition (SCADA) systems, so as to provide with more insight into real-time dynamic events occurring in power systems. A concept of supervisory protection is also explored to enhance the security of protection schemes by detecting misoperations. Published research findings facilitate critical evaluation regarding feasibility of these concepts.

Chapter 5 discusses the role of PMUs for dynamic model evaluation and correction for efficient state estimation as well as wide-area fault location.

Chapter 6 focuses on extracting useful information from a huge data (at least 1.5 TB in a month) to control smart grid operations such as fault detection and correction using hybrid multi-tier clustering approach.

Chapter 7 puts forth real-time self-healing protection methodologies through fault detection, classification and localization applicable to transmission lines having any configuration unlike the conventional algorithms applicable to specific configurations.

Chapter 8 deliberates on another aspect of self-healing of smart grid using maximal Lyapunov's exponent (MLE) to detect an unstable power swing and predict rotor angle instability. Based on vulnerable patterns, commensurate portioning control can be initiated to reduce overall system vulnerability by guarding against cascading failures.

Chapter 9 presents the application of synchrophasor measurements for online assessment of voltage stability, real-time load parameter estimation, and transient stability prediction. The existing techniques available in the literature for transient stability assessment have been reviewed. Application of MLE for PMU-based online transient stability detection is also discussed.

Chapter 10 explores the hybrid state estimation to integrate the advantages of synchronized phasor measurements with the conventional SCADA measurements. However, the presence of PMUs may cause ill-conditioning, leading to inaccurate estimates as well as to lack of convergence of estimates in extreme cases. In order to overcome such limitations in PMU, current phasors in different forms with uniform distribution of measurement errors are presented to overcome inaccuracy in estimates.

Chapter 11 explores the wide-area security employing decision tree (DT)-fuzzy-based data mining for catastrophic predictors. It has better transparency compared to black box models such as support vector machine (SVM). The transparency and interpretability of predictor ease its implementation and maintenance.

Dusmanta Kumar Mohanta
&
Maddikara Jaya Bharata Reddy

Preface to the second edition

Since the publication of the first edition in 2017, there has been a steady growth in the enactment of this technology in power systems, percolating to distribution systems and microgrids. Many countries have already implemented this enhanced technology, and some are in the pipeline to upgrade their networks with phasor measurement units (PMUs) and phasor data concentrators (PDC). With numerous encouraging responses from many researchers all over the world for the ***first edition*** of this book, the ***second edition*** was embarked to incorporate further application of this technology in the emerging area of energy management and cybersecurity with the integration of PMUs and wide-area monitoring systems (WAMS). Most of the authors also eagerly put their sincere endeavors to update their respective chapters. Certainly, it is hoped that this forthcoming second edition will have a significant impact on synchrophasor technology for students, researchers, and practicing power system engineers. In addition to the revised chapters, ***three new chapters have been incorporated into the present edition of this book***.

The modern electric power system is an amazingly complex system, having many synchronized generators swinging together almost at the same frequency, catering to consumers over a vast geographical area. Though phasor analysis is an effective theoretical tool for representation of AC power system, practically it was a daunting task to have real-time visualization of such widespread network due to lack of synchronized measurements of voltages and currents. The advent of global positioning system (GPS) paved the way for synchronized phasor measurements over large geographical sprawl and ushered the era for wide-area monitoring system (WAMS). Prof. A. G. Phadke and his team developed the phasor measurement unit (PMU) in Virginia Tech., USA during early 1980s. Subsequently synchronized phasor measurements using PMUs have emerged as the edifice of WAMS for monitoring and control of "Smart Grids."

The envisioned smart grids necessitate effective real-time monitoring, communication, and control at generation, transmission, and distribution levels so as to ensure quality and reliability of power. Synchronized phasor measurements have emerged as indispensable real-time sensors which form the basis of decision-making for different control algorithms of smart grids. A PMU installed at a bus can, not only measure voltage and current phasors at that bus, as well as at branches incident to that bus. Thus, PMUs located at optimal locations make the power system completely observable.

The synchronized phasor measurements have the ability to capture the dynamic behavior of the power system through effective real-time state estimation

(SE) to provide with enhanced situational awareness (SA) for initiating proactive remedial measures against catastrophic failures such as major blackouts. The real and reactive power imbalances manifest in terms of change of system frequency and voltage, respectively. The wide-area frequency and voltage monitoring have paved the way for improvised control mechanism.

Though there is a plethora of research contributions depicting the efficacy of synchronized measurements for smart grids, this book aims to provide a synthetic and synoptic view.

Chapter 1 begins with an interesting overview of synchrophasor technology and its pragmatic applications for enhancing real-time monitoring and control of the power grid. Many case studies supported by field data corroborate real-time event analysis and low-frequency oscillation monitoring for identifying critical incidences for initiating commensurate remedial measures. The experience from the implementation of the Indian URTDSM project has been incorporated in this edition.

Chapter 2 discusses the advantages of synchrophasor over SCADA systems for grid monitoring. It also emphasizes the importance of optimal placements of PMUs to enhance observability and redundancy. By applying linear programming and defining the optimal redundancy criteria, the authors were able to locate optimal PMU placement and evaluate configurations on IEEE Test Bus systems. This approach is validated by redundancy metrics to prove it can be scalable for larger grids. Case studies further illustrate synchrophasor applications, current challenges, and technological impacts in offering a comprehensive framework leading to smart grid visibility, control, and the potential of enhanced monitoring systems.

Chapter 3 proposes the adaptive distance relay for series compensated line using both ends of synchronized data. To improve security of the system during the stressed condition, the zone 3 element of the relay is adapted and tested under several conditions such as fault, power swing, and load encroachment. The adaptive alpha plane relaying is also discussed using synchronized data.

Chapter 4 postulates the feasibility of adding a new layer of disturbance analysis using PMU measurements. This layer would provide real-time detection and identification of major dynamic events taking place in power systems. Pattern creation and classification models, as well as physics-based models that underpin this layer, are described and evaluated. This chapter also assesses the feasibility of using PMU measurements for protection functions. Discussion includes the real-time constraints, quality of measurements, issues with field data, communication latencies, and finally identifying potential facets of power system protection that can benefit from PMU measurements. A new concept of supervisory protection to detect relay misoperations is explained.

Chapter 5 discusses the opportunities that PMUs bring to enhance system monitoring and protection. Three topics are included: (1) monitoring of power system operation based on SE techniques, (2) monitoring of power system models and parameter estimation for key components, and (3) wide-area fault detection and location based on PMU measurements.

Chapter 6 deals with a micro-PMU-based energy management system for maintaining supply–demand balance in the distribution system. The introduction of renewable energy sources and the ever-increasing demand for electric power present a constant threat to the stability of the system. To mitigate this, the demand and supply have to be matched. An energy management algorithm based on load shedding and dynamic restoration is discussed in this chapter.

Chapter 7 elaborates on the need for a resilient framework to provide ancillary power support to microgrids using plug-and-play sources. This framework relies on synchrophasor-based measurements for its operation. It encourages the sustainable use of plug-and-play sources, ensures economic viability, and ensures operational stability of microgrids while providing ancillary support.

Chapter 8 deals with a PMU-based online monitoring scheme developed for recognition of vulnerable patterns of rotor angle instability. The main idea is to calculate the maximum Lyapunov exponent along the system dynamic trajectory in order to predict the loss of synchronism of power systems when the system is in an oscillatory state. Also, the level of dynamic observability of the system is quantified based on the available PMUs.

Chapter 9 presents the application of synchrophasor measurements for real-time load parameter estimation, and generator parameter estimation. The existing techniques available in the literature for load and generator parameter estimation have been reviewed. Data preprocessing, bad data elimination, and the application of the dual Extended Kalman Filter and Ljung modification for parameter estimation are discussed.

Chapter10 explores the hybrid SE to integrate the advantages of synchronized phasor measurements with the conventional SCADA measurements. However, the presence of PMUs may be ill-conditioning, leading to inaccurate estimates as well as to lack of convergence of estimates in extreme cases. In order to overcome such limitations in PMUs, current phasors in different forms with uniform distribution of measurement errors are presented to overcome inaccuracy in estimates.

Chapter 11 discusses data-mining-based catastrophe predictors using PMU-based WASI features. The study validates the performance of the black box models such as SVM and RF through semi-transparent data-mining models such as DT and transparent models such as DT_Fuzzy. It is observed from the extensive studies of the developed data-mining-based catastrophe predictors that while switching from the black box solutions to transparent and interpretable solutions, there is an unavoidable tradeoff between accuracy, reliability, and security measures. The more transparent the predictor, the easier the implementation and maintenance by human actors. Overall, the fuzzy logic-based transparent solutions are preferred over black box solutions to ease the implementation with improved robustness and enhance their suitability for auditing process, even sacrificing the predictive performance indices.

Chapter 12 discusses enhancing SA in smart grids to preemptively detect and mitigate anomalies. It examines the role of PMUs, essential for real-time monitoring, generating substantial data requiring sophisticated, high-speed processing.

This chapter covers current machine learning and hybrid approaches for detecting anomalies in PMU data, highlighting the challenges of data integrity, availability, and addressing issues like missing data and concept drift.

Maddikara Jaya Bharata Reddy
&
Dusmanta Kumar Mohanta

Contributors

Praveen Kumar Agarwal
Formerly with Grid Controller of India Ltd (erstwhile Power System Operation Corporation Ltd), New Delhi, India

Sagnik Basumallik
New York Power Authority, Albany, NY, USA

Navin Bhatt
Grid Operations & Planning R&D Group, Electric Power Research Institute, USA

Sukumar Brahma
The Holcombe Department of Electrical and Computer Engineering, Clemson University, USA

Saikat Chakrabarti
Department of Electrical Engineering, National Institute of Technology Patna, India

Zakaria El Mrabet
NRG Energy, Inc., Houston, TX, USA

Evangelos Farantatos
Grid Operations & Planning R&D Group, Electric Power Research Institute, USA

Armina Foroutan
GE Vernova, Bothell, WA, USA

P. Gopakumar
Department of Electrical Engineering, National Institute of Technology (NIT) Calicut, India

Geza Joos
Department of Electrical and Computer Engineering, McGill University, Canada

Xiangqing Jiao
Department of Electrical and Computer Engineering, University of Kentucky, USA

Jennie Angela Jose Shirley
Department of Electrical and Electronics Engineering, National Institute of Technology (NIT) Tiruchirappalli, India

Innocent Kamwa
Faculty of Science and Engineering, Department of Electrical and Computer Engineering, Laval University, Canada

Prashant Khare
Department of Electrical and Electronics Engineering, National Institute of Technology (NIT) Tiruchirappalli, India

Sanjeev Kumar Mallik
Department of Electrical Engineering, National Institute of Technology (NIT) Patna, India

Pratim Kundu
Department of Electrical Engineering, Indian Institute of Technology Mandi, India

Yuan Liao
Department of Electrical and Computer Engineering, University of Kentucky, USA

Chen-Ching Liu
Department of Electrical Engineering and Computer Science, Washington State University, USA

Harini Manivelan
Department of Electrical and Electronics Engineering, National Institute of Technology (NIT) Tiruchirappalli, India

Sriram Prabhakara Rao
Center for Cyber Security Research (C2SR), University of North Dakota, USA

Ashok Kumar Pradhan
Department of Electrical Engineering, Indian Institute of Technology Kharagpur, India

Prakash Ranganathan
School of Electrical Engineering and Computer Science (SEECS), University of North Dakota, USA

Maddikara Jaya Bharata Reddy
Department of Electrical and Electronics Engineering, National Institute of Technology (NIT) Tiruchirappalli, India

Subhransu Ranjan Samantaray
School of Electrical and Computer Sciences, Indian Institute of Technology Bhubaneswar, India

Saumendra Sarangi
Department of Electrical Engineering, Motilal Nehru National Institute of Technology Allahabad, India

Sri Niwas Singh
Department of Electrical Engineering, Indian Institute of Technology Kanpur, India

Anurag K. Srivastava
Lane Department of Computer Science and Electrical Engineering, West Virginia University, USA

R. Sunitha
Department of Electrical Engineering, National Institute of Technology (NIT) Calicut, India

Tushar
GridAstra, Dallas, TX, USA

Ranganath Vallakati
Activision Blizzard Media, San Francisco, CA, USA

T.P. Vishnu
Department of Electrical Engineering, National Institute of Technology (NIT) Calicut, India

Guanqun Wang
Department of Electrical Engineering and Computer Science, Washington State University, USA

About the editors

Maddikara Jaya Bharata Reddy is a professor in the Department of Electrical and Electronics Engineering at the National Institute of Technology (NIT), Tiruchirappalli, India. He is a senior member of IEEE with 20 years of experience in teaching and research in the areas of digital relaying, power system protection, substation automation, wide-area monitoring and protection, and smart grid technologies. In 2010, he received the national-level IEI Young Engineer's Award and has published more than 110 journal and conference papers. He also served as a board member of several related journals.

Dusmanta Kumar Mohanta is a former professor in the Department of Electrical and Electronics Engineering of the Birla Institute of Technology, India. He is an IEEE senior member with more than 30 years in teaching and industry. He was an Electrical Engineer in the captive power plant of the National Aluminium Company (NALCO), Angul, India. He has been involved as a consultant for different industry problems and for testing different electrical equipment. His current research interests include power quality analysis, digital relaying, soft computing applications, smart grid, power system planning, and reliability analysis. He also served as a board member of several related journals.

Chapter 1

Synchrophasors for improving the performance of power grid

Praveen Kumar Agarwal[1]

1.1 Summary

In 2003, the National Academy of Engineering described the vast networks of interconnected wires and power plants, commonly known as the power grid, as the greatest engineering achievement of the twentieth century. Since the first power supply system of Pearl Street in North America, the electric power grid has expanded from a small cluster of customers served by localized power generators to a nation-wide system that delivers electricity to cities and towns of a country. Though the electric grid is like a single giant machine, it is owned and operated by several stakeholders in a coordinated way. A system operator coordinates the collective efforts of the stakeholders from a central control room to keep the power grid operational and live all the time. Conventionally, the central control room has a supervisory control and data acquisition (SCADA) system that acquires real-time values of electrical quantities from the current transformers (CT) and potential transformers (PT) installed in the fields. SCADA system typically acquires data once every 4–10 s. System operators, based on the values obtained by the SCADA system, take necessary control action to maintain the grid parameters within the acceptable limits. In the past, an update time of 4–10 s for the SCADA system was sufficient for the job of system operators as the electric grid was small with isolated clusters of fragmented geographical areas. However, the brisk pace of expansion of the generating capacities due to rapid industrialization resulted in interconnections of these isolated locations with long haulage transmission lines. Moreover, continuous pressure for resource optimization transformed the grid into a complex interconnected network of networks. System operators worldwide find the conventional SCADA system limited in its capability to manage such a large and complex grid.

The limitation of the SCADA system's ability has motivated grid operators worldwide to find a technology capable of handling complex and extensive inter-connected networks. Synchrophasor technology can capture field data much faster, unlike the SCADA system. Synchrophasor technology measures the phasor

[1]Grid Controller of India Ltd, formerly Power System Operation Corporation Ltd, New Delhi, India

quantities of the voltage and current at a very fast rate (25–30 times a second). Though the concept of phasor quantities has existed since the advent of the alternating current electricity generation system, its practical use could be achieved only in the early years of the twenty-first century, consequent to the advancement of information and communications technology. The invention of the phasor measurement unit (PMU) enabled the measurement of phasor values of voltage and current. Synchrophasor technology measures phasors in a time-synchronized manner from all the locations simultaneously; thus, the measurements are known as "Synchrophasor Measurements." The network of PMUs deployed all over the grid transfers the measurements to the phasor data concentrator (PDC) located at the central place over the fast communication link, which is called a "wide area measurement system" (WAMS) or wide-area monitoring (WAM) system.

Countries worldwide started deploying PMUs in early 1995. India began its journey to synchrophasor technology in 2009 by implementing a pilot project in the northern region. Later, more pilot projects were implemented, leading to a national-level pilot project in 2012. At that time, about 68 PMUs were deployed at various locations in the Indian grid. Consequently, a full-fledged large-size project was taken up, with about 1,700 PMUs and 38 PDCs covering the whole Indian power grid. System operators in India are utilizing synchrophasor data in many innovative ways to improve the performance of the Indian power grid.

This chapter introduces synchrophasor technology and its practical use in enhancing real-time monitoring and control of the electric grid. Many real-life examples have been illustrated with case studies. Real-time event analysis and low-frequency oscillation (LFO) monitoring are the most basic applications of synchrophasor data, which are possible with only a small number of PMUs deployed. Root cause analysis of critical incidents and applying corrective action are other areas where PMU data is utilized to improve grid performance.

1.2 Evolution of the power grid

Wikipedia defines the power grid as "An electrical grid is an interconnected network for delivering electricity from suppliers to consumers." It consists of generating stations that produce electric power, high-voltage transmission lines that carry power from distant sources to demand centers, and distribution lines that connect individual customers [1].

The evolution of the existing power grid started with the establishment of the first generating station at Pearl Street in the New York City, USA, in 1882. This was the first complete electric power system connecting a 100-volt generator that burned coal to power several hundred lamps in the neighborhood. Subsequently, many similar self-contained, isolated grid systems were built across United States. During this era, two broad power systems were developed: the alternating current (AC) and direct current (DC). Thomas Edison, who designed Pearl Street, propounded DC. In a DC, the electrons flow in a complete circuit from the generator through wires and devices and back to the generator. William Stanley, Jr. built the first generator that used AC. Instead of electricity flowing in one direction, the flow switches its direction back and forth [2]. AC has a significant advantage in that it can be transmitted over a long distance by converting to high voltage and then back to low voltage to serve end-users.

From the late 1800s onward, a patchwork of AC and DC grids cropped up across the country, directly competing with one another. Small systems consolidated throughout the early 1900s, and local and state governments began—together with regulations and regulatory groups. However, even with regulations, some business houses found ways to create elaborate and powerful monopolies. Public outrage at the subsequent costs began during the Great Depression. It sparked Federal regulations, as well as projects to provide electricity to rural areas, through the Tennessee Valley Authority and others [2].

AC had a decided advantage over DC for the transportation of electricity over long distances, as it was much easier and cheaper to "step up" and "step down" voltage. This is due to Ohm's law: $I = V/R$, where I = current, V = voltage, and R = resistance. By increasing voltage, the current is reduced. Since loss in the wire is proportional to the square of the current flow, less current value results in less loss. Also, because of IR loss, the voltage at the receiving end reduces if the distance is high. As a result, DC generators had to be located within a mile of the load, while AC generators could be constructed much further away.

Around the same time, economies of scale began to be applied to the electrical industry. It became increasingly apparent that a large centralized power plant was much more efficient (and cheaper) at providing electricity over a wide area than a small power plant was at providing electricity over a small area. Just before the turn of the century, Westinghouse built a hydroelectric power plant at Niagara Falls and, using AC technology, sent the electricity to Buffalo, NY, which was 20 miles away! In this way, the grid started expanding in the quest for reliability and availability, forming today's vast grid and complex network of generators, transmission lines, and transformers. The evolution of the electric grid took place in every country almost in a similar manner [3]. Today's power grid is big and complicated to operate.

1.3 Challenges of modern grid

The challenges in grid operation have increased manifold due to enlarged system size, the rapid pace of capacity addition, long-distance power flows, multiple players, increasing competition in the electricity market, emphasis on optimization, climate change, large-scale integration of renewable energy sources, and increasing customer expectations [4]. The quest for optimization and ever-growing demand for electricity are pushing the operation of the grid nearer to its operating limits. It is a critical issue for power system operators to keep the electricity grid stable and reliable. The ability of the system operators to make real-time decisions depends on the level of situational awareness derived from the data/information available to them in real time [4]. For this, the grid operators need reliable, fast, and accurate information-acquiring tools for interacting with the massively large and complex power system. Due to the ever-increasing penetration of variable energy resources into the grid, the characteristics of the electrical system keep changing randomly in response to variability due to variable generating, always changing load, and reconfiguration of transmission system due to outage and restoration. These challenges have forced the power system fraternity to rethink the power system operation.

Historically, SCADA systems have been used as a decision support tool for maintaining the parameters of the electric grid. Inherently, SCADA is an after-the-fact grid parameter-monitoring tool. The operator monitors various parameters, i.e., line/transformer loadings, bus voltages, grid frequency, and takes appropriate action to keep parameters within acceptable limits. Such actions are taken after the fact. However, there is always a time gap between the occurrence of grid events and the availability of affected parameters, except for some limit violation alarms or breaker opening alarms. This shortcoming limits the operator's ability to initiate appropriate action to prepare for the next immediate contingency, if any. The SCADA system's energy management system (EMS) applications include contingency analysis and security assessment. However, their outputs rely heavily on credible output from state estimators. These limitations of the SCADA system have been emerging as a bottleneck in maintaining the required level of performance of the power grid for some years.

1.4 SCADA system

The SCADA system is a tool utilities have used worldwide for decades to operate and manage the electric grid. Figure 1.1 shows a typical architecture of a SCADA system.

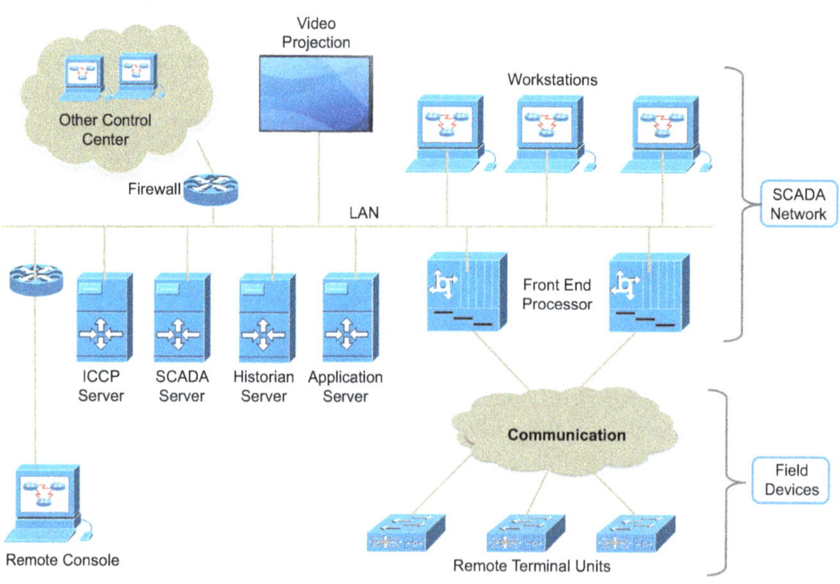

Figure 1.1 Typical architecture of a SCADA system

In the SCADA system, the values of grid parameters like voltage, frequency, MW, and MVAR are obtained from transducers connected to the CT and PT. The transducer is an electronic device that transforms electrical signals into equivalent current values ranging from 4 mA to 20 mA. The transformed current signal is fed to an analog-to-digital converter board installed in an remote telemetry unit (RTU) to convert it to digital values suitable for transmission on communication lines. A process running into the RTU scans cyclically (typically every 2–4 s) analog-to-digital boards for acquiring the values of the grid parameters. RTU then transmits acquired values to the SCADA computers in the central control room. The complete process of obtaining values from CT and PT and receiving them in the SCADA system takes about 4–6 s. Hence, we can say that in the SCADA system, data is updated every 4–6 s. However, any change in the close/open status of circuit breakers/isolators is immediately transmitted to the control room. In this way, the SCADA system maintains a real-time updated view of the entire power grid in the central control room and displays it on screens, helping the grid operators manage the grid. Figure 1.2 shows the data acquisition process in a SCADA system.

The large grid SCADA system may have multiple layers of control centers, each managing a particular grid area. The Indian power system is the third largest power system in the world. Its transmission network is spread all over India and has many cross-border connections with neighboring countries like Bhutan, Nepal, and Bangladesh. The Indian power grid is demarcated into five regional grids, viz. North, West, South, East, and Northeast. All five regional grids are connected synchronously. Each region has multiple control areas defined by state boundaries. The state and the regional grid have their own SCADA system and control center. All these SCADA systems are hierarchically connected to the National Control Centre SCADA system, as shown in Figure 1.3.

Data acquired through RTU is transferred to a direct-connected SCADA system and again moved to the SCADA system of the next level in the hierarchy. Hence, typically, each data point has to cross two to three levels of hierarchy before reaching the national control center. Due to this, the total delay between obtaining values from CT/PT and displaying data to the grid operator may reach 6–10 s.

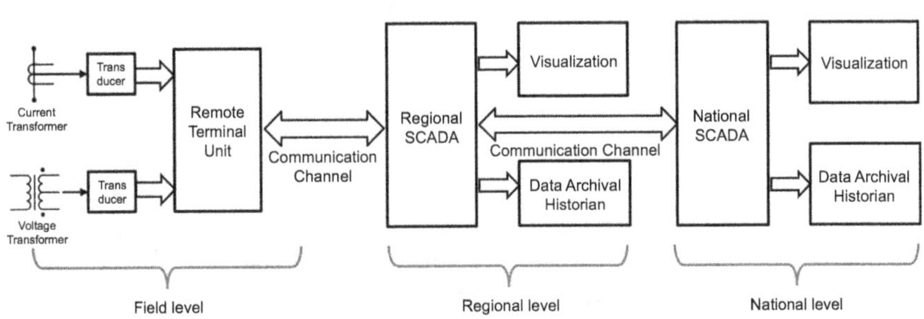

Figure 1.2 Data acquisition process in a SCADA system

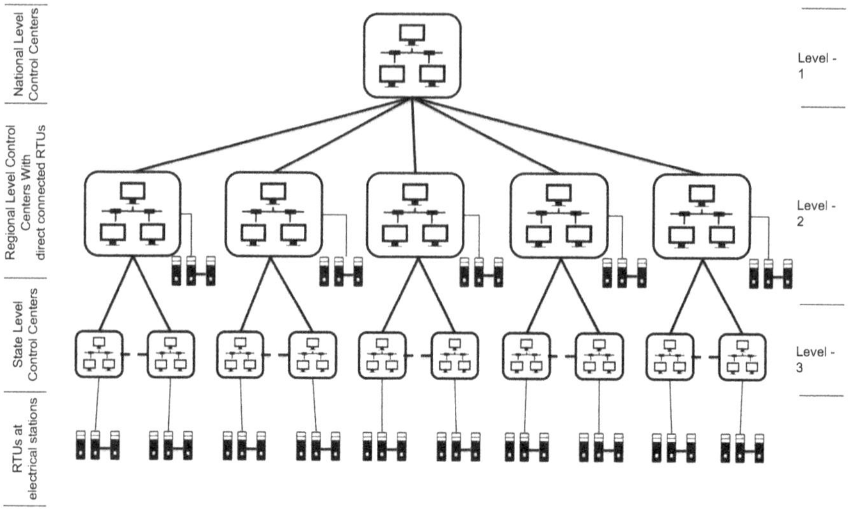

Figure 1.3 Hierarchical SCADA system for large grid

1.5 Synchrophasor technology

Synchrophasor technology has emerged as a niche, and new technology is being used increasingly worldwide to efficiently manage the electric power grid. Synchrophasor technology is where phasor quantities of power system parameters like voltage and currents are measured, unlike the SCADA system, which measures only the magnitude. In the synchrophasor system, all devices' phasor measurements are done simultaneously, and they are time synchronized with the global positioning system (GPS). Synchrophasor technology can potentially change the economics of power delivery by allowing increased power flow over existing transmission systems. Synchrophasor data could allow power flow up to the transmission line's dynamic limit instead of its worst-case limit. Synchrophasor technology will usher in a new process for establishing centralized and selective controls for the flow of electrical energy over the grid. These controls will affect large-scale (multiple states) and individual transmission line sections at intersecting substations. Transmission line congestion (over-loading) and protection and controls will, therefore, be improved on a multiple region scale (USA, Canada, Mexico) through interconnecting independent system operators [1].

1.5.1 Phasor measurement

In 1893, Charles Proteus Steinmetz presented a paper on a simplified mathematical description of the waveforms of AC electricity. Steinmetz called his representation a phasor. With the invention of PMUs in 1988 by Dr. Arun G. Phadke and Dr. James S. Thorp at Virginia Tech, Steinmetz's technique of phasor calculation evolved into the calculation of real-time phasor measurements that are synchronized to an absolute time reference provided by the GPS. We, therefore, refer to synchronized phasor measurements as synchrophasors. Early prototypes of the PMU were built at Virginia Tech, and Macrodyne built the first PMU (model 1690) in 1992 [1].

A phasor is a complex quantity representing the magnitude and phase angle of the sine waves found in electric systems. Phasor measurements that occur at the same time are called "synchrophasors." While it is a commonplace for the terms "PMU" and "synchrophasor" to be used interchangeably, they actually represent two separate technical meanings. A synchrophasor is the measured value, whereas the PMU is the measuring device. In typical applications, PMUs are sampled from widely dispersed locations in the power system network and synchronized from the common time source of a GPS radio clock. Synchrophasor technology provides a tool for system operators and planners to measure the state of the electrical system and manage power quality [1].

PMUs measure voltages and currents at principal intersecting locations (critical substations) on a power grid and can accurately output time-stamped voltage and current phasors. Because these phasors are truly synchronized, accurate comparison of two quantities is possible in real time. These comparisons can assess system conditions such as frequency changes, MW, MVARs, and volts. The monitored points are preselected through various studies to make extremely accurate phase angle measurements to indicate system (grid) stability shifts. The phasor data is collected on-site or at centralized locations in a PDC. The data is then transmitted to a regional monitoring system maintained by the local ISO. These ISOs will monitor phasor data from individual PMUs or as many as required PMUs. This monitoring provides an accurate means of establishing controls for power flow from multiple energy generation sources (nuclear, coal, wind, etc.) [1].

A PMU block diagram is shown in Figure 1.4. Each block is detailed below. A voltage or current level matched with the requirements of the converters is obtained by using appropriate measurement transformers. The analog AC waveforms are digitized by an analog-to-digital converter. The anti-aliasing filter placed before the signal sampler removes the signal components whose frequency is equal to or greater than one-half the Nyquist (sample) rate [5].

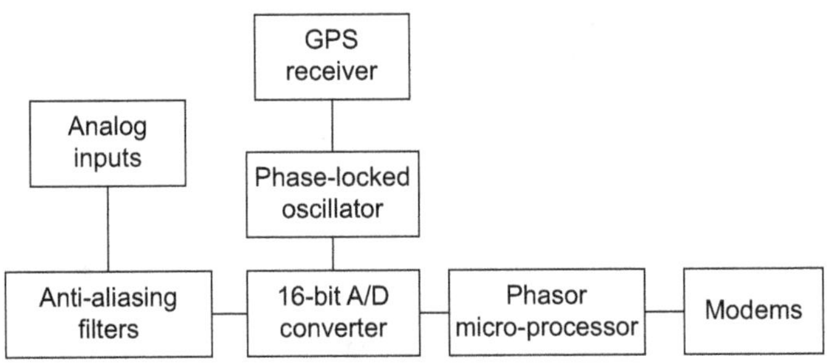

Figure 1.4 Block diagram of PMU

1.5.2 Time synchronized measurements

Since PMUs are located at different locations in the grid, time synchronization is essential to provide the reference point for the synchronized measurements. Every PMU produces accurate time-stamped values sent to a server for further processing. The measurements from different PMUs are combined and provide a comprehensive overview of the entire interconnection and grid stress; alternatively, they are used to trigger corrective actions to maintain stability [6]. To accurately provide a wide view of the geographically spread electric grid, the measurements at all the locations must be at the same instant, i.e., measurements must be time synchronized.

State estimation is one power system management system application, popularly known as "energy management system," which estimates the voltage and angle of each bus in the grid from various parameters measured by the SCADA system. The state estimator output is the fundamental input data to various other applications in EMS. Unsynchronized measurements introduce time-skewed data, leading to inaccuracies. Time-synchronized measurements provide the accurate and true state of the system.

Analyzing any grid events requires mapping and relating the sequence of events captured from various devices installed in the system. Time synchronization of all these devices is a critical requirement for accurate event analysis. Time-synchronized measurement provides a fast and accurate relation of the various sequences of events and, hence, timely and accurate assessment of the grid events.

The phasor quantities measurement process by PMU is shown in Figure 1.5. Here, unlike SCADA system measurement, PMU samples current and voltage directly from CT and PT. Additionally, each PMU is synchronized with a GPS for its time reference, providing better accuracy than 1 μs. PMU divides each second into equal parts for time-synchronized measurement based on the configured sample rate. The first measurement is taken at the start of each second and repeated for each part of the second. Since all PMUs follow the same philosophy and are time synchronized with GPS, they produce the measurements taken at the same instant of time from each location.

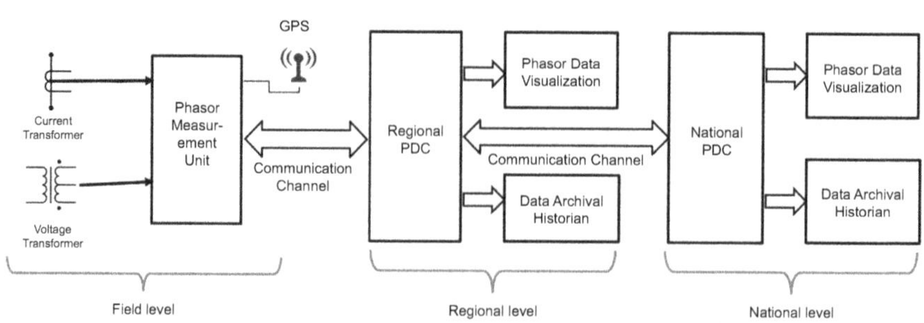

Figure 1.5 Data acquisition process in a WAM system

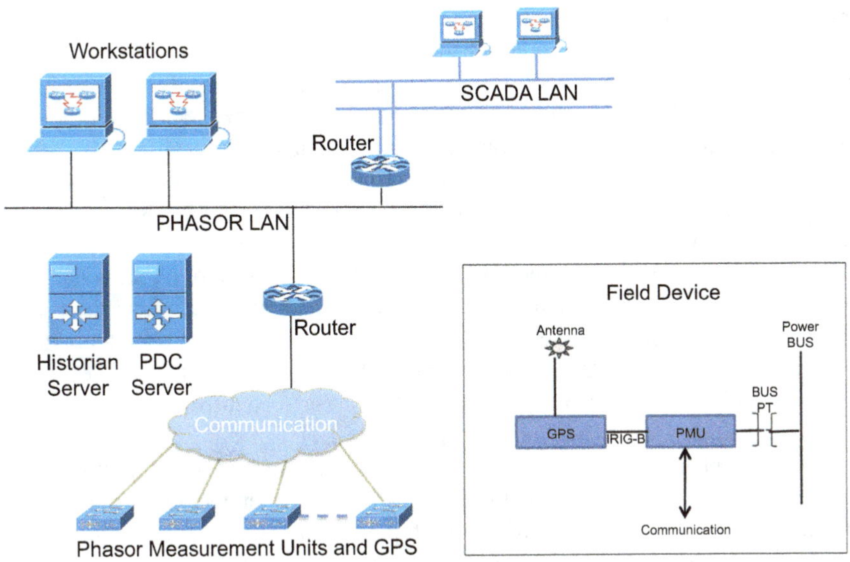

Figure 1.6 Architecture of a typical WAM system

1.5.3 Wide area measurement system

The power grid consists of many electrical stations, the measurements of which should be used for effective and efficient power system monitoring. Hence, PMUs are installed on nearly all the stations or at least on the minimum number of buses. So, voltage and angle calculations are also possible for stations not measured with PMUs. Typical installations of synchrophasor measurement systems have several PMUs connected to a central system known as PDC via various communication links. The PDC receives the time-stamped measurements from all PMUs. It consolidates the same in a packet with the same time stamp data and is made available to applications like visualization, oscillation monitoring systems (OMS), and data historians. This network of PMUs, PDCs, historians, visualization, and applications is known as the WAMS. Figure 1.6 shows a typical architecture of the WAM system.

1.6 SCADA versus WAMS

The SCADA system has been used by system operators worldwide to operate and monitor the power grid for decades. Historically, the power grid was a simple network of generating stations and transmission lines (mostly radials) and did not require an elaborate monitoring scheme. It was only necessary to evaluate the steady-state conditions on a few seconds time scale rather than continually observe the operating state on a millisecond time scale. However, the growth of demand and distributed energy sources such as renewable generation introduce more short-term and unpredictable fluctuations and disturbances. The emerging challenges and the pressing need to ensure stable and reliable operation of the power system motivated grid operators worldwide to implement emerging technologies for grid monitoring. Moreover, the synchrophasor technology is increasingly being used to supplement the conventional SCADA/EMS by providing refined and fast measurements capable of overcoming the challenges of managing increased variability and uncertainty in the grid.

The time resolution of SCADA measurements is generally 4–10 s, while for WAMS, it is 20–40 ms. The USP of synchrophasor technology or WAM system is the measurement of voltage and current phasor quantities, which gives the value of angular difference at any two points of the grid. The direct measurement of phasor values was impossible in the SCADA system except for some indirect estimation from the voltage and current measurements done by transducers. The Indian system operators are utilizing the estimated calculation of the angular difference between two pockets of the Northern Grid of India to avoid separation under skewed load dispatch or power swing conditions between these two pockets (see Figure 1.7). Since it was not possible to telemeter the phase angle separation between two distant nodes in the grid with a conventional SCADA system, the angular separation between them was calculated at the control center with the help of real-time power flow, bus voltages, and network reactance using standard equation $\sin \delta = P \times X / V1 \times V2$ [7]. A comparison of the SCADA system and the WAM system can be understood from Table 1.1.

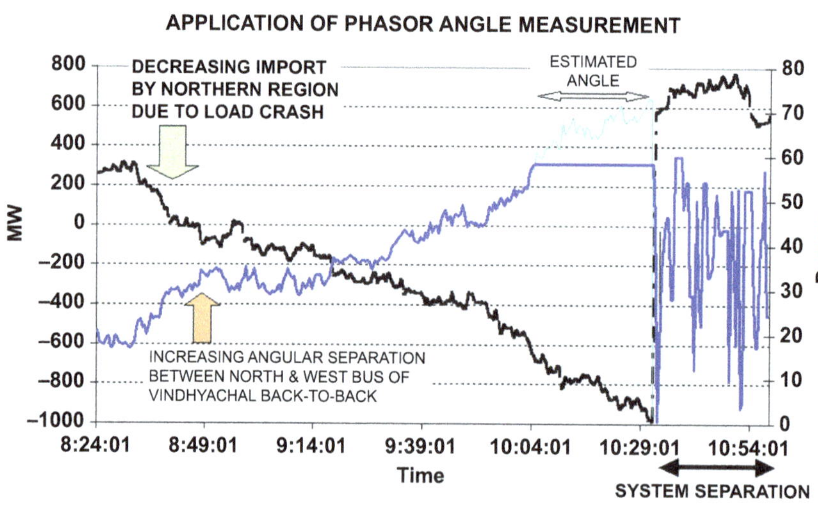

Figure 1.7 Phase angle measurement in a SCADA system

Table 1.1 Comparison of SCADA and WAMS

Description	SCADA	WAMS
Data updation	Slow, once in 4–10 s	Fast, once in 20 ms
State measurement	Steady sate measurement only	Dynamic state measurement also
Angle measurement	No direct measurement	Direct measurement
State estimation	Nonlinear and complex calculation	Linear, easy to calculate

1.7 Worldwide deployment of WAMS

The original use of PMUs began in parts of both the Western (such as the Bonneville Power Administration (BPA) and Western Area Power Administration) and Eastern Interconnections (such as New York Independent System Operator) in pilot projects during the 1990s. The use of synchrophasors has been increasing since 2004 when the US-Canada 2003 blackout investigation report recognized that many of North America's major blackouts had been caused by inadequate situational awareness for grid operators and recommended the use of synchrophasor technologies to provide real-time wide-area grid visibility [8].

1.7.1 WAMS in Americas

PMUs were first deployed by the BPA of the USA in its WAM system. The role of PMU data in uncovering the 2003 North American blackout renewed the DOE interest in PMU deployment. According to the North American Reliability Council (NERC), "A valuable lesson from the August 14 blackout is the importance of having time-synchronised system data recorders. NERC investigators labored over thousands of data items to synchronize the sequence of events. That process would have been significantly improved ... if there had been sufficient synchronized data recording devices." Today, over 1,700 production-grade PMUs are deployed across the USA and Canada, streaming data and providing almost 100% visibility into the bulk power system. In Brazil, a huge blackout in March 1999 revived the interest in PMU applications, derailed due to difficulties faced by the Brazilian economy during that decade, and the restructuring of the electric energy sector was delayed, mainly for dynamic performance analysis during disturbances.

1.7.2 WAMS in Europe

Switzerland and Italy started using WAM systems in 2003 as an additional power system operation and analysis tool. The Croatian transmission system operator HEP-TSO commissioned the first phasor measurement installation in 2003 and acquired the first dedicated PDC as part of a turnkey system that included five PMUs from the same manufacturer.

1.7.3 WAMS in Asia

China's power transmission grid is run by two major transmission system operators, the State Grid Corporation of China (SGCC) and the smaller China Southern Power Grid (CSG). Together, they deployed several PMUs as part of a nationwide mandate requiring that all substations in 500 kV networks and above and all generators larger than 100 MW be monitored by PMUs. As of 2013, approximately 2,400 PMU sets had been deployed in the power grid in China.

India started its synchrophasors initiative in 2009 by installing a pilot project consisting of eight PMUs and one PDC. As of January 2016, the central and state utilities had installed 68 PMUs. The first WAMS project was implemented in India in 2009. It was a simple pilot project with technical requirements that could be met with readily available products. The project was located in the Northern Region (NR). In this pilot project, four PMUs, each with a GPS, were installed at strategically selected locations, and one PDC was installed at the regional control center managing the northern power grid in India. The project became operational in 2010. Subsequently, five more PMUs were added to the NR project. The inputs to PMUs were taken from three-phase voltages of the 400 kV buses and three-phase currents of the feeders of the substations.

In 2011, an Indian-origin vendor developed an indigenously designed PDC and installed a demo project in the Southern Region (SR) with four PMUs. In this PDC, an oscillation monitoring application was also installed, providing small signals modal analysis of the grid on a real-time basis. Subsequently, one more demo project was commissioned in the Western Region (WR) using open PDC and two PMUs. After experiencing the utility of Synchrophasor data in managing and improving the grid's performance, similar pilot projects were taken up in other already operational regions.

Post-commissioning of regional pilot projects, which were isolated from each other, and as regional grids operate synchronously (other than the SR grid), the need for a unified availability and visualization of synchrophasor data from all the PMUs was felt. Consequently, all regional pilot projects were integrated, forming a national-level WAM project. Further, to improve observability, additional PMUs were installed.

A PDC was installed at the National Load Despatch Center in the national pilot project, and all regional pilot project's PDCs are integrated with this. Data from all the 68 PMUs across India was available at the national PDC. The deployment of PMUs in India is shown in Figure 1.8. Apart from this, some states of India have also started deploying PMUs. The state of Gujarat had planned to install 100 PMUs. Some PMUs have already been installed, and some were integrated into national PDC.

1.7.4 Indian WAMS (URTDSM) project

Based on the WAMS pilot project experience, a full-fledged synchrophasor project, known as the Unified Real Time Dynamic State Measurement System (URTDSM), was envisaged in 2012 and implemented in the Indian power system in two phases progressively by 2014–15 by the Power Grid Corporation of India, a deemed transmission licensee.

The URTDSM project envisaged the installation of 1,669 PMUs in two phases on all 400 KV and above substations, a generating station with a 220 kV switchyard, and all high-voltage direct current (HVDC) systems. Furthermore, all 765 kV and 400 kV lines will have PMUs installed at both ends, enabling the calculation of realistic line parameters. PDC will be provided at all state, regional, and national load dispatch centers. Nodal PDC will also installed if the number of PMUs is more than 40 nos in any one location. A research and development project was specially commissioned by POWERGRID to develop various applications for effectively utilizing the PMU data to improve further and maintain the performance of the Indian grid [9].

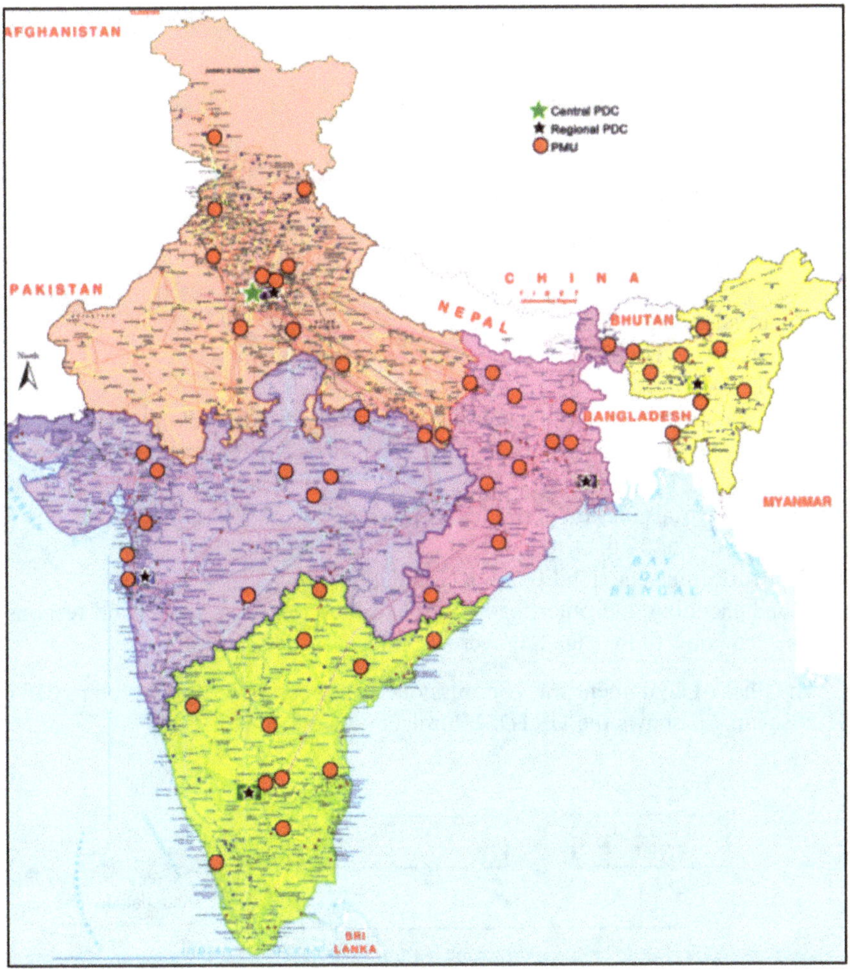

Figure 1.8. PMU Deployment in India.

Phase-I: Placement of 1,186 nos. PMUs at all lines in HVDC terminal stations, 400 kV and above voltage level S/s, generating station stepped up at 220 kV level and above where fiber optic (FO) cable along with communication equipment is either existing or being implemented by 2014–15 with the provision of expansion up to 50% capacity. Phase-I includes placement of nodal PDC (27 nos) at strategic substation, Master PDC (25 nos) at SLDCs, Super PDC (5) at RLDCs, 2 No PDC at Main and Backup NLDC, Remote console at each RPC (5), UT (3), Sikkim, NTMC (2), CEA (1), and NER States (4).

Phase-II: The 483 nos. PMUs envisaged in Phase-II to be placed at all HVDC terminal stations, 400 kV and above voltage level S/s, and the generating station will be stepped up to 220 kV and above, along with the provision of FO connectivity and communication equipment.

Development of analytics for various applications using PMU data shall be done in parallel, as given below:

1. Vulnerability test on relay characteristics.
2. Instrument transformer measurement validation.
3. Dynamic state measurements—wide Area measurement and control in regional transmission networks—linear state estimation.
4. Supervised Zone-3 protection scheme to prevent unwanted tripping of distance relays.
5. Schemes for controlling angular instability (i.e., out-of-step protection and smart islanding).
6. Emergency control schemes for controlling frequency and voltage instabilities.
7. Increase the power grid's reliability by detecting faults early, preventing local events that allow for isolation of the operative system, and preventing power outages.
8. Improved transmission corridor capability.
9. Adaptive islanding.
10. Network transient stability model validation.
11. Load shedding and other load control techniques, such as demand response mechanisms, to manage a power system.

The Phase-I equipment was commissioned progressively from the year 2018 to 2019. Figure 1.9 shows the URTDSM project's architecture.

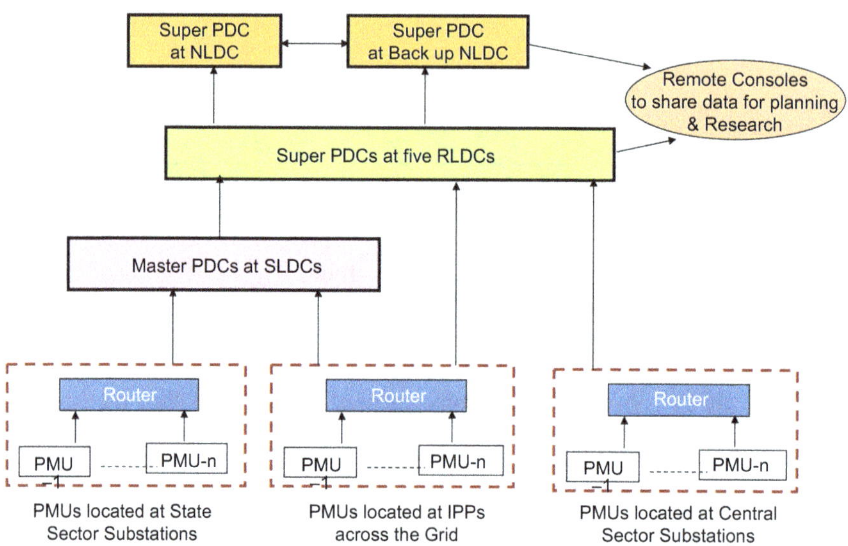

Figure 1.9 Unified Real Time Dynamic System Measurement Project Scheme Architecture

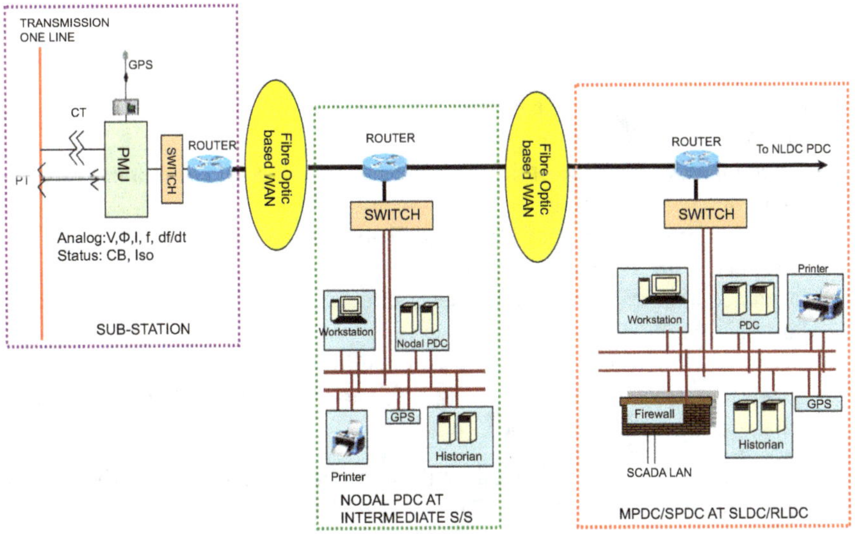

Figure 1.10 Data Flow Unified Real Time Dynamic System Measurement Project Scheme Architecture

PMUs under the URTDSM project were planned to report hierarchically to the 32 states, 5 regions, and 1 national control center. As of now, 1,200 PMUs are reporting at the National Control Center. Power system operators utilize data from these PMUs as an analytical tool for better real-time system operation. Figure 1.10 shows the data flow from PMU to different PDCs in the URTDSM project.

1.8 Applications of synchrophasors data

Synchrophasor data can provide a wealth of operational information on the electric grids that SCADA cannot dig out. Synchrophasor data can capture network dynamics, unlike the SCADA system that provides only steady state visibilities. These include fast monitoring of angular separation, oscillatory stability, voltage stability, and islanding detection. These solutions bring additional situational awareness for the grid operator [10]. Since the implementation of the first pilot project in India, grid operators have utilized synchrophasor data in many innovative ways for their real-time grid operation, as mentioned below:

1. Improvement of situational awareness
2. Detection of occurrence of transmission line tripping/revival within a flow gate by observing:
   (a) Real-time monitoring of trends of critical grid parameters like frequency, *df/dt*, and angular separation
   (b) Step change in angular separation
   (c) Step change in line current (MW/MVAR)

3. Detection of occurrence of generator tripping by observing:
   (a) Sudden decline in frequency
   (b) Increase in *df/dt* value
   (c) Change in angular separation
   (d) Decrease in voltage magnitude
4. Detection of occurrence of line circuit breaker auto-reclose event indicated by a kink in frequency trend
5. Detection of occurrence of load crash/load throw-off by observing
6. Sustained high frequency
7. Sustained abnormal phase angle separation
8. Sustained high voltage
9. Detection of islanding of any part of the grid
10. Subsystem synchronization during restoration using standing phase angle separation and phase sequence.

The various uses of synchrophasor data can be understood with the help of real-life examples and case studies. The synchrophasor systems have been operational in India since 2010, and Indian grid operators have been utilizing the synchrophasor data in many innovative ways since then. Grid Controller of India Ltd (erstwhile Power System Operation Corporation Ltd) has compiled such case studies and examples as a compendium and made these reports available to the public to benefit the power system fraternity. Since the author was also involved in the preparation of these reports, many examples and case studies have been adopted from these reports and have been presented in a lucid manner. The reports are available on the website http://www.posoco.in. Many examples have been presented here. The examples and case studies have been categorized into the following groups:

1. Fault detection, classification, and analysis
2. Low-frequency oscillation
3. Insight to coherency of generators
4. Electrical island detection and their resynchronization with the grid
5. Model validation.
6. Improving power plant performance
7. Monitoring of natural disasters.

1.8.1 Fault detection, classification, and analysis

The availability of synchrophasors has considerably enhanced the WAM and situational awareness of power system behavior under steady state and transient/dynamic conditions. The real-time trends of synchrophasor data at the control center have become firsthand information for the grid operators to view and analyze any transient phenomenon occurring in the grid. Various events that went unnoticed with the SCADA system can now be detected and analyzed in almost real time, opening up a new era in power system monitoring and control. With PMU data at the control center, the grid operators can observe the signature of events as firsthand information and then do further analysis by the SCADA alarm system.

Historically, analysis of any grid event is carried out with the help of relay flags and disturbance records (DR)/event logger (EL) outputs made available by the stakeholders after the occurrence of the fault, often after considerable time. However, challenges are often faced on account of issues such as non-availability/healthiness/failure of recording instruments, utility's inhibitions in data sharing, control area jurisdiction, time synchronization, portability/compatibility of DR/EL records, latency, and skewed timestamp of SCADA data. The availability of synchrophasor data has now become an effective tool for analyzing grid events and facilitated the preparation of an accurate "first information report" of an event occurring in the grid.

Figure 1.11 shows an event of multiple times faults in the HVDC line due to a storm in that area. A series of faults were observed from the display of trend data from PMU installed at the HVDC terminal in Gujarat, located in the western part of India. The HVDC terminal is a part of the HVDC line from Gujarat to Haryana (located in the northern part of India). This indicated the possibility of multiple faults near the HVDC location. The HVDC station operator also confirmed this. The system control room operator immediately directed the HVDC station to shift to the reduced voltage mode of operation, reducing the possibility of further faults.

Figure 1.12 shows the real-time display of voltage signals received from a PMU located at that Ultra Mega Power Project (UMPP) in Gujarat. A voltage dip and simultaneous fall in frequency were observed from the display. A maximum dip in this PMU's voltage was observed, indicating a phase-phase fault. The operator identified that a transmission line from this UMPP had tripped on a to-phase fault, and consequently, special protection scheme (SPS) operated, which backed down the generation. The operator informed the concerned constituent of the necessary load balance and monitored the line loadings in a parallel network for any instance of overloading.

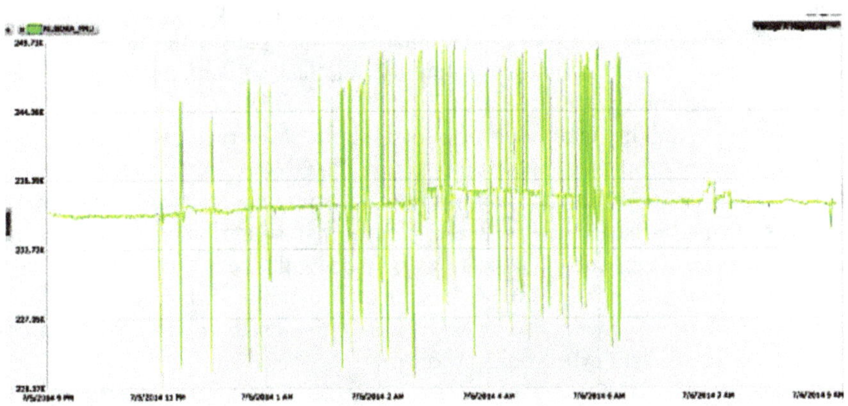

Figure 1.11 Detection of multiple faults in HVDC line

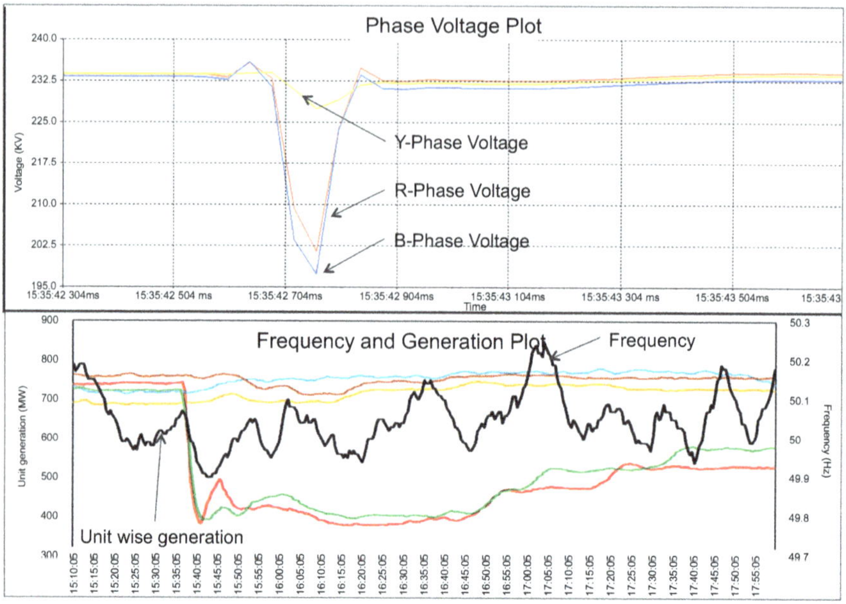

Figure 1.12 Real-time detection of fault and SPS action

1.8.2 Low-frequency oscillations

The power system is large and nonlinear, with many oscillation modes. The oscillations include electromechanical oscillations, control modes oscillations, and sub-synchronous resonance. The change in the electromechanical torque of a synchronous machine following a perturbation has two components. One component is called the synchronizing torque and determines the torque change in phase with rotor angle perturbation. The other component, damping torque, determines the torque change in phase with speed variation. Rotor angle stability depends on both components of torque. Lack of synchronizing torque causes the system's non-oscillatory instability or monotonic instability, and lack of damping torque results in oscillatory instability.

Rotor angle stability is of two types: small signal stability (small disturbance in the power system) and transient stability (large disturbance in the power system). Small signal stability is the ability of a power system to be in a steady state after a small disturbance. The instability due to this is mainly attributed to insufficient damping torque. Transient stability is associated with the ability of the power system to maintain synchronism when subjected to large disturbances like line faults, bus faults, and generator outages. However, the instability arising due to this results from insufficient synchronizing torque.

Small signal instability is due to insufficient damping torque, leading to LFO in the system, which is oscillatory in nature. If there are N generators in a system, then the total number of such LFO modes would be N-1. During LFOs, mechanical kinetic energy is exchanged between synchronous generators of the inter-connected system through tie lines. Most of these oscillatory modes in normal power system states are well damped. However, they get excited during any small disturbance, leading to oscillation in power system parameters like rotor velocity, angle, voltage, and current power flow. Due to oscillation in parameters, protection equipment may undesirably operate, leading to cascade tripping in the electrical system. Therefore, detecting such modes and initiating corrective actions is necessary to ensure system reliability and security. The generators' most important rotor velocity and the network's power flow are among these parameters. The rotor velocity variation causes fatigue to the mechanical parts of the turbine generator system. The power flow oscillations may amount to the entire rating of a power line when they are superimposed on the stationary line flow and would limit the transfer capability by requiring increased safety margins.

The LFO classification generally depends on the power system for which it has to be used based on Eigenvalue analysis based on the system model. The classification helps monitor the variables that may be participating globally or alone. The participation factor of each variable in one of the oscillatory modes helps decide the classification of modes. In general, LFO can be basically of four different types, as given below:

- Inter-area mode (0.1 Hz–0.7 Hz): Inter-area modes are associated with the swinging of a group of generators in one part of the system with a group of generators in other parts due to weak interconnecting lines between two power systems. These are also referred to as global modes.
- Local mode (0.7 Hz–2.5 Hz): It is either due to oscillation of one generator against the remaining of the system (very similar to the one-machine infinite-bus system) or oscillation of one generator against another, both located close to each other (two generators in the same power plant).
- Control mode: These are in the system due to poor design of controllers of AVR, HVDC, SVC, AGC, etc. These are also referred to as regulating modes.
- Torsional mode (10 Hz–40 Hz): These modes are associated with the turbine-generator shaft system and rotational components.

The ranges defined above may overlap others depending on the power system, but this is generally accepted in the power system fraternity.

With the current SCADA system, system operators cannot visualize LFOs due to the slow updating rate of data, which is once every 4–10 s (analog values). The oscillations at the generator level, i.e., intra-plant oscillations or local mode oscillations, were assumed as they appeared as hunting in the generators. In contrast, the inter-area modes were not visible to system operators apart from simulation studies. The SCADA data reporting rate is comparatively slow, which is not useful in detecting the oscillation or changes in the system in sub-seconds.

With synchronized phasor measurements available at a reporting rate of 25–50 frames per second from the PMU, the operator can now visualize such oscillations in the system. Tools and techniques like Wide Area Monitoring, Protection, and Control (WAMPAC) are also being developed to detect the source of such oscillations, analyze them in real time, and take corrective action before creating further system complexities. The detection of LFOs and their history is of great help in the planning and implementing damping controllers of HVDC, thyristor controlled series compensator (TCSC), etc. PMUs have enabled the operator to visualize such LFOs whose source can be tracked with the availability of the optimum number of PMUs, giving complete system observability.

Figure 1.13 shows the oscillations captured during the tripping of an EHV (extra high voltage) tie line. It can be seen that oscillations remained for quite some time but died down subsequently because of positive damping. The positive damping can be seen from the bottom graph of the figure, which shows the real-time calculation of damping indicated by a dot. The vertical axis shows the mode, i.e., frequency of oscillation. The red dot indicates the very low value of damping. All other dots indicate some value of positive damping. Since only one red dot converted to a yellow dot (positive damped), subsequently these oscillations were not dangerous to the grid.

Similarly, Figure 1.14 presents a case of a situation when LFOs were observed from the nearby PMU, synchronizing the 500 MW thermal unit in one of the

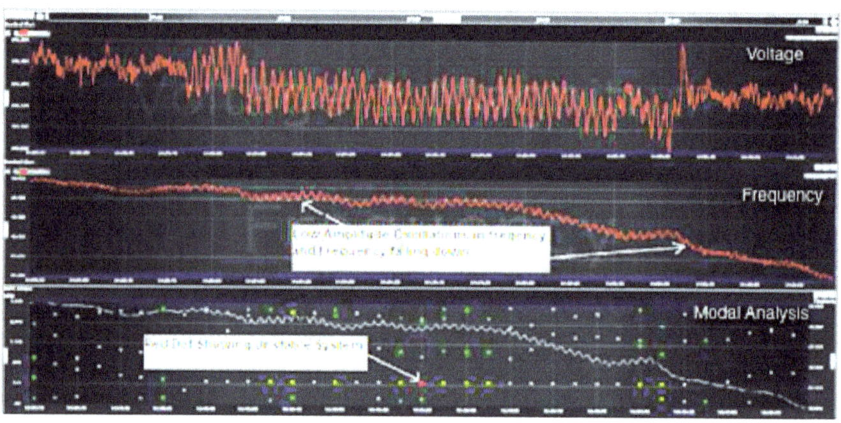

Figure 1.13 Low-frequency oscillation due to tripping of tie line

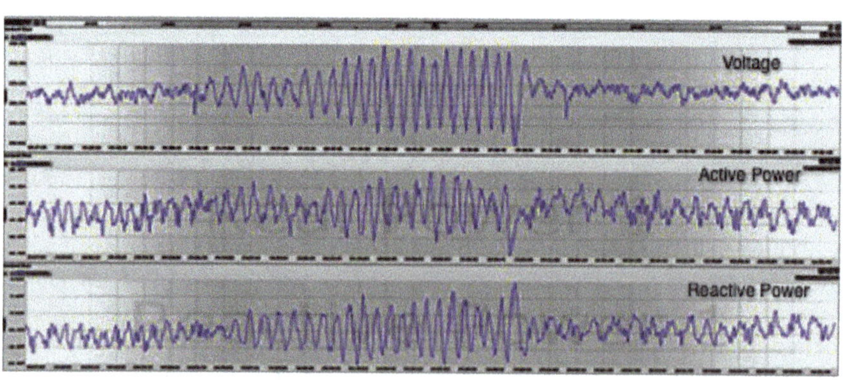

Figure 1.14 Low-frequency oscillation during synchronization of thermal unit

regional grids of India. The figure shows the oscillations observed in voltage, active power, and reactive power signals. Since oscillations were for a very small period (a few milliseconds) and had negative damping, they were harmless to the grid.

Oscillation monitoring system (OMS) is a fundamental application that is possible even with small numbers of PMUs installed. Since oscillations are like perturbations in a pond that start from a point and propagate all around, they are sensed by any PMUs in the grid. The small frequency oscillations are always present in the system due to the continuously changing condition of the grid; however, they are not harmful if they die down in a short time, i.e., they are not negatively damped, i.e., not of increasing amplitude. OMS provides this information to the grid operator. OMS does the modal analysis of signals captured by the PMUs and provides damping and other information to the operators.

LFO has been observed in the Indian system on several occasions. This case study presents LFO observed at only one location. No oscillation was observed in the rest of the system, as observed in the PMU installed across the different regions of the new grid. This indicates that this is a case of a local mode of oscillation.

Figure 1.15 shows the frequency plot of LFOs observed in the PMU installed at a thermal power station near New Delhi and not in other nearby PMUs. The observed mode of oscillations was 1.67 Hz, which was in the range of intra-plant oscillation, which was present only in that thermal power plant frequency signal. Figure 1.16 shows the zoomed view of the oscillation, which was of a growing nature. During this case, data was collected from all over India for oscillation. Once again, it was

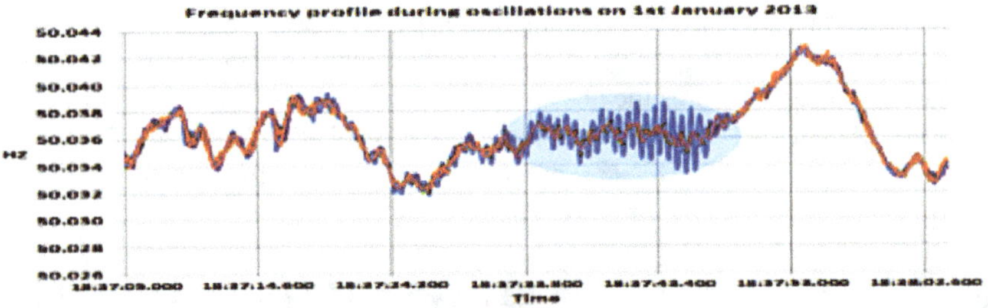

Figure 1.15 Low-frequency oscillations in thermal power plant frequency

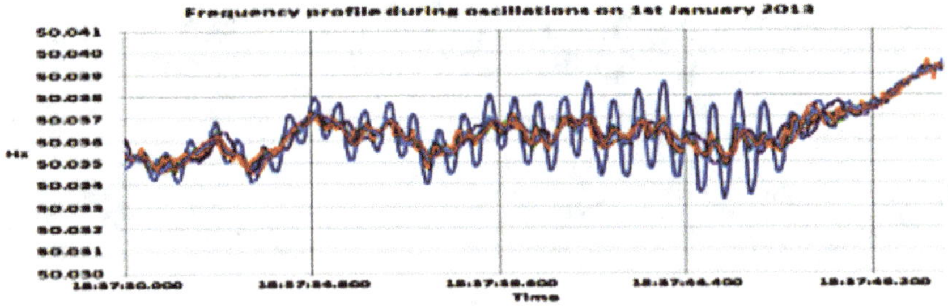

Figure 1.16 Zoomed view of oscillations

found localized only, as no oscillations were reported from other generators. During the investigation, it was learned that there was an event of malfunctioning of the electro-hydraulic governor (EHG) of a 490 MW unit of the thermal power plant. It was further clarified that LFOs were due to testing of the valve control system on the 490 MW thermal power plant unit, and the same unit was hunting from 350 MW to 470 MW. The oscillation frequency was 1.66 Hz.

1.8.3 Insight to coherency of generators

Oscillations captured by PMUs are very useful for giving insight into the coherency of generators in the system. The figure shows the coherent group of generators located in the eastern and western parts of the Western grid. The event that triggered the swinging of generators in the two areas is yet to be identified. However, for the system operator, having such a plot in real time gives enough warning that the system is exposed to inherent stress. This helps to avoid any nasty "surprises." The plot in Figure 1.17 reflects the stressed condition of lines connecting the eastern and western parts of the Western grid.

1.8.4 Remote detection of system separation

System separation is the phenomenon in the electric grid in which some part of the grid gets disconnected due to the tripping of connecting tie lines. The timely and early detection of such events is of utmost importance to grid operators. Since the separated part of the grid is always unstable due to distorted load generation balance, early detection and fast action may reduce the risk of further aggravation of the separated part. Measurement of angular separation with the help of synchrophasor data is very useful for detecting system separation conditions in the grid. Angular separation measures the phase angle difference between any two locations. WAMS display always needs a reference point to display the angular difference at various grid parts. If any part lose synchronization with the rest of the grid, WAMS

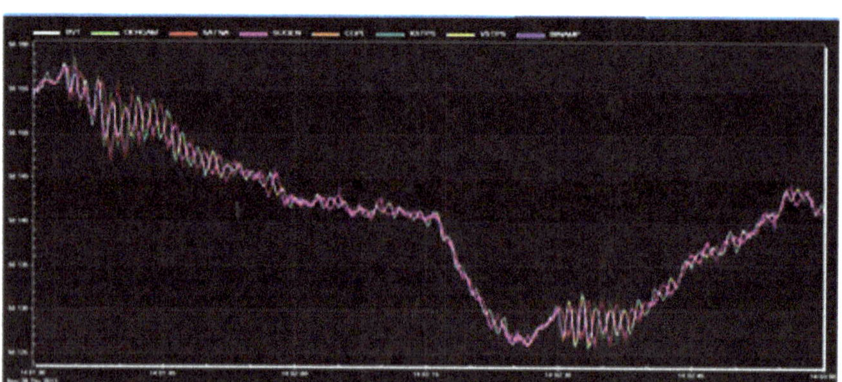

Figure 1.17 Zoomed view of oscillations

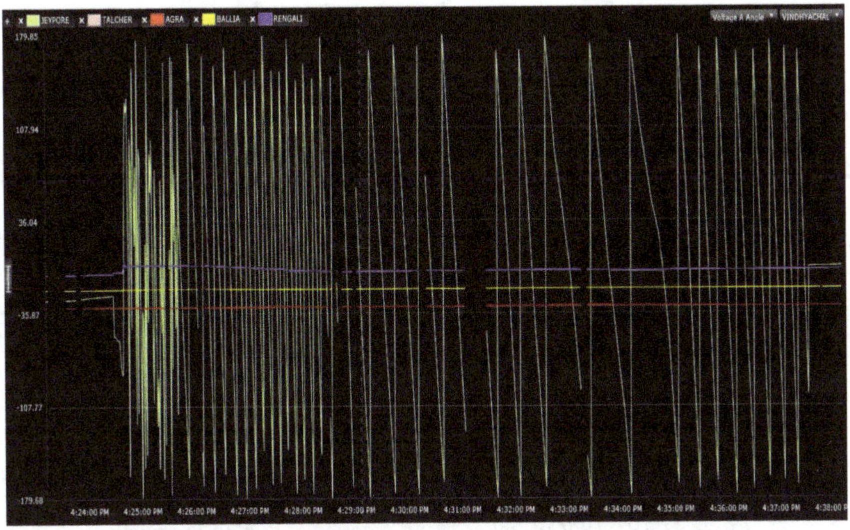

Figure 1.18 Angle trend after system separation

displays a pattern of angular difference varying between −180 degrees and +180 degrees. The variation frequency depends on the quantum of difference between the frequencies of the two systems. Figure 1.18 shows such an event in which the system of the south of Odisha (one of the Eastern states of India) separated from the rest of the grid due to the tripping of all tie lines connected to that area. The display shows a real-time angle of four locations with respect to the PMU located in the northern part of India. At about 1,625 h, the angle (shown in green color) of one of the PMUs installed at the substation of south Odisha, a state in the eastern part of India, started displaying values between +179.85 and 179.68. The angle of the other three PMUs (green, yellow, and violet) installed in northern India remained more or less the same. It depicted the system separation phenomena and prompted the operator to take corrective action, reducing the possibility of further outages in the grid.

Figure 1.19 shows the plot of voltage and frequency from the PMU located at the substation inside south Odisha. It gives a detailed account of the event before South Odisha's separation from the rest of the grid. It shows the fault in the R-phase in which voltage values shoot up to 100–300 kV at the time of separation.

1.8.5 Synchronizing two grids from remote

Synchrophasor data was found to be helpful in monitoring and controlling the synchronization of two grids from a remote control room. Generally, synchronizing an element with the grid is carried out from the substation by observing the synchroscope dials and closing the circuit breaker at the appropriate time. Using a synchroscope is convenient as far as the synchronization of a single element is concerned. However, synchronizing two grids required extensive monitoring and controlling of frequency difference and angular difference between the two grids, and the

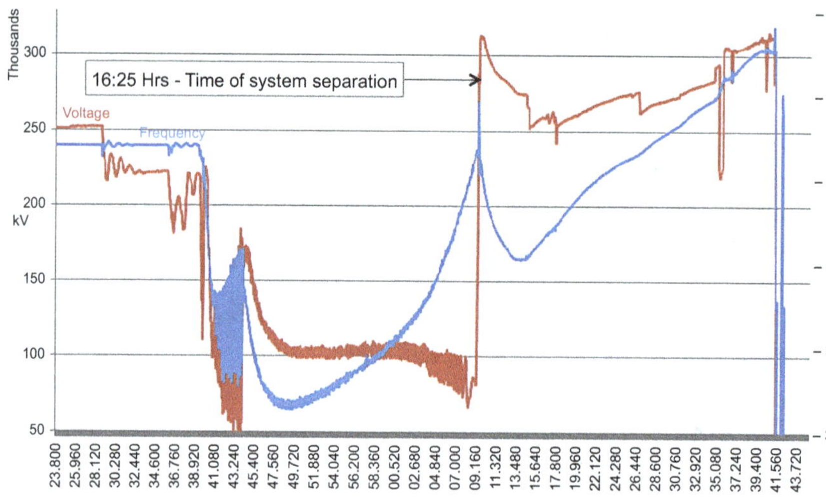

Figure 1.19 Plot of voltage and frequency during the event

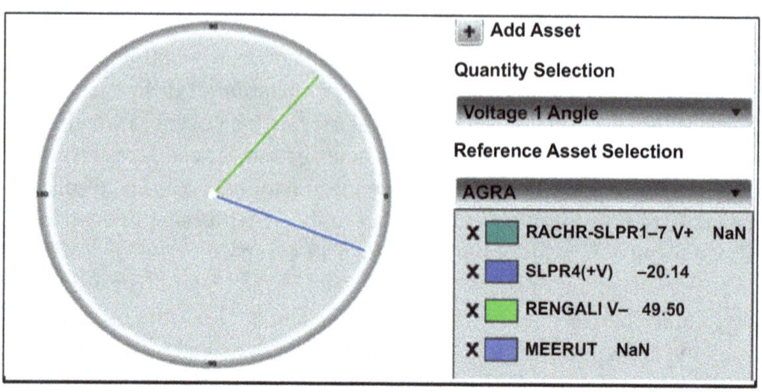

Figure 1.20 Synchronization of two grids

connecting breaker should be closed when both are within required limits. Since grid conditions always change dynamically, using a synchroscope may not be feasible.

On December 31, 2013, India needed to synchronize its Southern grid with the rest of the grid to realize its goal of "One Nation and One Grid." For this, synchrophasor data was used, and both grids synchronized successfully. Figure 1.20 shows the WAMS display used during the event. A synchroscope-like display was created for monitoring the phasors of both grids to get the moment when both are revolving at nearly the same speed with a minimum angle between them. Nearly, the same speed signifies that the difference in frequency between the two is minimal. The minimum angle difference between the two grids prevents the sudden large load flow between two grids to avoid unnecessary LFOs in the system.

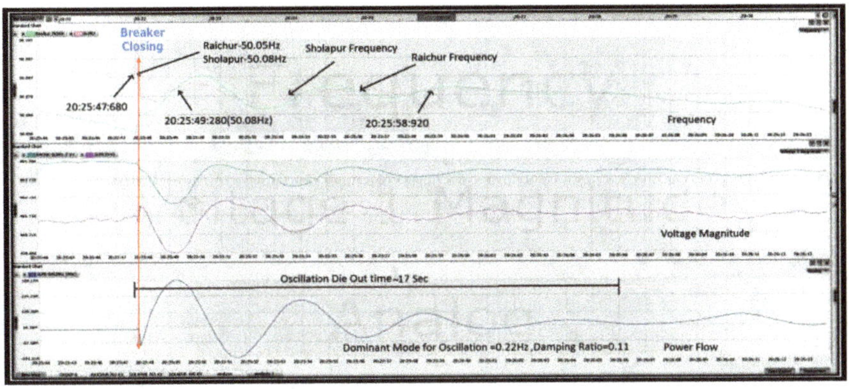

Figure 1.21 Synchronization of two grids

Figure 1.21 shows the real-time trend of frequency, voltage, and power flow of two stations located in two grids being synchronized. These real-time trends were very useful for matching the grid and breaker frequency when the frequencies were quite close.

1.8.6 Improving power plant performance

1.8.6.1 Ascertaining the performance of PSS

The power system stabilizer (PSS) is a supplementary control system of a generator's excitation system. The output signal of the PSS is injected into the summing junction of the exciter block in the generator to dampen LFOs of the power system. It is the most economical way to mitigate this kind of LFO. For the PSS to perform its role, it is crucial to tune optimally the internal parameters of the PSS, which are composed of lead-lag time constants and the gain. The tuning of the PSS and its application have been studied and applied to power systems worldwide since the 1960s. The underlying reasons for tuning PSS parameters are to compensate for the phase lag due to the electrical system, generator, and excitation system and to provide electrical torque in step with speed via the excitation system and generator [11].

PSS performance is often evaluated from the damping of the "local mode of oscillations," the generator swinging against the rest of the power system. This mode is usually at frequencies between 1 Hz and 3 Hz. Stronger network ties and lighter loading tend to give higher local mode frequencies, and weaker ties and heavier loading tend to give lower local mode frequencies. PSS performance must be designed to provide acceptable performance over various system conditions, which may result from different operating conditions (such as lines out-of-service and varying load levels). However, it is not straightforward to detect when the PSS needs tuning. However, the synchrophasor measurement could help determine the status of PSS tuning.

Figure 1.22 shows the oscillation experienced by a hydro plant in northern India due to PSS. Tehri Hydro Electric Power Plant (HEP), having 1,000 MW capacity, is located in the north of the NR of India. There are four 765 kV transmission lines

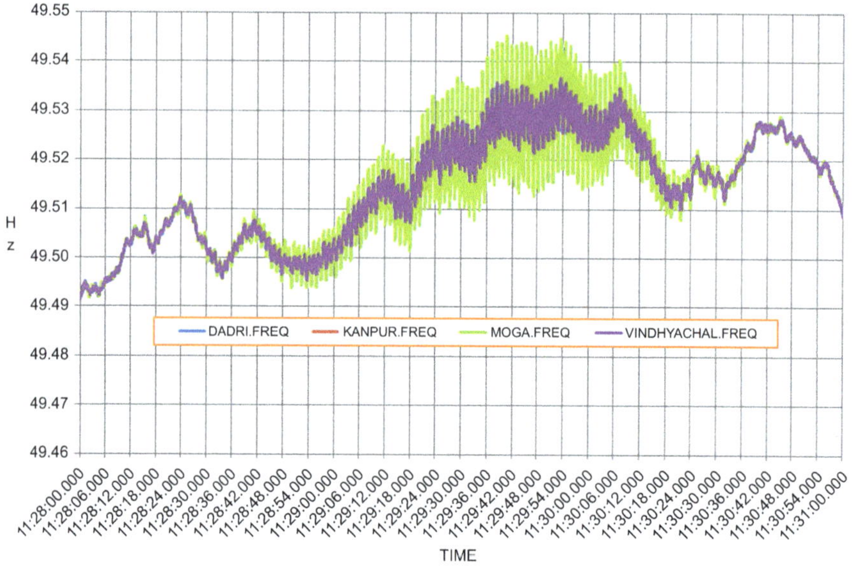

Figure 1.22 Oscillations due to the need for PSS tuning

emanating from this plant for evacuation of its generation. All four lines were charged at 400 kV at that time. The power station has complained about the experience of oscillation in the system. An exercise was carried out to examine the oscillation phenomena and ascertain the loadability of the line. The test was performed by keeping only two lines in service out of four lines (refer Figure 1.23), and generation at the Tehri HEP was increased in a controlled manner. At a certain power flow, the machine at the Tehri HEP started experiencing the local mode of oscillations. Oscillation disappeared as soon as the power flow on the lines decreased. Since the oscillations were in local mode and were appearing without any disturbances, it was concluded that the PSS of the machine needed tuning. After tuning the PSS, the oscillations were not experienced even with the flow increase.

1.8.6.2 Validation of performance of SPS

A hydroelectric power station, Karcham Wangtoo, is located in the north of northern India, having a capacity of 500 MW (2 × 250 MW). Two dedicated transmission lines were planned to evacuate its generation to a nearby pooling station at Abdullapur. For some reason, the commissioning of these lines was delayed, but the plant was ready to generate to its full capacity. To facilitate the evacuation of power generation, studies were carried out to explore the possibilities of the evacuation of the generation through existing transmission lines via the nearby power station Nathpa Jhakri. Since the evacuation of extra generations may lead to cascading tripping of other transmission lines during any $n-1$ contingency, a SPS was planned and implemented. The SPS would trip the machines at Karcham

Wangtoo in the case of an outage of any evacuation links, and the line loading of any of the remaining lines would increase by more than 800 MW. However, during the operation of SPS, system-wide oscillations were observed, as seen in the plot of PMU data. On investigation, it was found that there was an unintentional delay of about 10 s in the logic of tripping a unit through the operation of SPS as shown in Figure 1.24. The delay was unknowingly introduced. After the delay had been removed, oscillations were not observed again.

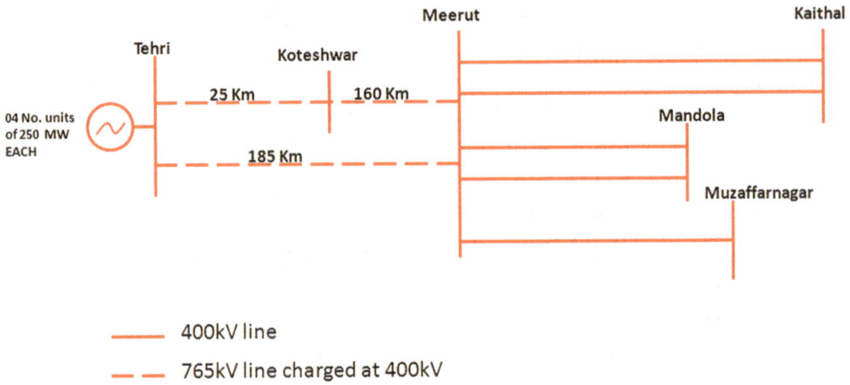

Figure 1.23 Evacuation from Tehri Hydro Plant

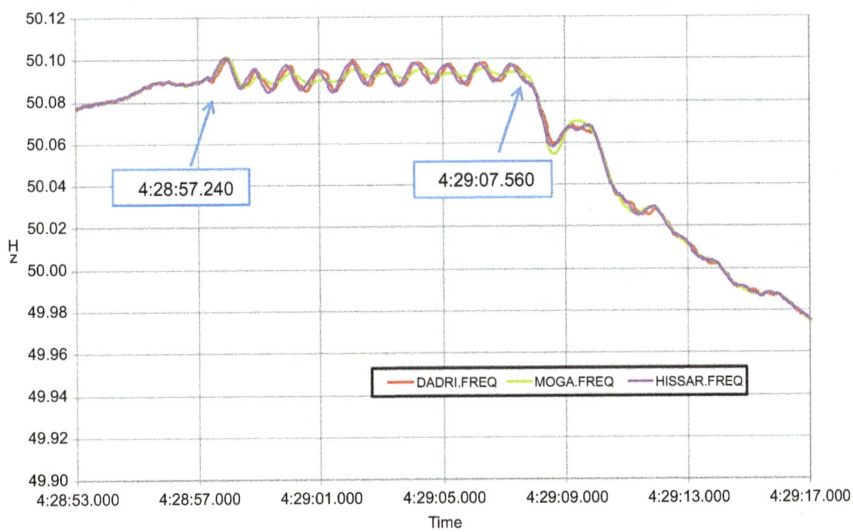

Figure 1.24 Oscillations due to the need for SPS delay

1.9 The experiences from URTDSM project

With the implementation of the URTDSM, the next level of challenges was faced during data utilization for the system operator's situational awareness. This has also provided an opportunity for the system operator to address the development of a new analytics tool for decision-making. This section explains a few of the challenges and experiences of using URTDSM for real-time operators.

1.9.1 Selection of reference node

Reference node plays a vital role in WAM as it provides stress monitoring, event detection visualization, and alarms to operators. Selection of the reference node is challenging in a large system as it requires baselining activities and validation. Before the implementation of URTDSM, a baselining activity for reference node based on the pilot project on all India basis was completed, and the criteria were defined as given below [12]:

Preferred locations for the reference node:

1. Node with strong interconnections and high fault level
2. Nodes with reliable communication
3. Node close to baseload generation complex
4. Nodes consistent with offline simulation/EMS studies
5. Node should be system-wide available and reliable.

Nodes to be avoided as reference node:

1. Nodes with low inertia, weak-tie lines, and low fault level
2. Nodes located in oscillation area (based on real-time experience)
3. Nodes near to HVDC bus (varying power order during the day-to-day operation)
4. Nodes near to varying generation complex.

1.9.2 Selection of wide-area node pair for angular separation monitoring

One of the major advantages of the WAMS system is the availability of angular separation between any two nodes in the system. It helps identify the system's healthiness and stress at the various operating points. Baselining activities play a major role in determining the wide-area node pairs. In the Indian power system, node pairs were selected based on the following criteria [12]:

1. Generation complex to load complex
2. Power corridors/flow gates
3. Power corridors from one region to other
4. Maximum angular separation in the grid (wide distant node).

1.9.3 Event detection setting and alarms segregation

Early warning situational awareness can be achieved through quick event detection on synchrophasor data. It mainly helps take preventive action immediately after events to avoid its cascading consequences. The event can be detected from the sudden variation in monitored parameters like frequency, positive sequence voltage magnitude, rate of change frequency (ROCOF), and angle difference. Based on analysis of several events and their baselining, it is observed that the rate of change in frequency and positive sequence voltage, ROCOF, and adjacent node. Angle difference signals can be configured for event detection globally through common settings for all PMUs. However, in the case of MW, MVAR signals individual event detection settings must be configured [12].

1.9.4 Classification of low-frequency oscillation modes

The fine-resolution time-synchronized data with higher granularity helps the power system observe LFO in real time. However, to provide correct situational awareness of the oscillation, the baselining activity of oscillation and tuning of the OMS has to be completed. This exercise was completed during the synchrophasor pilot project, during which various LFO events were analyzed and documented to provide good insight. Few reports have been published on the LFO documenting the whole exercise, which helped in the classification and damping criteria. The above analysis has tuned the OMS in URTDSM [12].

1.9.5 Visualization display for operator situational awareness

Various kinds of displays are available for the system operator, including the geographical display, single line diagram, trend and dial display, and contour display. These displays help the operator in observing the power system with better visibility [12].

1.9.6 Training of system operator

This is also one vital aspect in improving the utilization of URTDSM across the system operator at various levels of control centers. To improve the utilization of WAMS at various control centers across the country, a training module has been designed with various use cases for the Indian power system. This has helped enhance the utilization of URTDSM in the Indian power system.

1.10 Way forward

Since implementing the first pilot project in India and realizing its utility in managing and improving the power grid's performance, many separate pilot projects were implemented and integrated with a national-level WAM system. This national-level WAM system has been utilized as the mainstream system by the grid operators and no longer has pilot status. Consequently, based on the experience gained in the last 5–6 years, India had ventured into a full-scale WAMS popularly known as the URTDSM project, probably the single largest project in the world wherein 1,600 PMUs and 32 nos PDC have been installed and integrated in a hierarchical manner. Many applications for utilizing data from the URTDSM project were envisaged. Some of the applications had already been developed and deployed. Other applications are in development or being tested with actual data.

Acknowledgments

The views expressed in this chapter are of the authors only; Grid Controller of India Ltd (erstwhile Power System Operation Corporation Ltd) does not subscribe to the same in any way.

References

[1] Phasor measurement unit. Available from: https://en.wikipedia.org/wiki/Phasor_measurement_unit [Accessed 15 Apr 2016].

[2] BURN – An Energy Journal. The electricity grid: A history. Available from: http://burnanenergyjournal.com/the-electricity-grid-a-history/ [Accessed 16 Apr 2016].

[3] JS. How electricity grew up? A brief history of the electrical grid … . Available from: https://power2switch.com/blog/how-electricity-grew-up-a-brief-history-of-the-electrical-grid/ [Accessed 25 Oct 2016].

[4] Agrawal, V. K., Agarwal, P. K., and Rathour, H. Experience of upscaling and integration of regional level synchrophasors pilot projects to a national level project. *D2 colloquium, CIGRE, 13–15* November 2013, Bangalore, India.

[5] Hurtgen, M. and Maun, J.-C. Applications of PMU measurements in the Belgian electrical grid, Report, 9 May 2012.

[6] Stastny, L., Franek, L., and Z. Bradac. Time synchronised low-voltage measurements for smart grids, *Procedia Engineering*, vol. 100, 2015, 1389–1395, http://dx.doi.org/10.1016/j.proeng.2015.01.506.

[7] Soonee, S. K., Narsimhan, S. R., Porwal, R., and Pandey, V. Application of phase angle measurement for real time security monitoring of Indian electric power system – an experience, CIGRE Paris Session, *CIGRE*, August 2008, Paris, France.

[8] SmartGrid.gov. Synchrophasor technologies and their deployment in the recovery act smart grid program, Report, US Department of Energy, 2013.

[9] Central Electricity Regulatory Commission, New Delhi. CERC order date 07-July-2022 against Petition No. 678/TT/2020. Available from: https://cercind.gov.in/2022/orders/678.pdf.

[10] Giri, J. and Parashar, M. Transitioning from wide area monitoring to wide area management, *IEEE Smartgrid Newsletter*, 2016. Available from: https://smartgrid.ieee.org/bulletins/april-2016/transitioning-from-wide-area-monitoring-to-wide-area-management.

[11] Shin, J., Nam, S., Lee, J., Baek, S., Choy, Y., and Kim T. A practical power system stabilizer tuning method and its verification in field test, *Journal of Electrical Engineering & Technology*, vol. 5, no. 3, 2010, 400–406.

[12] Agarwal, P. K., Anumasula, R., and Kumar, C. Implementation experience of URTDSM project in Indian power system, *2019 CIGRE Conference, September 16–19, 2019*, Montréal, Canada.

Chapter 2

An optimal rcliability criterion (ORC) index for optimal placement of phasor measurement units (PMU) for full observability of power grid

Ranganath Vallakati[1], *Prakash Ranganathan*[2] *and Sriram Prabhakara Rao*[3]

2.1 Introduction

The dependency of human lives on electrical power is increasing exponentially. Our lives have become so reliant on electricity that a brownout situation in the grid could mean life and death conditions. Thus, a stable power grid is important. Electricity possesses two important qualities: reliability and continuity. The goal of an electric utility is to supply power continuously and reliably. We need reliable electricity rather than electricity without reliance. Continuity and reliability go hand in hand; one can be achieved only if the other is achieved. With increased injection of renewable energy sources or distributed generations (DGs), various problems are showing up more often than normal. The problems include a deterioration in grid characteristics such as power quality, supply efficiencies, and reliability. Rerouting of excess power becomes a concern due to lack of energy transportation infrastructure. This is because existing transmission and distribution infrastructures are designed only for unidirectional power flows rather than bidirectional power flows. This means that power flow starts from high-voltage lines to medium-voltage lines and then to low-voltage distribution lines. As penetration of DGs increases, the current unidirectional flow infrastructure does not work due to the power following into the low-voltage lines from the DGs. Thus, effective management, monitoring, and forecasting methods are needed to maintain the stability of the grid. A smarter grid provides management solutions for a stable operation.

A smart grid (SG) is simply a conventional power grid that is integrated with intelligent sensing and software technologies. SG contains automated substations, switch gears, sensors for intelligent monitoring and control, technologies for

[1]Activision Blizzard Media, San Francisco, CA, USA
[2]School of Electrical Engineering and Computer Science (SEECS), University of North Dakota, USA
[3]Center for Cyber Security Research (C2SR), University of North Dakota, USA

improvising generators, and load efficiency that ultimately keeps the power system in a stable condition. SG framework also contains conventional sources such as coal-fired power plants and DGs such as wind farms and solar panels. There are intelligent monitoring sensors on transmission lines, buildings, and homes at DG locations. Sensors can make decisions on their own and need to be well connected for better monitoring and situational awareness. Data sensed in an SG are collected for various functionalities such as real-time monitoring, offline analysis, load-flow studies, generation, and load-shedding schemes.

The electric utilities currently use supervisory control and data acquisition (SCADA) for controlling and monitoring operations in a power system. There are several drawbacks to the existing SCADA system. In a SCADA system, data samples of voltage, current magnitude, and phase angle parameters are estimated rather than measured. The magnitudes are not time-synchronized, so it is difficult to relate the occurrence and timing of an event. The limitations of SCADA systems were overcome with the invention of phasor measurement units by Dr. Arun G. Phadke and Dr. James S. Thorp in 1988. Synchrophasors are modern-day measurement tools that gather key sensor parameters such as voltage, frequency, and phase angle information, which can be utilized to monitor the power grid [1,2]. With these sensors, every detail of the grid is observable, which would assist system operators in supplying electricity reliably to consumers.

Synchrophasors provide time-synchronized measurements of the power system such as voltage and current phasors used to calculate active power, reactive power, and frequency. The data are then utilized for applications such as state estimation (SE), transient analysis, capacitor bank monitoring, load-shedding schemes, inter-area oscillations, and microgrid operations [3,4]. Typical microgrid operations include islanding, anti-islanding, and identifying fault locations on transmission lines [5–7].

2.2 Synchrophasors

To avoid blackouts, parameters such as frequency, voltage, current, and phase angle are analyzed concerning time and location of occurrence. The key advantage of a synchrophasor is its ability to timestamp the measured data with a high data rate. Synchrophasors utilize global positioning system (GPS) time clocks to time stamp their measurements. Synchrophasors helped independent system operators (ISOs) in realizing the wide-area measurement systems (WAMSs). WAMSs are an essential upgrade for SCADA systems for they use synchrophasors to monitor power system conditions over large geographical areas to stabilize the grid. The recent model of synchrophasors has a sampling rate of around 240 samples per second. This level of high granularity is much required because the quantities such as frequency and phasor values vary at a sub-second time interval. ISOs utilize synchrophasor data collected from different geographical locations at the same time and make critical decisions in real time to maintain the grid.

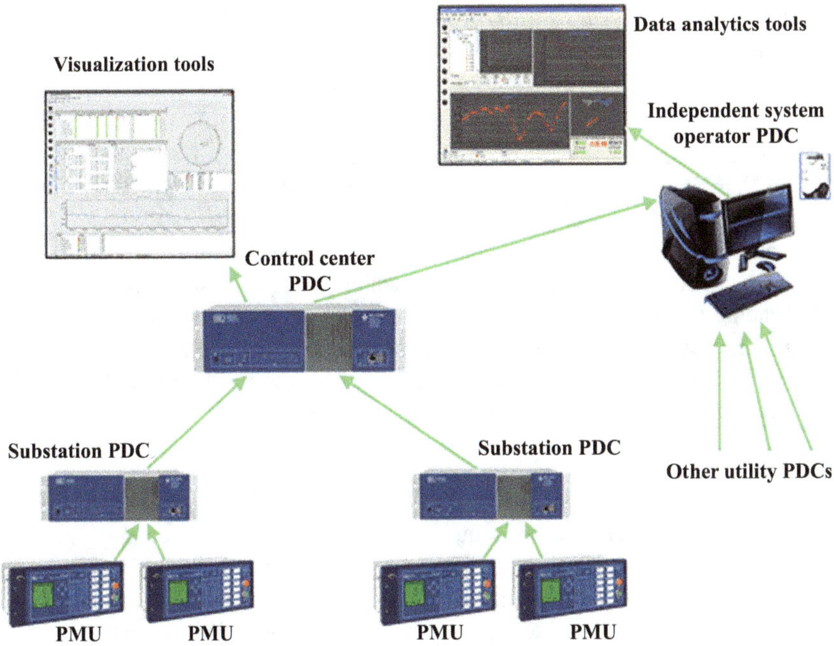

Figure 2.1 Wide area monitoring system (WAMS)

Different components form a WAMS network as shown in Figure 2.1. They are as follows:

1. Phasor measurement unit (PMU)
2. Phasor data concentrator (PDC)
3. GPS clock

PMU uses the most common technique for determining the phasor representation of a signal, that is, by taking the samples from the waveforms and then applying the discrete Fourier transform to compute the phasors [8]. The AC waveforms are digitized by an analog-to-digital converter for each phase. A phase-lock oscillator along with a GPS source provides the needed synchronization for the samples.

Waveforms are obtained through a line of components such as transformers and filters, and these measured signals are time-synchronized using the GPS clock. The GPS clocks allow the synchronization of signals into a Universal Time Frame (UTC) to assist the wide-area system operations. The time-synchronized phasor measurements are aggregated in PDC as per IEEE Standard C37.118 [9]. PDC acts as a centralized main-stream data server that collects and stores the streaming information from various synchrophasors across the grid as per IEEE Standard C37.244 [10]. The data stored are then extracted and used for various applications.

2.2.1 Advantages of synchrophasors over SCADA measurements

SCADA provides information about voltage magnitude, active and reactive power flows, load measurements, generator outputs, the breaker status, and performances of capacitor banks and transformers as calculated estimates from SE. When the power system state is changing quickly, conventional SE fails to capture certain situations making the SCADA scan inefficient.

In addition, analog measurements are proportional to the time differences between the measurements and the rate at which states of system change [11,12]. With synchronized measurements, variables such as phase angles need not be estimated as PMU captures all measurements that are synchronous to UTC. Also, the quickly changing states can be captured for accurate SE. Figure 2.2 provides a comparative analysis of SCADA and PMU measurements.

2.2.2 Challenges with synchrophasor measurements

1. Economics of synchrophasor units
   PMUs are deployed across utilities to improve the SE of power systems. However, the major constraint for utilities is the cost of PMUs. In real-world applications, utilities need to set up several things before PMUs can be placed on the grid. This includes, but is not limited to, the construction of communication infrastructure, procurement of PMUs, PDC for data aggregation, and synchronization clocks. Thus, it is an expensive issue for utilities to upgrade their monitoring systems. Table 2.1 shows the cost of installing a synchrophasor

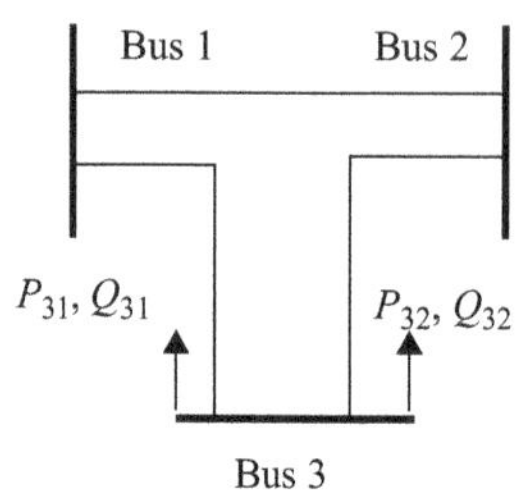

	Measured parameters at Bus 1, Bus 2, and Bus 3	Calculated
SCADA	$\|V_1\|,\|V_2\|,\|V_3\|, P_{31},Q_{31},P_{32},Q_{32}$	$\|\delta_{31}\|, \|\delta_{32}\|$
PMCU	$\|V_1\|,\| \delta_1\|,\|V_2\|,\|\delta_2\|,\|V_3\|,\| \|\delta_3\|$	–

Figure 2.2 SCADA Vs PMU measurements [11]

Table 2.1 Typical approximation of synchrophasor cost

Type of equipment	**Cost (US$)**
PMU with protection, automation, and controls	~15,000
PDC (up to 40 PMUs)	~8,000
Synchronization clocks	~2,000
Digital equipment (firewalls, cables, and routers)	~5,000

system [13]. The total cost is approximately US$30,000 excluding any infrastructure, operational, and labor costs. However, devices such as PDC may collect data from several PMUs. Therefore, a small or mid-size utility can install only a few PDCs as a typical PDC can handle data up to 40 PMUs.

However, there is a solution. By applying the fundamental Ohm's law on the transmission lines, a PMU placed on a bus allows for observation of the neighboring buses, that is, adjacent bus voltages are easily computable. Thus, the placement of PMUs in a grid became a challenge for the utilities to reduce the cost as well as improve the power system situation to maintain reliability and continuity in the system. This problem of reducing the cost factor as well as increasing the reliability of the grid using PMU placement is addressed in this chapter by using the method of linear programming (LP).

2. Reliability of synchrophasor data
The main objective of PMU placement problems is to minimize cost. Different optimization methods try to reduce system redundancy by making the buses measured the least number of times. During contingency situations such as the failure of a PMU or a communication line, the system loses its observability and the system becomes unobservable affecting the operator's ability in decision-making. Thus, optimal methods for PMU placement problems should include the concept of redundancy in its measurements to keep the system stable during contingencies. This research proposes a criterion for including such a concept of redundancy in the PMU-placement algorithm.
3. Utilization of synchrophasor data
Synchrophasors generate around 30–120 samples per second that contain information about the state of the power system. Thus, ISO is bombarded with nearly 108,000 samples/h, which equates to 2,592,000 samples/day or 77,660,000 samples a month. These numbers grow exponentially, and approximately around 1.5 TB of data are accumulated in a month. With all such big data accumulated, ISO makes use of these data for meaningful applications. It is not possible to monitor every data point unmonitored. A visualization technique had been developed using a unit-circle representation of phasor data. Unit-circle visualization type will help ISO monitor the incoming data with ease of understanding. The data representation is stored in the database (db) before data mining tasks. ISO uses synchrophasor data for predictive analytics to forecast the system behavior. Any forecasting task requires knowledge of the history of data. Data mining is the process of extracting useful information to understand more meaningful interpretations of the system. A density-based clustering algorithm called density-based spatial clustering of applications with noise is used to cluster the synchrophasor data. The results of clustering are stored in the db after the initial stage of visualization.

2.3 Optimal placement of phasor measurement units

PMUs provide time-synchronized voltage and current measurements when placed on a bus. These time-synchronized values are critical for the continuous operation of WAMS in SGs. It is not required to place a PMU on every bus in the system to get full observability. Using Ohm's law, a PMU placed on a specific bus can be used to measure the voltage and current phasors at that point. It can also be used to calculate the voltage and current phasors on the lines that are connected to the placed bus. Also, it is not feasible for the utilities to afford a PMU for every bus in their grid due to its cost. Thus, there is a need to find a solution for the optimal placement problem (OPP) for PMUs.

2.3.1 Need for OPP for synchrophasors

Case 1: In Figure 2.3(a), all four bus points are accompanied by PMUs for complete observability. The situation is shown in Table 2.2. There exist many solutions for redundant observabilities than required. Thus, it would be useless to have a PMU on every bus.

Case 2: Figure 2.3(b) shows how three PMUs on buses 1, 2, and 3 are sufficient to cover the full observability of the system. This situation is given in Table 2.2. There is a reduced redundancy in the measurements, but the system is yet fully observable.

Case 3: Figure 2.3(c) shows a four-bus system with only two PMUs placed on buses 1 and 4. By analyzing the case 3 column of Table 2.2, we can say that the

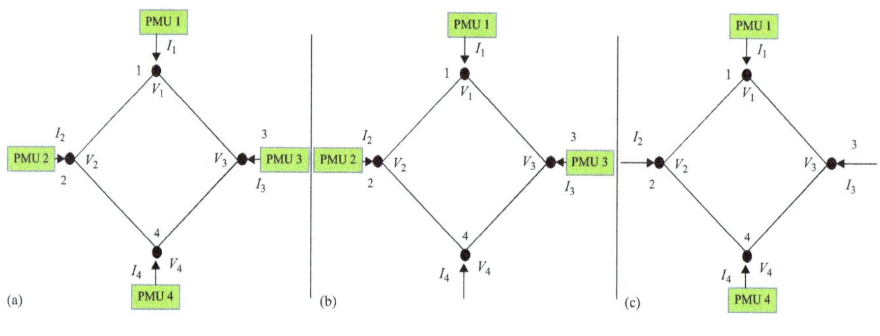

Figure 2.3 Various cases of placing the PMUs on a four-bus system

Table 2.2 Case studies of PMU placement on four-bus system

		Case 1	Case 2	Case 3
Bus 1	Measured	PMU 1	PMU 1	PMU 1
	Calculated	PMU 2, PMU 3	PMU 2, PMU 3	–
Bus 2	Measured	PMU 2	PMU 2	–
	Calculated	PMU 1, PMU 3	PMU 1, PMU 3	PMU 1, PMU 4
Bus 3	Measured	PMU 3	PMU 3	–
	Calculated	PMU 2, PMU 4	PMU 2	PMU 1, PMU 4
Bus 4	Measured	PMU 4	–	PMU 4
	Calculated	PMU 2, PMU 3	PMU 2, PMU 3	–

values at buses 2 and 3 without the PMU can be calculated using the line parameters (R; X) and measured parameters at buses 1 and 4. In this way, even with two PMUs, the system remains fully observable.

Thus, we can conclude that placing PMUs using the known topology in a certain manner could provide complete observability of the system with cost savings that include operational costs and hardware (equipment and infrastructure) costs. In addition, this research includes the zero-injection buses (ZIBs) of the power systems to further reduce the number of PMUs required for full observability. ZIBs are similar to transshipment nodes in the OPP algorithm that reduce the minimum number of PMUs. Details about ZIBs are explained in a later section of the chapter. An index, optimal redundancy criterion (ORC) is proposed to address the issue of reliability which is very important for the power systems. ORC assists in identifying the best solution for OPP while enforcing some conditions to maintain redundancy and full observability of critical buses.

2.4 PMU placement problem formulation using linear programming

In this research, the OPP of PMU is solved using the LP technique. In any LP method, the problem is defined by having an objective function that maximizes or minimizes a linear function subject to a set of linear constraints. The constraints include a combination of equalities or inequality constraints. Our goal is to reduce the minimum number of PMUs required for a power system without losing the complete observability of the system.

2.4.1 Problem formulation

For a system with N buses, the OPP of synchrophasors can be formulated as shown in (2.1).

$$\begin{aligned} &minimize\sum_{i}^{N} W_i * X_i \\ &Subject\ to\ f(X) \geq \hat{1} \end{aligned} \tag{2.1}$$

where X_i is a binary decision variable vector, whole entries are defined using the binary variables as follows:

$$X_i = \begin{Bmatrix} 1 & if\ a\ PMU\ is\ installed\ at\ bus\ i \\ 0 & otherwise \end{Bmatrix} \tag{2.2}$$

$\hat{1}$ is a vector whose entries are all ones

W_i is the cost associated with the installation of PMU at that bus i. Figure 2.4 shows a single-line diagram of an IEEE 14 bus system. There are five generation points at buses 1, 2, 3, 6, and 8. The generators at bus 1 and 2 generate both active

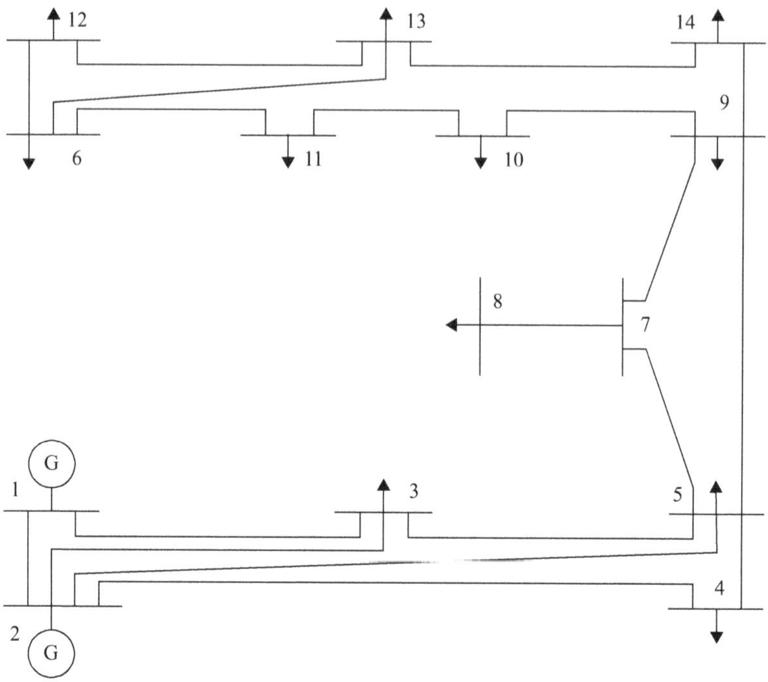

Figure 2.4 IEEE 14 bus system [14]

and reactive power, while generators at 3, 6, and 8 generate reactive power. There are 11 load points connected at bus 2, 3, 4, 5, 6, 9, 10, 11, 12, 13, and 14 that consume the active and reactive power. Using the LP approach, the system is evaluated to find the PMU placement locations.

The two cases of the system that are considered to determine the OPP solution are as follows:

1. By considering a regular test-bus system with any ZIB
2. By considering ZIB as special constraints.

2.4.2 System with no zero-injection buses

Assuming all of the 14 buses in the IEEE 14 bus system are regular ones having at least a load or a generation point. There is a requirement to measure or observe all the buses as changes can happen at each of them due to the generation and load points. A binary connectivity matrix is formed to indicate links between the buses. The entries of $A_{i,j}$ are defined as follows:

$$A_{i,j} = \left\{ \begin{matrix} 1 & if\ i = j \\ 1 & if\ i\ and\ j\ are\ connected \\ 0 & if\ otherwise \end{matrix} \right\} \tag{2.3}$$

Matrix $A_{i,j}$ can be directly obtained from the bus admittance matrix by transforming its entries into binary form as shown in (2.4):

$$A_{i,j} = \begin{bmatrix} 1 & 1 & 0 & 0 & 1 & 0 & 0 & 0 & 0 & 0 & 0 & 0 & 0 & 0 \\ 1 & 1 & 1 & 1 & 1 & 0 & 0 & 0 & 0 & 0 & 0 & 0 & 0 & 0 \\ 0 & 1 & 1 & 0 & 0 & 0 & 0 & 0 & 0 & 0 & 0 & 0 & 0 & 0 \\ 0 & 1 & 1 & 1 & 1 & 0 & 1 & 0 & 1 & 0 & 0 & 0 & 0 & 0 \\ 1 & 1 & 0 & 0 & 1 & 0 & 0 & 0 & 0 & 0 & 0 & 0 & 0 & 0 \\ 0 & 0 & 0 & 1 & 0 & 1 & 0 & 0 & 0 & 0 & 1 & 1 & 1 & 0 \\ 0 & 0 & 0 & 0 & 0 & 0 & 1 & 1 & 1 & 0 & 0 & 0 & 0 & 0 \\ 0 & 0 & 0 & 1 & 0 & 0 & 1 & 1 & 0 & 0 & 0 & 0 & 0 & 0 \\ 0 & 0 & 0 & 0 & 0 & 0 & 1 & 0 & 1 & 1 & 0 & 0 & 0 & 1 \\ 0 & 0 & 0 & 1 & 0 & 0 & 0 & 0 & 1 & 1 & 1 & 0 & 0 & 0 \\ 0 & 0 & 0 & 0 & 0 & 1 & 0 & 0 & 0 & 1 & 1 & 0 & 0 & 0 \\ 0 & 0 & 0 & 0 & 0 & 1 & 0 & 0 & 0 & 0 & 0 & 1 & 1 & 0 \\ 0 & 0 & 0 & 0 & 0 & 1 & 0 & 0 & 0 & 0 & 0 & 1 & 1 & 1 \\ 0 & 0 & 0 & 0 & 0 & 0 & 0 & 0 & 0 & 1 & 0 & 0 & 1 & 1 \end{bmatrix} \tag{2.4}$$

Here, a zero in the connectivity matrix derived from the admittance matrix implies that there is no connection between buses i and j. However, a value of one in the matrix implies that there is a connection between the buses. Once the binary connectivity matrix is formed, the LP constraints for all the 14 buses from the 14 connections are shown in (2.5)–(2.18):

$$\text{Bus 1}: x_1 + x_2 + x_5 \geq 1 \tag{2.5}$$

$$\text{Bus 2}: x_1 + x_2 + x_3 + x_4 + x_5 \geq 1 \tag{2.6}$$

$$\text{Bus 3}: x_2 + x_3 + x_4 \geq 1 \tag{2.7}$$

$$\text{Bus 4}: x_2 + x_3 + x_4 + x_5 + x_7 + x_9 \geq 1 \tag{2.8}$$

$$\text{Bus 5}: x_1 + x_2 + x_4 + x_5 \geq 1 \tag{2.9}$$

$$\text{Bus 6}: x_6 + x_{11} + x_{12} + x_{13} \geq 1 \tag{2.10}$$

$$\text{Bus 7}: x_4 + x_7 + x_8 + x_9 \geq 1 \tag{2.11}$$

$$\text{Bus 8}: x_7 + x_8 \geq 1 \tag{2.12}$$

$$\text{Bus 9}: x_4 + x_7 + x_9 + x_{10} + x_{14} \geq 1 \tag{2.13}$$

$$\text{Bus 10}: x_9 + x_{10} + x_{11} \geq 1 \tag{2.14}$$

$$\text{Bus 11}: x_6 + x_{10} + x_{11} \geq 1 \tag{2.15}$$

$$\text{Bus 12}: x_6 + x_{12} + x_{13} \geq 1 \tag{2.16}$$

$$\text{Bus 13}: x_6 + x_{12} + x_{13} + x_{14} \geq 1 \tag{2.17}$$

$$\text{Bus 14}: x_9 + x_{13} + x_{14} \geq 1 \tag{2.18}$$

The operator (+) in the formulation is a logical "OR" operation, while the inequality operator ($\geq$) ensures that at least one of the variables appearing in the sum will be non-zero. For example, consider the following constraints (2.5) and (2.6) associated with bus 1 and bus 2. Equation (2.5) indicates that at least one PMU must be placed on at least one of the three buses: bus 1 (x_1), bus 2 (x_2), or bus 5 (x_5). Similarly, (2.6) implies that at least one PMU should be installed on any one of the buses 1, 2, 3, 4, or 5 to make bus 2 observable.

2.4.3 System with zero-injection measurements

ZIBs are buses in which no currents are neither injected nor drawn into/from the system (i.e., buses that have neither load nor generation units). In the power systems, there are few bus locations where there is no requirement for measuring. The use of Kirchhoff's current law to include ZIBs in modeling the OPP further reduces the number of required. Each zero-injection node in a system creates an additional constraint. Thus, the number of PMUs required for complete observability of the system can be reduced. For a ZIB i, let A_i indicate the set of adjacent buses to bus i. Let $B_i = A_i \cup \{i\}$. Let the number of zero injections be Z. As there is no change in current (i) values at a ZIB, the variables related to the ZIBs are eliminated from all the constraining equations. The model formulation in an integer linear programming (ILP) framework, therefore, selectively allows some of the buses to be unobservable. However, there are additional constraints that need to be included for an observable system. They are as follows:

- The unobservable buses must be adjacent to ZIBs.
- The number of unobservable buses in the set (ZIB and its adjacent buses) is mostly as one (1) in the matrix.

In the IEEE 14 bus system shown in Figure 2.4, bus 7 is a ZIB. Thus, the new constraining equation can be written as

$$\text{Bus}\,1: x_1 + x_2 + x_5 \geq 1 \tag{2.19}$$

$$\text{Bus}\,2: x_1 + x_2 + x_3 + x_4 + x_5 \geq 1 \tag{2.20}$$

$$\text{Bus}\,3: x_2 + x_3 + x_4 \geq 1 \tag{2.21}$$

$$\text{Bus}\,4: x_2 + x_3 + x_4 + x_5 + x_8 + x_9 \geq 1 \tag{2.22}$$

$$\text{Bus}\,5: x_1 + x_2 + x_4 + x_5 \geq 1 \tag{2.23}$$

$$\text{Bus}\,6: x_6 + x_{11} + x_{12} + x_{13} \geq 1 \tag{2.24}$$

$$\text{Bus}\,8: x_4 + x_8 + x_9 \geq 1 \tag{2.25}$$

$$\text{Bus}\,9: x_4 + x_8 + x_9 + x_{10} + x_{14} \geq 1 \tag{2.26}$$

$$\text{Bus}\,10: x_9 + x_{10} + x_{11} \geq 1 \tag{2.27}$$

$$\text{Bus}\,11: x_6 + x_{10} + x_{11} \geq 1 \tag{2.28}$$

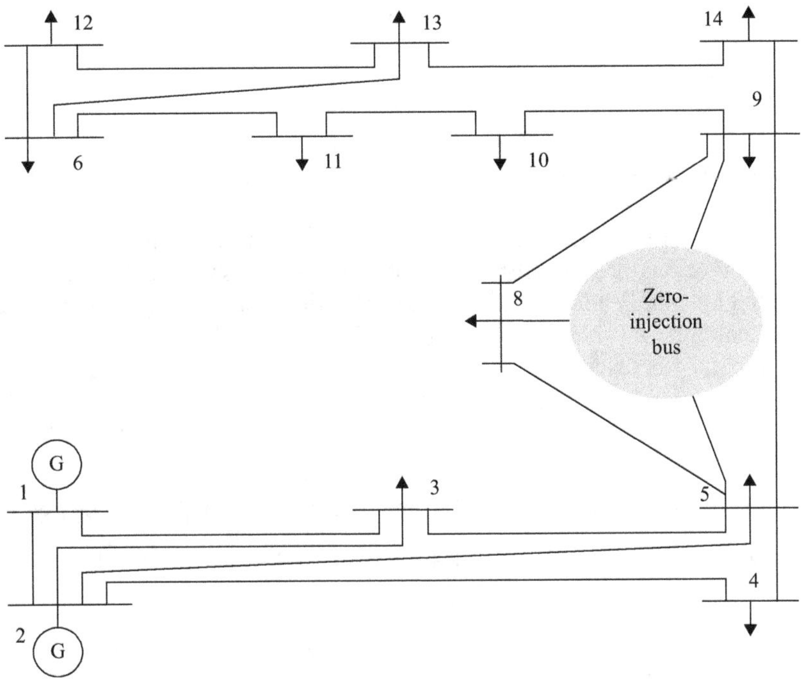

Figure 2.5 IEEE 14 bus system with zero-injection bus

$$\text{Bus } 12: x_6 + x_{12} + x_{13} \geq 1 \tag{2.29}$$

$$\text{Bus } 13: x_6 + x_{12} + x_{13} + x_{14} \geq 1 \tag{2.30}$$

$$\text{Bus } 14: x_9 + x_{13} + x_{14} \geq 1 \tag{2.31}$$

The constraints (2.19)–(2.31) are the updated (or modified) connections obtained from the bus connectivity matrix formed from the modified 14 bus system as shown in Figure 2.5. There are only 13 constraints, as bus 7, which is a ZIB, is removed and thereby reduces the number of PMUs. In this way, the number of PMUs required to maintain complete observability is reduced, if a system has zero-injection nodes.

2.5 System observability redundancy index

Identifying an optimal solution is challenging as there can be multiple solutions for an OPP. To solve this problem, a system observability redundancy index (SORI) is used [15]. If a bus i is observed by a PMU "n" several times, then the SORI (γ) is given by (2.32):

$$\gamma_i = \sum n_i \tag{2.32}$$

The SORI will help in choosing an optimal solution that will provide a higher redundancy enabling complete observability. In general, larger redundancy values yield higher SORI values. Thus, maximizing the SORI is a good measure of success for the OPP as this approach helps keep the system observable during contingencies. For example, let us consider the IEEE 14 bus system in Figure 2.4. The system is made completely observable with 4 PMUs either being placed at 2, 6, 7, and 9 bus locations or at 2, 6, 8, and 9 bus locations. In both cases, all 14 buses are fully observed. Table 2.3 provides a comparison of bus-observabilities for both solutions. Using (2.32), the SORI is calculated by simply adding all bus observabilities in a system. In the first case (second column of Table 2.3) when the PMUs are placed at 2, 6, 7, and 9 locations, the SORI is calculated as follows:

$$SORI = 1 + 1 + 1 + 3 + 1 + 1 + 2 + 1 + 2 + 1 + 1 + 1 + 1 + 1 = 18 \tag{2.33}$$

While in the second case (third column of Table 2.3), the SORI index will be:

$$SORI = 1 + 1 + 1 + 2 + 1 + 1 + 2 + 1 + 1 + 1 + 1 + 1 + 1 + 1 = 16 \tag{2.34}$$

Thus, even though with the same number of PMUs, full observability is obtained in both cases, it is recommended to place the PMUs at buses 2, 6, 7, and 9 as this would yield better results and observability. In addition, a complete redundancy is attained, which is critical for contingency scenarios or events.

Table 2.3 Observability index for various placement solutions in the IEEE 14 bus system

	PMU placement at buses: 2, 6, 7, 9	**PMU placement at buses: 2, 6, 8, 9**
Bus 1	PMU at bus 2	PMU at bus 2
Bus 2	PMU at bus 2	PMU at bus 2
Bus 3	PMU at bus 2	PMU at bus 2
Bus 4	PMU at buses 2, 7, and 9	PMU at buses 2 and 9
Bus 5	PMU at bus 2	PMU at bus 2
Bus 6	PMU at bus 6	PMU at bus 6
Bus 7	PMU at buses 7 and 9	PMU at buses 8 and 9
Bus 8	PMU at bus 7	PMU at bus 8
Bus 9	PMU at buses 7 and 9	PMU at bus 9
Bus 10	PMU at bus 9	PMU at bus 9
Bus 11	PMU at bus 6	PMU at bus 6
Bus 12	PMU at bus 6	PMU at bus 6
Bus 13	PMU at bus 6	PMU at bus 6
Bus 14	PMU at bus 9	PMU at bus 9

2.6 Optimal redundancy criterion

Although the OPP yields solutions for full observability, there is a likelihood of uncertainty during contingency situations, which may result in partial or zero observability of the system. Contingency here refers to a situation in the power system where,

- There is a loss of PMU measurements from the observing bus. This would result in the loss of PMU measurements from all the connected buses.
- There is a loss in the communication channel that is transmitting information from the observing bus.
- There is a loss of PMU measurements from the observable bus.

The above contingency cases can be caused by physical damage to equipment and infrastructure, as well as cyber-physical damages. This means, in certain cases, it is better to include a certain number of buses to be observed by multiple PMUs. Thus, the loss of PMU measurements from one device will not affect the observability of any other bus. The factor of percentage or observability weights of buses can be considered in different ways. Some literature studies consider high-voltage buses to be more critical than other buses, whereas others consider all generating buses to be critical buses [16]. We propose a new index called ORC that is based on the number of connections to each bus. ORC is defined as the ratio of the total number of connections between buses to the total number of buses in the system. We can select the buses that have more connections or links by adapting the following criterion (2.35):

$$ORC = \frac{\sum_{i=1}^{n}\sum_{j=1,i\neq j}^{n} B_{i,j}}{n} \tag{2.35}$$

where i and j represent the buses, $B_{i,j}$ represent connections between i and j, and n is the total number of buses in the system. The ORC conditions are enforced as follows:

Step 1: ORC is calculated using the formula provided in (2.35).
Step 2: Based on ORC, the critical bus constraints are selected from all the constraints.
Step 3: Increase the minimum observability requirement of the selected critical buses.

Let us consider constraints (2.5)–(2.18), where each constraint represents a bus in the IEEE 14 bus system. Here, the total number of connections for all 14 buses is the sum of their connections.

$$B_1 + B_2 + \ldots B_{14} = 56 \tag{2.36}$$

The total number of buses in the test system are 14. Thus, ORC for the 14 bus system is given by

$$ORC = \frac{56}{14} = 4 \tag{2.37}$$

Therefore, from the constraints formed earlier, we consider (2.6), (2.8), and (2.13), and increase the weights, as these buses have a greater number of connections than ORC values (five, six, and five connections, respectively). The weights in constraints appear on the right-hand side of the equation. Now, the updated constraints would be as shown in (2.38)–(2.40).

$$\text{Bus}\,2 : x_1 + x_2 + x_3 + x_4 + x_5 \geq 2 \tag{2.38}$$

$$\text{Bus}\,4 : x_2 + x_3 + x_4 + x_5 + x_7 + x_9 \geq 2 \tag{2.39}$$

$$\text{Bus}\,9 : x_4 + x_7 + x_9 + x_{10} + x_{14} \geq 2 \tag{2.40}$$

For full observability, a "1" is placed on the right-hand side of the constraint equation of each critical bus, indicating that each bus is observed at least once. Based on system and individual utility needs or requirements, one can adjust the observability weights. For our test system, we chose to make the critical buses observable twice. This will provide sufficient redundancy and help increase the reliability of the system.

2.7 OPP on standard test systems

This section discusses the results obtained for the OPP using LP approach on Standard IEEE test cases. To validate the approach, the following test cases have been considered:

1. IEEE 14 bus system
2. IEEE 30 bus system
3. IEEE 57 bus system
4. IEEE 118 bus system
5. SRIPG 208 bus system
6. IEEE 300 bus system

All the required data for IEEE test cases have been taken from the University of Washington webpage [14, p. 14]. The data for the SRIPG test system have been considered from [17]. A Mathematical Programming Language (AMPL) integrated development environment (IDE) is used to develop LP models. IBM ILOG CPLEX Optimization Studio solver is used to solve the formulated linear constraints obtained from the test cases.

2.7.1 A Mathematical Programming Language

AMPL is an algebraic modeling language for mathematical programming. It was designed and implemented in 1985 and has been evolving ever since [18]. This is a sophisticated modeling tool that supports the entire optimization modeling life-cycle: development, testing, deployment, and maintenance. AMPL integrates

1. Modeling language for describing optimization data, variables, objectives, and constraints.
2. A command language for browsing models and analyzing results and a scripting language for gathering and manipulating data and for implementing iterative optimization schemes. All use the same concepts and syntax for streamlined application-building.

The main advantages using AMPL are as follows:

1. It has a general and easy to use syntax for arithmetic, logical, and conditional expressions. It uses familiar conventions for summations and other iterated operators.
2. A wide range of solvers such as CPLEX, Gurobi, Xpress (Linear and convex quadratic solvers), CONOPT, Ipopt, KNITRO, MINOS, SNOPT (nonlinear solvers), Bonmin, Couenne, KNITRO (mixed-integer solvers) are available. In addition, open-source solvers can be hooked to AMPL.

AMPL is notable for the similarity of its arithmetic expressions to customary algebraic notation and for the generality and power of its set and subscripting expressions. AMPL also extends algebraic notation to express common mathematical programming structures such as network flow constraints and piecewise linearity. AMPL has a flexible interface that enables several solvers to be available at once so a user can switch among solvers and select options that may improve and optimize the solution. All of the general set and arithmetic expressions of the AMPL modeling language can also be used for displaying data and results; a variety of options are available to format data for browsing, printing reports, or preparing input to other programs. Using AMPL, mathematical algebraic problems can be directly converted into a small system of inequalities and an objective function. The availability of an algebraic modeling language makes it possible to emphasize the kinds of general models that can be used to describe large-scale optimization problems.

The model (.mod) files are fed into the AMPL program that works like a compiler. The model and the input data are put into a format that is understood by the solver. The solver actually finds an optimal solution to the problem by reading the intermediate file produced by AMPL and applying an appropriate algorithm. Based on the topology information, the constraint formulations can be directly embedded into the AMPL. In this work, the methodology of formulating the algebraic topological constraints in AMPL and CPLEX linear optimization method is used to solve the problem.

2.7.2 Zero-injection buses

Table 2.4 shows the various ZIBs in the test cases considered. It provides various bus numbers that act as ZIBs in the test systems to model the zero-injection constraints for the OPP.

Table 2.4 Zero-injection buses

Bus systems	Zero-injection buses	Number of zero injections
14	{7}	1
30	{6,9,11,25,28}	5
57	{4,7,11,21,22,24,26,34,36,37,39,40,45,46,48}	15
118	{5,9,30,37,38,63,64,68,71,81}	10
300	{4,7,12,16,19,24,34,35,36,39,42,45,46,60,62,64,69,74,78,81,85,86,87, 88,100,115,116,117,128,129,130,131,132,133,134,144,150,151, 58,160,164,165, 66,168,169,174,193,194,195,210,212,219,226, 237,240,244,1201,2040,9001,9005,9006,9007,9012,9023,9044}	65

Table 2.5 Optimal number of PMUs without zero injections

Test systems	ILP method	Other methods
IEEE 14 bus	4 {2,6,7,9}	4
IEEE 30 bus	10 {1,5,6,9,10,12,15,18,25,27}	10
IEEE 57 bus	17 {2,6,12,19,22,25,27,32,36,39,41,45,46,49,51,52,55}	17
IEEE 118 bus	32{1,5,9,12,15,17,21,23,29,30,36,40,44,46,50,51,54,62,64, 71,75,77,80,85,86,90,94,101,105,110,115,116}	32
SRIPG 208 bus	55{5,6,7,12,22,25,27,28,30,33,47,48,50,54,58,60,63,67,73, 75,88,89,92,95,99,104,107,110,111,115,117,118,120,122,127, 128,131,132,136,138,145,146,152,154,157,159,163,165,181, 183,188,190,191,196,201}	58
IEEE 300 bus	87{1,2,3,11,12,15,17,20,23,24,26,33,36,39,43,44,49,55,57,61,62,63, 70,71,72,74,77,81,84,86,97,102,104,105,108,109,114,119,120,122, 130,132,133,134,137,139,140,143,153,154,159,160,164,166,173,178, 181,184,189,194,204,208,210,211,214,217,221,225,229,231,233,234, 237,238,240,244,245,249,9003,9004,9005,9006,9006,9006, 9006,9006,9533}	87

The LP formulations are carried out for all the test systems in AMPL. First, solutions are obtained with normal test cases without considering any ZIBs. Second, zero-injection constraints are added as additional constraints to study the number of PMUs required. Then, an ORC condition is enforced on both the scenarios (i.e., with and without zero injection) and bus constraints to determine the reliability of the model. Table 2.5 shows the number of PMUs required for the test systems without zero-injection constraints. Column two in Table 2.5 identifies the number of PMUs required for each of the test systems. The table also lists the bus

Table 2.6 Optimum number of PMUs while considering zero-injection constraints

Test systems	**ILP method**	**Other methods**
IEEE 14 bus	3 {2,6,9}	3
IEEE 30 bus	7 {2,3,10,12,19,24,27}	7
IEEE 57 bus	10 {1,13,18,25,29,32,38,41,51,54}	11
IEEE 118 bus	27{2,8,11,12,19,22,27,28,32,34,40,45,49,53,56, 62,70,76,77,80,85,87,91,94,101,105,110}	28
SRIPG 208 bus	N/A	N/A
IEEE 300 bus	68{1,2,3,11,15,21,23,26,33,43,44,49,55,57,61,63,664,70, 72,73,77,7071,97,102,104,105,108,109,7166,114,119,120, 122,126,137,139,140,143,152,155,167,175,178,183,184,188, 198,204,205,214,217,223,225,229,231,233,234,238,245,249, 7017,7039,9002,9003,9004,9006,9006,9006}	N/A

locations, on which the PMUs needed to be placed, and column three shows the optimal number of PMUs obtained in various literatures.

By comparison of columns two and three, the minimal set of PMUs required for the test cases yielded similar results that have been previously proposed in the literature [19]. However, the SRIPG system seems to provide a better result of 55 PMUs when compared to the results provided in [20]. Table 2.6 provides the results obtained for the OPP when ZIBs are considered. Generally, the number of PMUs required decreases with zero-injection considerations due to the increase in the constraints in the LP approach and this can be seen in Table 2.6.

While considering ZIBs for solving the OPP for PMUs, there is a significant decrease in the number of PMUs required. This can be seen by comparing Tables 2.5 and 2.6. Using the LP formulations, the number of PMUs required for complete observability is smaller than the solutions provided in the literature (IEEE 57 and 118 bus systems). To the best of the authors' knowledge, no study has solved the OPP for the IEEE 300-bus system by considering ZIBs. This is carried out as an additional task in this research. This research has successfully implemented LP formulations in AMPL on the IEEE 300 bus system to obtain the minimal set of PMUs required.

2.7.3 Enforcing optimal redundancy criterion

Another case was to find the optimal number of PMUs with ORC conditions enforced. Table 2.7 shows the result when the ORC condition is enforced. This is done simply by modifying the right-hand side of the inequality constraint. Here, the redundancy of the critical buses is set to 2. By enforcing ORC, all critical buses are observed at least twice. This means that any failure in PMU measurements will not affect the system's observability. For more clear observation of ORC's performance, consider the results in Table 2.8. Column two from Table 2.8 shows the minimum number of PMUs required, and column three shows the number of PMUs required with ORC enforced. Column four gives us an idea of the percentage

Table 2.7 Optimal number of PMUs enforcing ORC

Test systems	Without zero injections with ORC	With zero injections with ORC
IEEE 14 bus	5 {1,4,8,10,13}	4 {4,5,10,13}
IEEE 30 bus	10 {1,2,6,9,10,12,15,18,25,27}	8{2,3,10,12,13,19,23}
IEEE 57 bus	19{1,4,6,9,10,11,15,20,24,25,28,32, 36,38,41,46,49,53,57}	13{3,6,13,16,17,20,25,29,32,38, 41,51,54}
IEEE 118 bus	37{3,5,9,12,15,17,19,22,23,26,28,34, 37,40,45,49,52,56,59,61,66,68,71, 75,77,80,85,86,90,93,94,100,104, 107,109,110,114}	29{3,8,12,15,19,21,27,29,32,34, 42,45,49,53,56,57,62,70,75,77,80, 85,86,90,94,100,101,105,110}
SRIPG 208 bus	60{1,6,9,12,20,24,25,27,28,30,33, 47,48,50,55,57,58,60,63,67,73,75, 83,88,89,92,95,99,102,103,107,110, 111,115,117,118,120,122,127,128, 132,136,138,142,145,146,152,153, 154,157,159,163,165,181,184,188, 190,191,196,201}	N/A
IEEE 300 bus	96{1,2,3,11,12,13,15,17,21,23,24,26, 323,39,40,41,42,43,44,55,57,61, 62,63,69,70,71,72,74,77,84,88,94, 102,104,105,108,109,110,117,119, 120,122,127,130,132,133,134,137, 139,145,146,148,153,154,160,164, 166,173,175,177,184,188,189,194, 196,205,206,210,211,214,217,219, 221,225,229,231,232,236,237,238, 240,245,246,249,7049,9003,9004, 9005,9006,9006,9006,9006,9006, 9006,9006}	76{1,2,3,11,13,15,17,21,23,26, 37,41,43,44,55,57,61,63,664, 70,71,72,73,77,94,102,104,105, 108,109,110,7166,119,120,121, 125,126,137,139,140,145,148, 153,156,163,167,173,176,180, 184,188,196,198,201,206,214, 217,221,225,229,231,233,234, 238,245,247,249,7039,7049, 9002,9003,9004,9006,9006, 9006,9006}

Table 2.8 Number of PMUs with ORC enforced without zero-injection buses

Bus system	Number of PMUs for full observability	Number of PMUs with ORC	Percentage of increase in PMUs with ORC
IEEE 14 bus system	4	5	25
IEEE 30 bus system	10	10	0
IEEE 57 bus system	17	19	11.764
IEEE 118 bus system	32	37	15.625
SRIPG 208 bus system	55	60	7.272
IEEE 300 bus system	87	96	10.344

increase in the number of PMUs when ORC is enforced. It is observed that with around a 10% increase in the number of PMUs, more redundancy in measurements is seen.

A similar table is prepared for test cases while considering ZIBs. Table 2.9 shows the percentage of increase in the number of PMUs considering zero-injection constraints.

2.7.4 SORI

Table 2.10 shows the SORI obtained for various test systems considering both with and without zero-injection constraints. It also shows a comparison of the results obtained through the proposed approach and other methods used in the literature [21]. It is seen that the LP-formulated SORIs are higher than the ones in the existing literature. This indicates that using the LP approach, the test case systems are much more redundant and the full observability of the buses is higher, compared to the values indicated in other literature.

2.7.5 PMU placement costs

One of the main objectives of the PMU placement problem is maintaining the reliability in measurements of system parameters with the least cost to utility companies. Table 2.11 lists the approximate cost of placing a PMU in the system. Based on the approximate costs, the costs associated with the PMU placement for test systems are estimated. Table 2.11 provides the cost information for a utility to

Table 2.9 Number of PMUs with ORC enforced with zero-injection buses

Bus system	Number of PMUs or full observability	Number of PMUs with ORC	Percentage of increase in PMUs with ORC
IEEE 14 bus	3	4	25
IEEE 30 bus	7	8	14.29
IEEE 57 bus	10	13	30
IEEE 118 bus	27	29	7.41
SRIPG 208 bus	N/A	N/A	N/A
IEEE 300 bus	68	76	11.76

Table 2.10 System observability redundancy index

Bus system	SORI			
	Without zero injections	With zero injections	Without zero injections	With zero injections
	ILP method		Other methods	
IEEE 14 bus	18	14	19	15
IEEE 30 bus	51	46	N/A	N/A
IEEE 57 bus	72	52	72	61
IEEE 118 bus	167	182	164	152
SRIPG 208 bus	320	NA	N/A	N/A
IEEE 300 bus	427	355	N/A	N/A

Table 2.11 Costs associated with PMU placement

Test system	Without zero injections (in US$)	Without zero injections with ORC (in US$)	With zero injections (in US$)
IEEE 14 bus	96,000	96,000	74,000
IEEE 30 bus	228,000	228,000	162,000
IEEE 57 bus	382,000	426,000	228,000
IEEE 118 bus	712,000	822,000	602,000
SRIPG 208 bus	1,226,000	1,336,000	NA
IEEE 300 bus	1,938,000	2,136,000	1,512,000

Table 2.12 System observability redundancy index with ORC

Test bus systems	Without ORC	With ORC
IEEE 14 bus system	18	18
IEEE 30 bus system	51	51
IEEE 57 bus system	72	85
IEEE 118 bus system	167	215
SRIPG 208 bus system	320	358
IEEE 300 bus system	427	490

place the PMUs in a grid. Column two shows the dollar amount required for placing PMUs without considering zero injections, whereas the third column indicates the amount required without considering zero injections and ORC. The fourth column shows the dollar amount required for systems considering zero injections. By observation, using zero-injection constraints in the LP formulations, the total cost could be reduced significantly. The ORC increases the cost of PMUs by a fraction of 9 (for larger test systems). Nevertheless, the cost increase is judiciously justified by trading reliability for full observability of critical infrastructure in the power grid.

2.7.6 Justification of ORC index

ORC increases the total cost of an electric utility. However, it also increases the redundancy in observability of the critical buses required to keep a system stable. Table 2.12 shows an increase in SORI values with ORC enforced. For example, the IEEE 57 bus system requires 17 and 19 PMUs for full observability with and without enforcing ORC respectively as seen in Table 2.8. Compared to Table 2.12, the SORI rose from 72 to 85 in the same system when ORC was enforced. This indicates that with just an increase of two PMUs in the system, full observability is increased to around 13 buses in the 57-bus system, thus increasing the observability for critical buses in the system. The other test cases also show positive results with an increased value in SORI. Table 2.8 shows that the number of PMUs required for full observability with and without ORC index for an IEEE 118 bus system is 32 and 37, respectively. Now, let us consider Table 2.13 for some contingency conditions in the IEEE 118 bus system.

Table 2.13 Contingency cases in IEEE 118 bus system with and without ORC

Scenario	Critical buses	PMU failures (number of unobserved buses)	# of Unobservable buses	Percentage of un-observability without ORC	PMU placed buses with ORC enforced (critical buses are observed at least twice)	Percentage of un-observability with ORC
1	37	Case 1: PMU at bus **40**: {37,39,40,41,42}	5	4.237	Bus 37 is observed thrice :{34,37,40}	0
2	37,59	Case 1: PMU at bus **40**: {37,39,40,41,42} Case 2: PMU at bus **54**: {49,53,54,55,56,59}	11	9.32	Bus 37 is observed thrice :{34,37,40} Bus 59 is observed thrice :{56,59,61}	0
3	37,59,105	Case 1: PMU at bus **40**: {37,39,40,41,42} Case 2: PMU at bus **54**: {49,53,54,55,56,59} Case 3: PMU at bus **105**:{104,105,106,107,108}	16	13.55	Bus 37 is observed thrice :{34,37,40} Bus 59 is observed thrice :{56,59,61} Bus 105 is observed twice :{104,108}	0
4	37,59,92,105	Case 1: PMU at bus **40**: {37,39,40,41,42} Case 2: PMU at bus **54**: {49,53,54,55,56,59) Case 3: PMU at bus **105**:{104,105,106,107,108} Case 4: PMU at bus **94**:{92,93,94,95,96,100}	22	18.6	Bus 37 is observed thrice :{34,37,40} Bus 59 is observed thrice :{56,59,61} Bus 105 is observed twice :{104,108} Bus 92 is observed twice :{93,94}	0
5	5,37,59,92,105	Case 1: PMU at bus **5**:{3,4,5,6,8,11} Case 2: PMU at bus **40**: {37,39,40,41,42} Case 3: PMU at bus **54**: {49,53,54,55,56,59} Case 4: PMU at bus **105**:{104,105,106,107,108} Case 5: PMU at bus **94**:{92,93,94,95,96,100}	28	23.72	Bus 5 is observed twice :{3,5} Bus 37 is observed thrice :{34,37,40} Bus 59 is observed thrice :{56,59,61} Bus 105 is observed twice :{104,108} Bus 92 is observed twice :{93,94}	0
6	5,12,37,59,92, 105	Case 1: PMU at bus **5**:{3,4,5,6,8,11} Case 2: PMU at bus **12**:{2,3,7,11,12,14,16,117} Case 3: PMU at bus **40**: {37,39,40,41,42} Case 4: PMU at bus **54**: {49,53,54,55,56,59} Case 5: PMU at bus **105**:{104,105,106,107,108} Case 6: PMU at bus **94**:{92,93,94,95,96,100}	36	30.5	Bus 5 is observed twice :{3,5} Bus 12 is observed twice :{3,12} Bus 37 is observed thrice :{34,37,40} Bus 59 is observed thrice :{56,59,61} Bus 105 is observed twice :{104,108} Bus 92 is observed twice :{93,94}	0
7	5,12,17,37,59, 92,105	Case 1: PMU at bus **5**:{3,4,5,6,8,11} Case 2: PMU at bus **12**: {2,3,7,11,12,14,16,117} Case 3: PMU at bus **17**:{15,16,17,18,30,31,113} Case 4: PMU at bus **40**: {37,39,40,41,42} Case 5: PMU at bus **54**: {49,53,54,55,56,59} Case 6: PMU at bus **105**: {104,105,106,107,108} Case 7: PMU at bus **94**:{92,93,94,95,96,100}	43	36.4	Bus 5 is observed twice :{3,5} Bus 12 is observed twice :{3,12} Bus 17 is observed twice :{15,17} Bus 37 is observed thrice :{34,37,40} Bus 59 is observed thrice :{56,59,61} Bus 105 is observed twice :{104,108} Bus 92 is observed twice :{93,94}	0

Scenario 1: During normal operating conditions, a PMU at bus location 40 in the IEEE 118 bus system will observe buses 37, 39, 40, 41, and 42 . If the PMU at bus 40 fails for some reason (weather, planned, or unplanned failure) to observe bus 37, then this failure will affect the observability of seven other buses that are interconnected to bus 37. This can be visually understood with the help of case 1 in Figure 2.6. This is the main reason why bus 37 is selected as a critical bus when ORC is enforced. On application of ORC, the critical bus 37 will be observed by three other PMUs as shown in Figure 2.7 and Table 2.13.

Scenario 2: Similarly, under normal conditions, PMUs at buses 40 and 57 will observe buses 37, 39, 40, 41, 42 and 49, 53, 54, 55, 56, 59, respectively. However, if any of these PMUs fails, there will be an unobservability in the IEEE 118 bus system including two critical buses 37 and 59 that are interconnected to 14 other buses. This counts toward 9.32% unobservability in the total system. But when ORC is enforced, this unobservability is prevented. Both the PMUs at 37 and 59 will observe the critical buses redundantly zeroing down the unobservability from 9.32%. Several case studies are conducted using the IEEE 118 bus system as shown in Figure 2.6 in which the reliability of 100% is obtained by enforcing ORC in our formulations. There are many other scenarios in which similar contingency cases can happen, and applying the ORC rule will prove useful and reliable. The contingency scenarios are listed in Table 2.13 with 1, 2, 3, 4, 5, 6, and 7 number of PMU failures. Table 2.13 provides the following information:

1. What if certain PMU(s) are lost and which buses are getting affected?
2. If there is a loss of a critical bus, which buses are unobservable?
3. Percentage of unobservability due to the loss of PMU.
4. Reliability results with ORC are enforced.
5. Percentage of unobservability during the loss of PMU(s), and when ORC is enforced.

This is another example that demonstrates the effectiveness of ORC index in assisting the utilities to keep the full observability of critical buses.

2.7.7 Time computation using LP

In addition, the LP run time is calculated by measuring the computational run time (CPU) for each of the test cases. Table 2.14 shows the time taken by the LP method using AMPL and CPLEX solvers. The computational results of the LP approach with and without ORC were compared with existing works of literature [15]. From Table 2.14, it is seen that it takes virtually 0 s to compute the optimal solutions for all test cases. The other methods take some measurable amount of time to find the optimal solutions. All the test cases ran on the following configured PC system:

- Processor: Intel Xeon W3503 @ 2.4 GHz
- Installed memory: 8 GB
- Hard drive: 465 GB

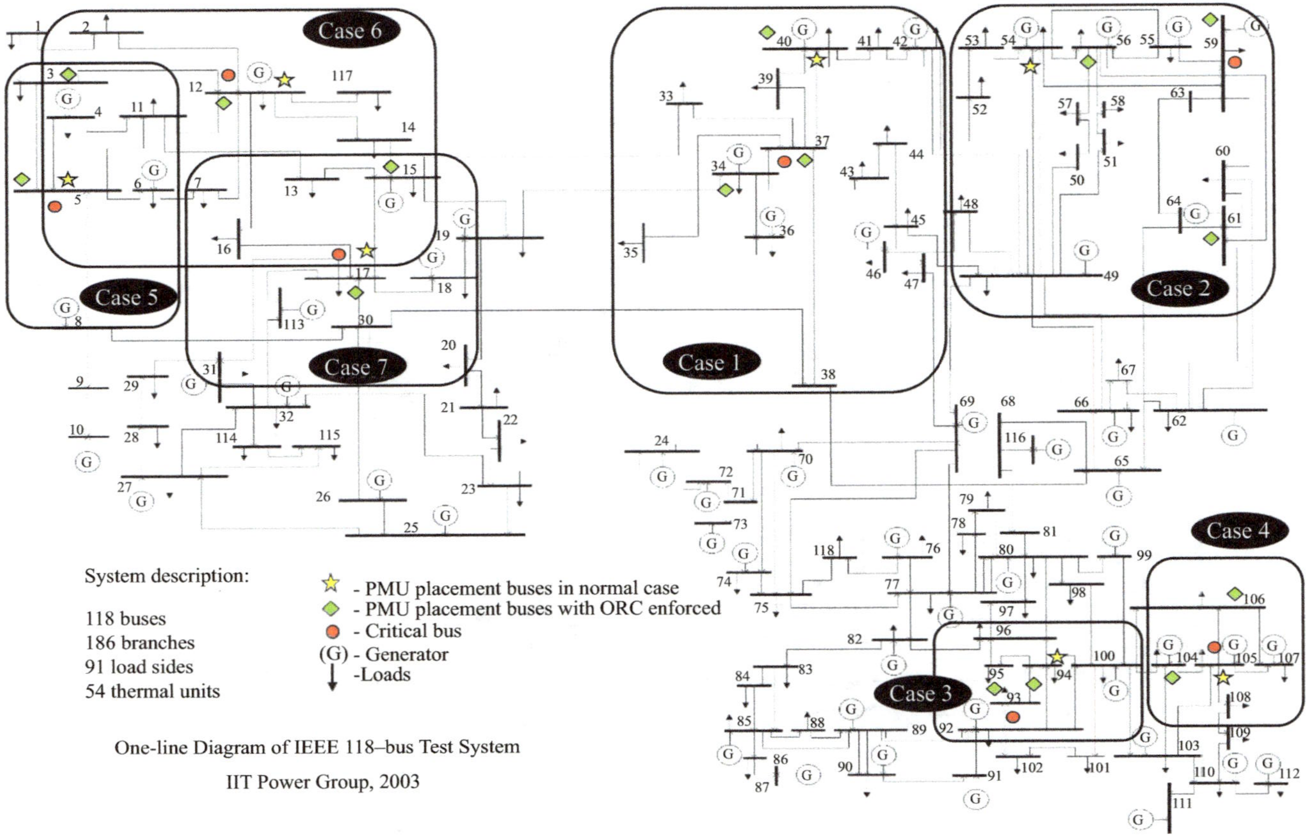

Figure 2.6 IEEE 118 bus system with contingency cases

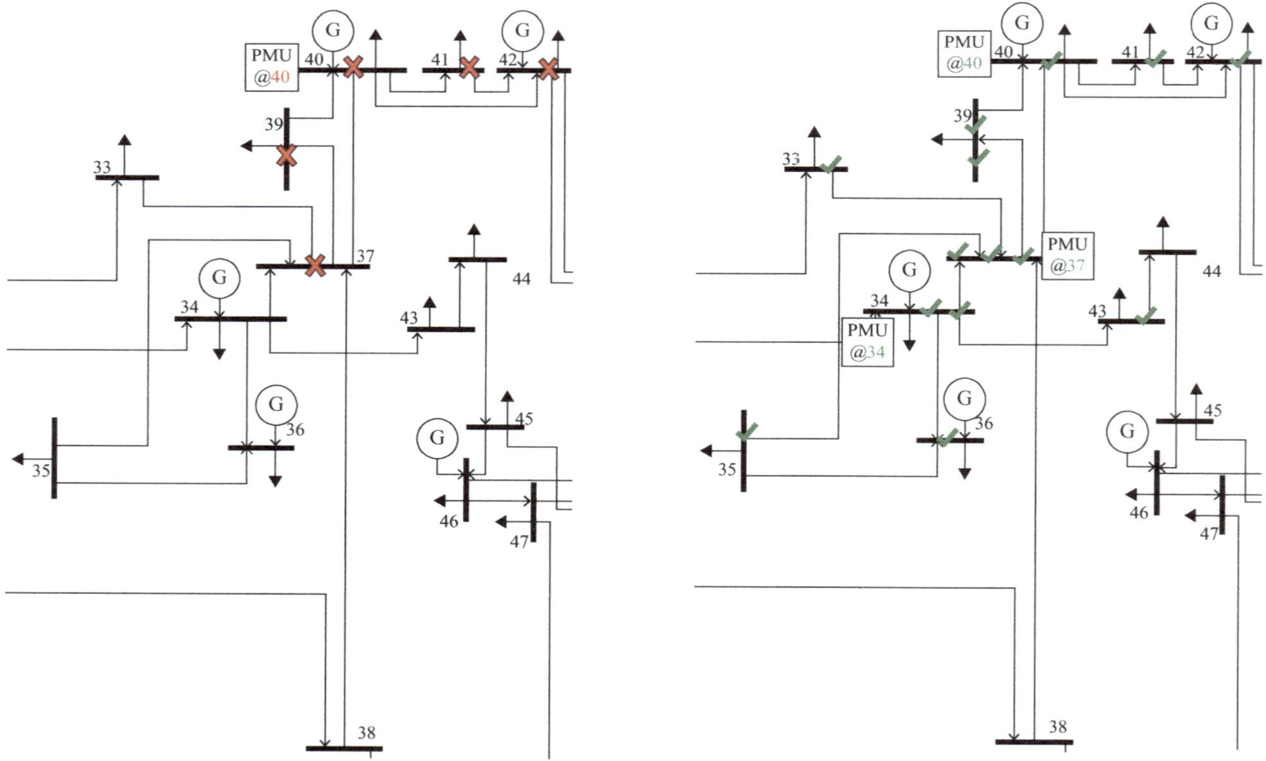

Figure 2.7 *Case (1): loss of PMU at bus 40 (left) and with ORC enforced (right)*

Table 2.14 Computation time for LP method

Test bus systems	ILP method (milliseconds)	ILP method with ORC (milliseconds)	Other methods [22] (milliseconds)
IEEE 14 bus	0.00	0.00	660
IEEE 30 bus	15.6	15.6	830
IEEE 57 bus	15.6	31.2	870
IEEE 118 bus	15.6	46.8	1,340
SRIPG 208 bus	15.6	46.8	–
IEEE 300 bus	15.6	46.8	–

2.7.8 Forecasting the scalability of OPP using linear regression

In [22], the authors have concluded that only one-fourth to one-third of the system buses need to be provided with PMUs to make a system fully observable. Based on our LP results, the estimation of the number of PMUs required for random bus systems was done by utilizing the well-known forecasting linear regression (LR) method. LR is a statistical approach for modeling the relationship between the dependent and independent variables. It is a simple, yet powerful approach to predict the value of a variable based on the value of another variable.

Using the technique of LR with least squares method, the number of PMUs required for different bus systems is estimated using (2.41)–(2.43):

$$Y = \alpha + \beta * X \tag{2.41}$$

where

$$\alpha = \overline{Y} - \beta * \overline{X} \tag{2.42}$$

$$\beta = \frac{\sum_{i=1}^{n} \left(X_i - \overline{X}\right) * \left(Y_i - \overline{Y}\right)}{\sum_{i=1}^{n} \left(X_i - \overline{X}\right)^2} \tag{2.43}$$

$\overline{X}$ and $\overline{Y}$ are the averages of X and Y variables, respectively. The values of X and Y have been considered from our results in Tables 2.5 and 2.6. X refers to the number of buses in the systems, whereas Y refers to the number of PMUs required. Table 2.15 provides the estimation of the PMUs required concerning the number of buses in the system.

Column two shows the number of PMUs estimated through LR, whereas column three provides the actual results obtained through the ILP approach. From column four, the percentage of full observability in any system is around 30%, which

Table 2.15 Estimation of number of PMUs required

No. of buses in a system (X)	Minimum observability		Percentage of observability (no. of PMU placed buses in the system)
	Estimated no. of PMUs using linear regression (Y)	Through ILP approach	
7	2	2	28.5
14	4	4	28.5
30	9	10	30
57	16	17	28.07
100*	28	*	28
118	33	32	27.11
200*	56	*	28
208	58	55	26.44
300	84	87	29
500*	140	*	28
1000*	280	*	28
1500*	421	*	28.06
2000*	561	*	28.05
3000*	842	*	28.06

* No bus systems are practically available, so there are no solutions.

indicates that approximately 30% of the buses in a system need to be equipped with PMUs to make the system fully observable.

2.8 Latest research and developments

The recent study [23] in the field of ORC-based optimal PMU placement has provided new methodologies/algorithms. For example, a branching optimal PMU placement method yielding feasible solutions that are not unique for the ideal lines is proposed in [24] to foster power system observability. Similarly, a method based on nonlinear function programming has been proposed and recommended to determine the optimal PMU placement with contingency considerations [23]. The authors in [25] present a novel strategy using conventional measurements to visualize the phasor placement strategy within a power system. One of the great challenges in determining the state of a power system is due to these multi-level noise signals that occur during changing patterns of power flow [26]. It was tackled by finding the PMU placements with reliabilities set levels [27] using a genetic algorithm (GA).

2.8.1 Case studies

This [24] study offers a unique positioning of PMUs in the grid systems to ensure system monitoring and redundancy. It is based on a topology-centric approach that focuses on the binary connectivity matrix model to investigate all potential solutions. This method is tested in IEEE 14 and 30-bus systems with different conditions. A few best solutions for PMU placement were found that leaf room for the system operator to choose from. The proposed method has much distinct ability to provide multiple multi-objective expanded optimal solutions in one shot, unlike meta-heuristic algorithms. For the IEEE 14 bus system, 6,181 feasible solutions are achieved, while in the case of the IEEE 30 bus system, the test results provide 126,786,561 viable points. It outperformed conventional mathematical optimization and meta-heuristic methods by yielding an extended range of multiple feasible solutions.

While using conventional measurements, the innovative approach for PMU placement guarantees global observability of the power system provided by the authors in [25]. They made a novel criterion for observability as well as a network conversion technique that evolved to improve the solutions of PMU placement. They have used this approach through binary integer linear programming to solve the problem of OPP. The proposed method is capable of identifying the best locations for PMUs where other techniques may fail. As such, it delivers complete network visibility and can be combined with other OPP approaches that consider diverse constraints. Case studies on IEEE 14, 118, and 2736 bus systems are used to demonstrate the proposed heuristic, which is capable of avoiding non-observable solutions with an acceptable computation time. This method was established to be faster and more reliable in execution time and valid compared to other OPP methods.

The work in [28] considers the problem of better localization of PMUs to increase accuracy in SE results. It uses two different mathematical programming approaches, which are mixed-integer linear programming (MILP) and non-linear programming (NLP), to represent power grid observability and solve the online OPP. MILP and NLP also solve the OPP problem with similar performance, whereas MILP requires less CPU time in general. This work extends prior methods to propose novel NLP for zero-injection monitoring. The MILP and NLP methods can reduce the number of PMUs required for the inclusion of power flows and zero injection. More PMU is required in case of either restricted communication or failure of any one type, and this may be necessary in some situations. The suggested zero-injection formulation of NLP with a different number of PMUs is analogous to the MILP, which can be solved with a minimum amount. The effectiveness of the proposed methods was demonstrated with simulations on IEEE bus systems and a 2383-bus large-scale Polish system.

In the [29] study, they use a GA to optimize PMU placement for outage detection. It assesses different placement strategies of PMUs in terms of heat maps and empirically observed probabilities. In addition, the optimal placement of PMUs is important for effective line outage detection and identification. This placement produced by a GA provides a similar performance as full PMU placement for most of the outages. The various types of GA-based placements showed improved detection and classification results over the control random, tree-structured, and degree-based locations. The success rate of GA positioning was almost the same as that achieved by a full deployment, although some lines were not pinpointed with high resolution due to no PMUs being close. The study also observes a point of diminishing returns in the number of PMUs, suggesting that performance falls below an acceptable level below a certain threshold.

The study [23] tried to improve the placement of Distribution-PMUs (D-PMUs) to facilitate intelligent monitoring applications over secondary distribution networks. The authors utilized a multi-objective optimization model, related to minimizing D-PMUs dependent investment cost, maximizing redundancy, and also guaranteeing observability under contingencies. NSGA-III (non-dominated sorting GA) has proved also an effective optimum solution approach. This trade-off represents a balanced middle ground between the investment cost and reliability of the observation system. It considers changes in network topology and potential single points of failure. Experimental results show that the proposed method provides more accurate and reliable measurements using fewer D-PMUs compared with other available methods. When applied in Nansha, China, it shows impressive results on the average bus observability and network measurement redundancy detection.

The study [27] proposes a GA for optimal PMU placement and its number to maintain the observability and reliability of the system. They offer a method of considering both large power system observability and minimal PMUs for expected levels of reliability. The scheme considers ZIBs and radial buses (RBs) for better network coverage. The method outperformed other techniques when applied to IEEE 14, 30, and NE 39 bus systems and identified optimal PMU placement in different systems. This would increase the system reliability and false discharge rates with a minimal number of PMUs. NE 39 bus system observability is increased from 74.64 to 91.18, indicating the better reliability of the proposed method than without optimal running situations.

In [30], the Adam Eve Genetic Algorithm (AEGA) was investigated and its outcome was compared with a simple genetic algorithm (SGA). The human-implemented SGA possessing a fixed-size population and static cross-over/mutation rates was compared with the AEGA which dynamically alters both population size and probabilities. According to the simulation results, the AEGA has a better chance of converging to the global optimal solution even if the convergence rate is slower than the behavior space.

While considering ZIBs and future network contingencies, the optimum number of PMUs along with optimal locations in the network to provide full observability is determined using the binary dragonfly algorithm (BDA) in [31]. PMU or line failures might disrupt the BDA solution, but it has demonstrated its resilience in solving the OPP issue. This capability is very important in terms of keeping the topology observable during non-standard conditions, and the algorithm considers that ZIBs can affect system observability along with network contingencies. BDA has shown its efficacy in various test systems (9–14, 30, and 39, all the way up to 57 bus BPA systems) where multiple PMUs might be required for full observability when compared with other approaches. One example of this is shown with the 57-bus IEEE system, where BDA needed one to three fewer PMUs compared to other methods. These results demonstrate that the BDA can significantly reduce the investment cost of PMUs while maintaining full observability of the power network.

In [32], it is pointed out that the maintainability of observability in large power grids is becoming an issue with the installation of renewable energy and aging infrastructure. This paper introduced a type of meta-heuristic algorithm—binary accelerated particle swarm optimization (BAPSO) to identify the optimal placement of PMUs. This BAPSO method combines the local and global search strategies to escape from the primary problem of sticking to the local optima as with the standard BPSO algorithm. The BAPSO algorithm can converge faster and provide solutions within a configurable amount of time as compared to the general BPSO algorithm. The method further comprises a scheme for the integration of RBs to reduce PMU usage. The proposed BAPSO method emerges as the best among all other meta-heuristic algorithms in required computational time and several PMUs needed for complete observability of the power grid if the observability of the grid is maintained with a minimum number of PMUs, which can be located on ideal locations to attain maximum utilization for various possible forms of bus systems.

In [33], the two-tier structure is utilized to minimize both the cost and network vulnerability of PMU placement and to research optimal positions and costs using evolutionary algorithms. This work has been implemented in the IEEE bus environment using MATLAB®. The strategy assigns PMU placements to contingencies with a higher probability of enhanced remedies. This research investigates the problem from an antithetical point of view compared to the other methods. It considers minimizing PMUs but not installation costs. This made the desired method far more performant, with computation times of under 2 seconds. Then, it leads a step pragmatic approach for PMU placement, which takes into account the cost and observability of system. The technique may be implemented as one tier (one-stage setup) or in a many tiers (multi-tier arrange).

The authors in [34] examined the traditional, heuristic, and metaheuristic techniques from a number of IEEE (142,430) and New England (3,987,118) testbeds. Consequently, PMU technology, in general, is essential for enabling high-resolution monitoring of power systems to prevent blackouts. The optimal location of PMUs is important because PMUs are expensive and fewer numbers can measure the nearby buses. The paper [35] presents a survey of the most advanced approaches to optimal PMU placement by using both traditional and non-traditional optimization techniques. To evaluate these strengths and weaknesses, optimization algorithms like particle swarm optimization, LP, and GA are done in existing research on PMU placement. A categorization of optimization techniques is presented to classify algorithms. Specifically, so much as it offers an analysis of distinguished algorithms and optimization solvers within the compare-contrast framework, in addition to discussing doable testbed structures. It highlights important research gaps about PMU placement as well.

2.8.2 Challenges and solutions

Fault observability becomes challenging with the evolution from unidirectional power flow to bidirectional power flow in distribution networks, driven by continued capability and scale expansion. The shift makes fault detection and localization highly complex, which is troublesome for the already deployed measurement devices.

The problem is further amplified by signal noise that makes the fault detection less precise and demands higher measurement precision [26]. Although PMUs have been shown to serve as an indispensable measurement for attaining system-wide fault observability, the high expense and deployment restriction of such equipment poses difficulty [35]. There is thus a sophisticated optimization going between the need for observability versus constraints on PMU numbers. The research highlights the minimization of PMU count with accurate fault localization under a wide range of engineering requirements [26,27]. Moreover, a comparison of GAs such as the Adam-Eve algorithm and simple GA presents tradeoffs in terms of global optimality versus convergence speed, thus reiterating that an efficient optimization approach is needed for optimal PMU placement to ensure good fault observability [30].

2.8.3 Impact of advancements in technology

1. **Enhanced reliability and fault tolerance**
   Positioning the PMUs such that they are deployed in critical and susceptible contingencies-prone regions of the power networks is one of the key strategies that go a long way to enhancing the efficiency concerning remedial actions. Therefore, this proactive approach increases the fault tolerance of the network by increasing both preventive and corrective measures. Several studies [36] show a significant enhancement of network reliability achieved with PMU applications. They greatly contribute to enhancing the reliability of power systems and remarkably improved outcomes are witnessed, especially in instances where multiple PMUs have been deployed.
2. **Cost optimization**
   Recent progress in the technology of ideal PMU placement seeks to balance the tradeoff between network vulnerability and PMU installation cost. There are others with a similar aim, based on a bi-level framework to select sites and installation costs using evolutionary algorithms [33].
3. **Improved observability**
   For the optimal PMU placement technology, it aims to keep the full network observability, which means determining "where," but with how many PMUs we should separate on sites. This is achieved by using optimization algorithms such as the BDA which considers network contingencies and ZIBs [31].
4. **Outage detection and identification**
   Advancement in PMU placement technology has a considerable influence on the efficiency of outage detection and identification (D&I) strategies. As pointed out by a study [29], placement of PMU at an optimal location improves the performance of the D&I strategy, and rounded results are observed for both single-line line outages.

2.8.4 Future trends

Researchers are approaching exploiting ORC in searching for Pareto-optimal PMU placement. Fundamentally, in the field of optimal PMUs placement research, which involves optimization and redundancy concepts; significant directions to look up for are rest upon. Because of the global increase in power grids and distributed energy resources, new ORC algorithms and techniques are considered essential. This extension and integration make the problem of PMU placement more complicated and might require new optimized approaches, i.e., distributed algorithms or metaheuristics.

Integrating cybersecurity considerations into ORC models was also considered as a crucial factor. Given this increasing dependence on PMU data for grid monitoring and control, the protection of communications with these devices is essential. In that aspect, more than one North American utility contacted by Network Engineering Services has mentioned it as an issue. ORC may help to place PMUs strategically for high data observability and low-security risk.

Applying ORC to co-optimize PMU placement with ultra-precise phasor measurement units (micro-PMUs) is a problem that also has significant modeling issues. For example, IEDs could be placed instead of any sensor, but if tackled within the broader framework, these should cancel each other out in the analysis. However, the problem of finding optimal combinations and placements of sensors across variant assets is unaddressed. One work proposes the use of new data sources such as phasor measurement data or SCADA readings for improving ORC objectives functions. This might improve measurement redundancy and the elimination of bad data.

One question of interest is to assess whether the ORC-based PMU placement can be resilient against grid disturbances and variations in time. Optimization should ideally be carried out for a wide spectrum of operating conditions and topologies. The final research question should include adaptive ORC strategies for policy development. Those could be able to provide updated advice on the optimal PMU locations dynamically as a grid changes (using synchrophasor data analytics and machine learning techniques). This multi-pronged model may ultimately lead to better, safer management of the power grid.

2.9 Conclusion

This chapter provides an ORC for the placement of synchrophasors and their latest developments with challenges and future aspects. Test cases were evaluated for IEEE systems for scalability, and the criteria provided promising results for the placement problem.

Acknowledgments

The authors gratefully acknowledge financial support from the Division of Research & Economic Development at the University of North Dakota (UND-21418-4010-02294) and National Science Foundation award number 1537565. Dr. Prakash Ranganathan also thanks his graduate students Ranganath Vallakati and Sriram Prabhakara Rao for their contributions.

References

[1] "How Smart Grid Works." [Online]. Available: http://www.nature.com/news/2008/080730/images/454570a-6.jpg [Accessed 01 Jan 2015].

[2] A. G. Phadke, "Synchronized phasor measurements in power systems," *IEEE Comput Appl Power*, vol. 6, no. 2, pp. 10–15, 1993.

[3] B. Mohammadi, A. Rabiei, and S. Mobayen, "Online inter-area oscillation monitoring in power systems using PMU data and Prony analysis," *International Journal of the Physical Sciences*, vol. 6, no. 22, pp. 5267–5272, 2011.

[4] K. S. Shim, S. T. Kim, J. H. Lee, E. J. Choi, and J. H. Choi, "Detection of low-frequency oscillation using synchrophasor in wide-area rolling blackouts," *Int J Electr Power Energy Syst*, vol. 63, pp. 1015–1022, 2014.

[5] K. P. Lien, C. W. Liu, C. S. Yu, and J. A. Jiang, "Transmission network fault location observability with minimal PMU placement," *IEEE Trans Power Deliv*, vol. 21, no. 3, pp. 1128–1136, 2006.

[6] A. Reddy, "PMU based real-time short term voltage stability monitoring – Analysis and implementation on a real-time test bed," in *North American Power Symposium*, Pullman, Washington, 2007.

[7] L. Huang, Y. Sun, J. Xu, W. Gao, J. Zhang, and Z. Wu, "Optimal PMU placement considering controlled islanding of power system," *IEEE Trans Power Syst*, vol. 29, no. 2, pp. 742–755, 2014.

[8] J. De La Ree, V. Centeno, J. S. Thorp, and A. G. Phadke, "Synchronized phasor measurement applications in power systems," *IEEE Trans Smart Grid*, vol. 1, no. 1, pp. 20–27, 2010.

[9] K. E. Martin, G. Brunello, M. G. Adamiak, *et al.*, "An overview of the IEEE standard C37.118.2—synchrophasor data transfer for power systems," *IEEE Trans Smart Grid*, vol. 5, no. 4, pp. 1980–1984, 2014.

[10] P. S. R. Committee and Others, "IEEE guide for phasor data concentrator requirements for power system protection, control, and monitoring," IEEE, Piscataway, NJ, USA, 2013.

[11] E. O. Schweitzer and D. E. Whitehead, "Real-time power system control using synchrophasors," in *2008 61st Annual Conference for Protective Relay Engineers*, IEEE, 2008, pp. 78–88.

[12] M. V. Mynam, A. Harikrishna, and V. Singh, "Synchrophasors redefining SCADA systems," *J Int Assoc Electr Gener Transm Distrib*, vol. 26, no. 2, pp. 22–28, 2013.

[13] "SynchroPhasor Products." [Online]. Available: https://www.selinc.com/synchrophasors/products/ [Accessed 08 Aug 2014].

[14] "IEEE 14 bus system." [Online]. Available: http://www.ee.washington.edu/research/pstca/pf14/pg_tca14bus.htm [Accessed 08 Aug 2014].

[15] S. Dambhare, D. Dua, R. K. Gajbhiye, and S. Soman, "Optimal zero injection considerations in PMU placement: An ILP approach," in *16th PSCC Glasgow*, July 14, vol. 18, 2008.

[16] A. Pal, G. A. Sanchez-Ayala, V. A. Centeno, and J. S. Thorp, "A PMU placement scheme ensuring real-time monitoring of critical buses of the network," *IEEE Trans Power Deliv*, vol. 29, no. 2, pp. 510–517, 2013.

[17] P. Gopakumar, G. S. Chandra, M. J. B. Reddy, and D. K. Mohanta, "Optimal redundant placement of PMUs in Indian power grid—northern, eastern and north-eastern regions," *Front Energy*, vol. 7, pp. 413–428, 2013.

[18] R. Fourer, D. M. Gay, and B. W. Kernighan, "AMPL: A mathematical programming language," *Manag Sci*, vol. 35, no. 5, pp. 519–544, 1990.

[19] M. Alvarez, F. Sellschopp, and E. Vazquez, "A PMUs placement methodology based on inverse of connectivity and critical measurements," *Int J Electr Power Energy Syst*, vol. 68, pp. 336–344, 2015.

[20] P. Gopakumar, G. S. Chandra, and M. J. B. Reddy, "Optimal placement of phasor measurement units for Tamil Nadu state of Indian power grid," in *2012 11th International Conference on Environment and Electrical Engineering*, IEEE, 2012, pp. 80–83.

[21] D. Dua, S. Dambhare, R. K. Gajbhiye, and S. Soman, "Optimal multistage scheduling of PMU placement: An ILP approach," *IEEE Trans Power Deliv*, vol. 23, no. 4, pp. 1812–1820, 2008.

[22] T. L. Baldwin, L. Mili, M. B. Boisen, and R. Adapa, "Power system observability with minimal phasor measurement placement," *IEEE Trans Power Syst*, vol. 8, no. 2, pp. 707–715, 1993.

[23] Y. Sun, W. Hu, X. Kong, Y. Shen, and F. Yang, "Multi-objective optimal D-PMU placement for fast, reliable and high-precision observations of active distribution networks," *Appl Sci*, vol. 12, no. 9, Art. no. 9, 2022, doi:10.3390/app12094677.

[24] M. K. D. Chejarla and S. K. Matam, "Multiple solutions for optimal PMU placement using a topology-based method," *J Inst Eng India Ser B*, vol. 102, no. 2, pp. 249–259, 2021, doi:10.1007/s40031-020-00532-y.

[25] Z. Jin, Y. Hou, Y. Yu, and L. Ding, "Optimal PMU placement in the presence of conventional measurements," *Int Trans Electr Energy Syst*, vol. 2022, p. e8078010, 2022, doi:10.1155/2022/8078010.

[26] Z. Dong, Z. Han, F. Shi, and Y. Liu, "Review of fault observability analysis and optimal PMU placement for distribution networks," in *2023 International Conference on Power System Technology (PowerCon)*, IEEE, 2023, pp. 1–7.

[27] S. S. Ritu and T. Aziz, "Genetic algorithm based optimal PMU placement for enhanced power system observability and reliability," in *2022 2nd Asian Conference on Innovation in Technology (ASIANCON)*, IEEE, 2022, pp. 1–7.

[28] A. Almunif and L. Fan, "Optimal PMU placement for modeling power grid observability with mathematical programming methods," *Int Trans Electr Energy Syst*, vol. 30, no. 2, p. e12182, 2020.

[29] X. Yang, N. Chen, and C. Zhai, "Optimal placement of limited PMUs for transmission line outage detection and identification," in *2020 International Conference on Probabilistic Methods Applied to Power Systems (PMAPS)*, IEEE, 2020, pp. 1–6.

[30] M. Shahraeini and R. Soltanifar, "Performance comparison between simple and Adam—Eve-like genetic algorithms in optimal PMU placement problem," in *2022 8th Iranian Conference on Signal Processing and Intelligent Systems (ICSPIS)*, IEEE, 2022, pp. 1–6.
[31] C. Patel, T. Tailor, S. Shah, and S. Srivastava, "Binary dragonfly algorithm based optimal PMU placement considering contingency constraints," *Int J Adv Res Eng Technol IJARET*, vol. 11, no. 4, pp. 34–43, 2020.
[32] R. Babu, S. Raj, and S. Mahapatra, "Allocation of synchronized phasor measurement units for power grid observability using advanced binary accelerated particle swarm optimization approach," *J Electr Syst Inf Technol*, vol. 10, no. 1, p. 42, 2023.
[33] S. Almasabi and J. Mitra, "A fault-tolerance based approach to optimal PMU placement," *IEEE Trans Smart Grid*, vol. 10, no. 6, pp. 6070–6079, 2019.
[34] M. A. A. Sufyan, M. Zuhaib, and M. Rihan, "Optimal PMU placement for smart grid: A technical case study," in *2020 IEEE 17th India Council International Conference (INDICON)*, IEEE, 2020, pp. 1–7.
[35] M. M. Ahmed, M. Amjad, M. A. Qureshi, K. Imran, Z. M. Haider, and M. O. Khan, "A critical review of state-of-the-art optimal PMU placement techniques," *Energies*, vol. 15, no. 6, p. 2125, 2022.
[36] B. Jimada-Ojuolape, "Reliability enhancement of synchrophasor-based DTR system considering N-1 contingency for PMU placement," in *2021 IEEE International Future Energy Electronics Conference (IFEEC)*, IEEE, 2021, pp. 1–6.

Chapter 3
Secured relaying using synchronised data

Pratim Kundu[1], *Saumendra Sarangi*[2] *and*
Ashok Kumar Pradhan[3]

3.1 Introduction

Industries today depend heavily on electric power. Reliability of electric service is the major concern to electric power industry. Fault in a power system component disrupts the power flow and if not cleared quickly can lead to component damage and other associated problem in the system. Protective relaying is concerned with the design and operation of equipment that detects abnormal power system condition and initiates action as quickly as possible to isolate the faulted component [1,2]. A wide-area disturbance/blackout is the power outage over large geographical region that affects the power system reliability, causing interruption of the electric supply to thousands of users. Analysis of major disturbances has revealed maloperation of protection schemes as the most important reason [3–5].

Wide area measurement system (WAMS) is a promising technology in smart grid initiative. To be able to monitor, operate and control power systems in wide geographical area, WAMS combines the functions of metering devices (i.e. new and traditional) with the abilities of communication systems [6]. Phasor measurement unit (PMU) otherwise known as synchrophasor is a monitoring device that provides accurate phasors, frequency and rate of change of frequency with time tagged. Data from PMUs with dedicated communication system are recorded at a central location known as phasor data concentrator (PDC). Such a measurement system, WAMS, provides a dynamic view of the system and helps deriving accurate protection decision. With use of PMUs, improved monitoring, protection and control of power network are being achieved [7–11]. Over the last decade applications of PMU data in power system have gained importance [12].

Discriminating a fault condition from other stressed power system events such as power swing, load encroachment and voltage instability condition, especially for zone 3 of relay, is a challenging issue. This discrimination helps in blocking relays

[1]Department of Electrical Engineering, Indian Institute of Technology Mandi, India
[2]Department of Electrical Engineering, Motilal Nehru National Institute of Technology Allahabad, India
[3]Department of Electrical Engineering, Indian Institute of Technology Kharagpur, India

and prevents maloperation. Detecting a fault during such events is also essential to clear fault as quickly as possible. Protection scheme operation depends on the collective operation of various components such as relays, circuit breakers, etc. Failure of any component can disrupt the functioning of protection schemes. This leads to inability of the protection scheme to clear the fault and thus affecting the power system stability. Identification of the failed component online can enhance the backup protection to act faster and will prevent mal-function of protection schemes due to unidentified component error.

Distance relay used for series compensated line finds the apparent impedance modulated in accordance with the level of compensation and the operation of metal oxide varistor (MOV) [13]. Sometimes, the series capacitor may be bypassed for maintenance or any other reason. Information on level of compensation is necessary for correct setting of distance relay [14]. Further, the asymmetric MOV operation during a fault introduces coupling between the sequence networks, limiting the application sequence component analysis for computation of the apparent impedance during fault. Further, it is to be noted that the settings of the distance relays can be changed in accordance with the system operating condition [15,16]. Though differential current principle using both ends data improves the selectivity, the channel delay and charging current limit the application of current differential relay (CDR) in protecting a transmission line. Instead of difference of current, the ratio of currents of both ends current in alpha plane relay (APR) provides an alternative to detect the fault. The phase angle and magnitude modulation during high resistance fault and CT saturation cause maloperation of such scheme. However, both ends synchronised data that is collected through dedicated communication channel can be applied in improving the performance of the APR.

In this chapter, various solutions have been provided to prevent maloperation of distance relays under different circumstances using WAMS data. Assessment of a system during stressed condition using information obtained from various system parameters is carried to prevent zone 3 maloperation due to infeed. In this context, features derived from WAMS data are applied to estimate the infeed current and adaptive setting is verified for various other conditions such as power swing and load encroachment. Performance of protection scheme is affected due to malfunctioning of any of its component such as relay, circuit breaker, etc. A method for online checking of protection operation using WAMS data is developed. In addition, identification of any protection component failure is achieved. WAMS data-based adaptive distance relay setting of series compensated and adaptive APR is also provided. The algorithms are tested for various fault and system conditions of power systems.

3.1.1 WAMS architecture

Synchrophasor data concentrators or PDCs are the key building blocks of the WAMS architecture. The primary function of the PDC is to communicate with phasor measurement units (PMUs) and other PDCs, acquire and time align synchrophasor data from various PMUs and transmit an output stream to several clients which can be other PDCs or a super PDC. Figure 3.1 shows the tiered WAMS architecture starting from a remote substation PMUs to control centre super PDCs.

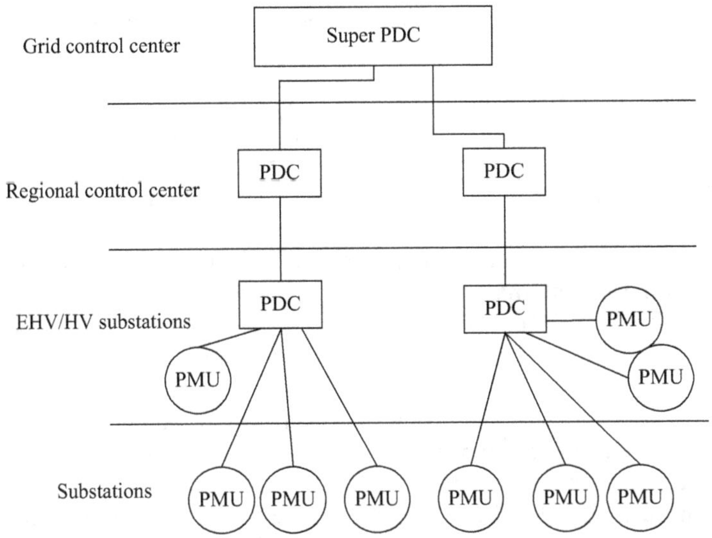

Figure 3.1 Wide area measurement system (WAMS) hierarchy

A synchrophasor system makes measurements throughout a power system. In order to analyse and respond to events occurring in the system, the measured states of the system need to be aligned in time. This function of time alignment is performed by a device known as a PDC. A PDC should be able to communicate with multiple PMUs, align all received data by time stamp, identify missing data, archive the data for post event analysis, aggregate and re- transmit, and decimate and filter the output streams as required by the receiving application. This function is facilitated by a PMU's ability to identify events based on frequency, voltage, current, power and status changes. The PDC system must provide a mechanism to enable visualisation of the information for the purposes of installation, commissioning, diagnostics and post-event analysis. Operator visualisation occurs at the energy management level.

3.1.2 WAMS communication

Smart grid of the future will have PMU data available across the grid. To meet the latency requirements and to handle the huge amounts of data, a real-time information infrastructure was suggested. Because of the huge amount of data generated at each substation, not all the data can be sent to one central location. Separating out application servers will also help to tag packets for latency purpose. The middleware system to handle this distributed data base and to provide the latency and other quality of service (QoS) is one of the major goals of the North American SynchroPhasor Initiative (NASPI) and some research initiatives like Gridstat [17,18].

3.1.2.1 Communication delay causes

Although more and more control systems are being implemented in a distributed fashion with networked communication, the unavoidable time delays in such systems impact the achievable performance. Delays due to the use of PMUs and the communication link involved are due primarily to the following reasons:

1. Transducer delays: Voltage transducers (VT) and current transducers (CT) are used to measure the rms voltages and currents, respectively, at the instant of sampling. Window size of the discrete Fourier transform (DFT): Window size of the DFT is the number of samples required to compute the phasors using DFT.
2. Processing time: The processing time required in converting the transducer data into phasor information with the help of DFT.
3. Data size of the PMU output: Data size of the PMU message is the size of the information bits contained in the data frame, header frame and the configuration frame.
4. Multiplexing and transitions: Transitions between the communication link and the data processing equipment leads to delays that are caused at the instances when data are retrieved or emitted by the communication link.
5. Communication link involved: The type of communication link and the physical distance involved in transmitting the PMU output to the central processing unit can add to the delay.
6. Data concentrators: Data concentrators are primarily data collecting centres located at the central processing unit and are responsible for collecting all the PMU data that is transmitted over the communication link.

This research work focuses on the issues of maloperation of distance relays at various system conditions like power swing and protection component failure. It proposes the use of WAMS data to support protection schemes during such conditions. The latency associated with communication scheme restricts the application of WAMS data for network protection only to backup schemes (slower protection). In this chapter, the maximum latency associated with WAMS data is considered to be 200 ms. Methods developed enhances the backup relay decision during various condition such as fault, changes in system operating condition and any dynamic condition. However, a blackout in a large area may happen due to failure of a protection component. Following such an event detection of such a failure of the component is essential. Hence, method for identification of failure of protection components is obtained using WAMS data. To test the techniques, simulations were carried out for different systems with EMTDC/PSCAD. The sampling rate considered in the study is 1 kHz and phasors are estimated using discrete Fourier DFT technique to obtain synchronised data from the system.

3.2 Adaptive protection using synchrophasor data

Relays in power systems are set according to the certain assumption on normal operating conditions of power system. As the system conditions changes, the relay settings may not be appropriate at that instant which causes its maloperation. The changing system condition requires adaptive setting for reliable operation. With the

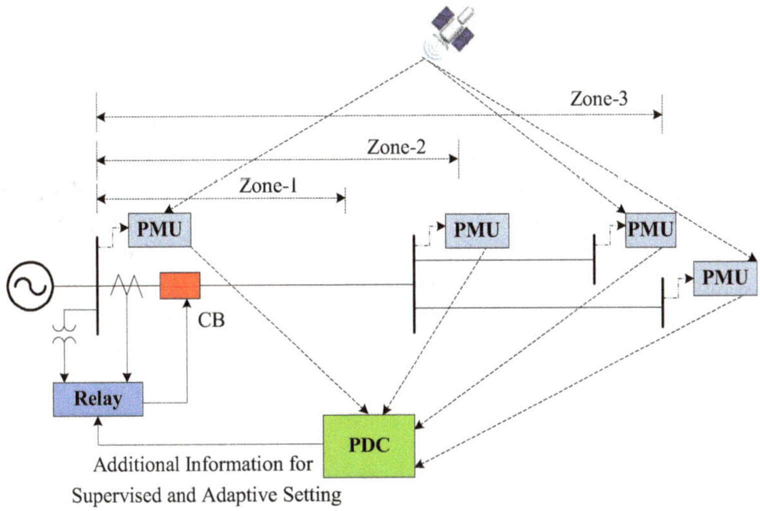

Figure 3.2 Adaptive protection scheme using synchrophasor data

advent of the digital relays, protection functions can be changed to make the settings more appropriate and this is called as adaptive setting. However, the adaptive protection can be applied to improve both the dependability and security of the relay.

As shown in Figure 3.2, distance relay protecting the line, the zone 2 and zone 3 which provides backup operates with time delay, i.e. 15 and 50 cycles, respectively. This delay provides ample time to record the changes during any disturbance using PMUs data collected at the PDC, and using this information relay maloperation can be prevented. However, the communication delays are the major obstacle to implement such type of adaptive protection scheme. Further, if any changes in the network are recorded, such a data can be applied to adapt the setting of relay. With large reach of zone 3, settings of distance relay protecting the series compensated line can be adapted according to the changes in the compensation level. Further, the infeed current may cause the under reaching of the zone 3 relay and the fault remains undetected if the primary fails to detect it. Hence, adaptive zone 3 setting to ensure the security of the system is essential and this can be achieved by using synchronised data available at the PDC.

Unit protection schemes such as CDR and APR which uses both ends data are applied to improve the selectivity. Failure of such a scheme may cause delayed tripping during fault leading to oscillation in the system. To improve the security of the power system under stressed condition prevention of maloperation of both ends of data-based scheme is inevitable and this can be achieved using additional information collected using synchrophasors from both ends of the line. It is to be noted that data collected before and after the fault can be applied to security decision. Information with data synchronisation gathered at PDC or available to the relay, can fetch accurate system condition which can be applied for both instantaneous and delayed protection schemes as shown in Figure 3.2.

3.2.1 Adaptive zone 3 operation

The change in operating conditions of the system affects the relay reach. A protection boundary setting adjustment method is proposed which uses an equivalent Thevenin's model to obtain relay settings required for prevailing system conditions. The proposed method aims at improving security of vulnerable relays during a stressed situation by managing protection operation with relay characteristic adjustment or line loading change to avoid undesirable operations. The method provides the necessary steps to be taken to avoid blocking of the relays, which helps to reduce errors in the decision of the fault during stressed conditions.

A step-distance protection scheme is illustrated in Figure 3.3. The P, S and Q buses of the system are connected to synchronous generators and a renewable-based source is connected at the T-bus. As a backup for relay at the bus S for SR line, the zone 3 setting of relay for the line SR is given by (3.1), where the fault is considered at the end of line RQ.

$$Z_{reach} = Z_A + 1.2 \times Z_{RQ} \times \left(1 + \frac{I_{PR} + I_T}{I_T}\right) \tag{3.1}$$

The apparent impedance of relay R_{SR} depends on the current from buses P and T. Similar calculations are done considering faults at ends of lines TR and PR. The selected setting is the highest value obtained among all the faults considered. Due to infeed, the reach of zone 3 setting becomes greater as compared to a simple two-line configuration consisting of line SR and any one of adjacent line. This results in reduced loadability for the relay. To calculate the current contribution

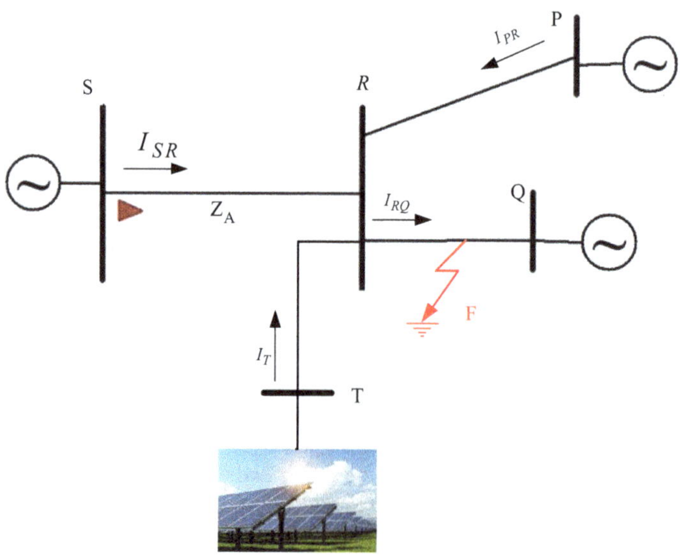

Figure 3.3 Single line diagram of distance relay

from buses P and T, the equivalent source behind the buses is required. The equivalent source voltage (E) and impedance value can be written as follows:

$$E_{TH} = \mathrm{V} + Z_{TH} \times \mathrm{I} \tag{3.2}$$

The source impedance may be changing time to time as per the number of generators connected at any time. Furthermore, weather conditions and converter control methods determine the infeed current from the renewable-based source connected at the T bus. Real-time computation of the infeed current and identification under reaching of zone 3 requires real-time measurement of the system. With synchrophasor data collected at central location, the equivalent source voltages and capacity can be computed. Using this data, the infeed current from all other buses can be computed for a three-phase fault at the end of the reach. However, the infeed current from the inverter-based sources connected may be changed in accordance with the weather condition. Further, the infeed fault current may be limited to 2–2.5 pu. Consider the rated thermal capacity of the converters.

To identify the underreaching effect due to the infeed, the prefault condition is identified by using a search-based technique that uses topology information of the system [19]. The transmission lines serve as the branches and the buses as the nodes in a network graph. The status of the line or lines linked to each bus node is displayed. Information about breaker operation is used to establish the line's status. The buses for which the underreaching must be computed are chosen based on the graph. The node to which the relay is linked is the first bus. We designate the next node as the primary node after proceeding along the branch from the chosen node, where the relay is linked. Only when the branch that connects a node to the primary node is active are nodes that are connected to the primary node picked from the graph. The steps for calculating the reach are as follows:

1. Select set of buses using the graph search approach
2. Obtain measurement data for selected buses
3. Calculate the equivalent source voltage using (3.2)
4. Z_{reach} is calculated using (3.1) considering fault close to reach of the selected bus
5. Repeat step (3.4) for all the selected buses and select the highest value as the setting for the relay.

3.2.1.1 Case study

The initial setting for the relay at bus 28 is obtained considering infeed from two lines 25–26 and 29–26 in 39 bus New England network as shown in Figure 3.4. If one of the lines are disconnected, the reach of the relay changes, and the revised characteristic is shown in Figure 3.5.

Disconnecting line 29–26 alters the relay's reach in the event of a three-phase breakdown in line 26–27 near bus 27. The zone 3 reach for the relay at bus 28 goes beyond the nearby line when the infeed from line 29–26 is not present. As seen in Figure 3.6, the apparent impedance enters the relay operating characteristic for a defect in line 27–17, which should ideally be outside the zone 3 reach of the relay at bus 28.

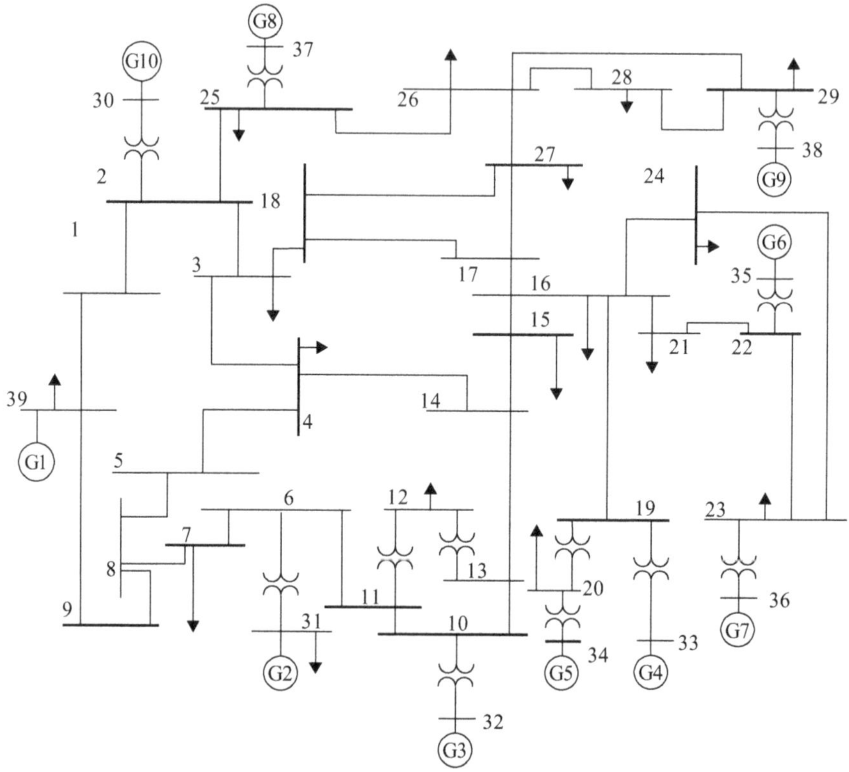

Figure 3.4 Modified 39 bus New England test system

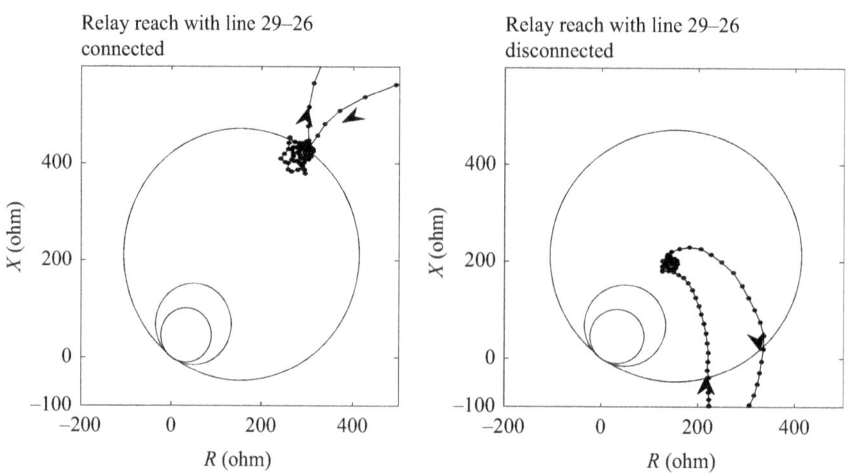

Figure 3.5 Effect of infeed from line 29 to 26 on the reach of relay at bus 28

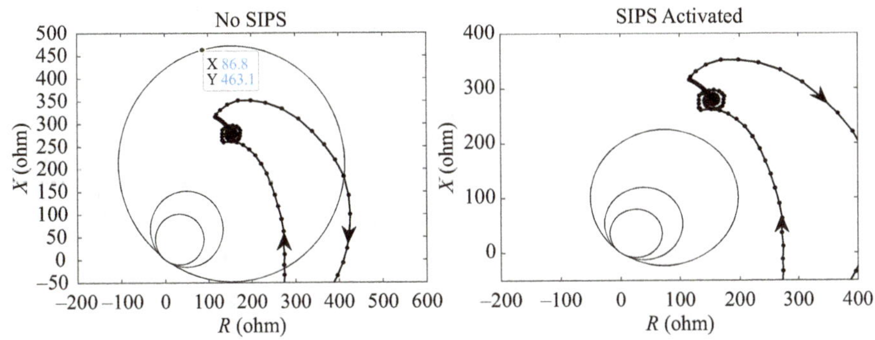

Figure 3.6 Relay reach at bus 28 for fault in line 27–17

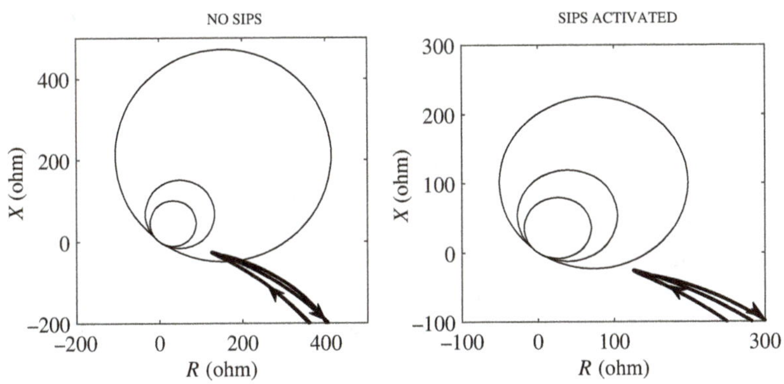

Figure 3.7 Improvement in security of relay at bus 28 using the proposed method

This can cause coordination error between backup relays at buses 26 and 28 in the system. With the proposed adaptive setting of the zone 3 relay is activated in the system, the adjustment is done based on the online calculation. With the relay reach adjusted, the fault in line 27–17 does not affect the relay operation at bus 28. This helps with the coordination of zone 3 operation of relays. The change in reach of the relay also helps to improve the security of the relay. The system's mismatch between generation and load is what causes the power swing. In these situations, relays' overreaching backup zones are typically at risk. In this instance, a power swing condition where the apparent impedance observed by the relay at bus 28 for line 28–26 oscillates is caused by clearance of line 29–26 for the system depicted in Figure 3.6. As seen in Figure 3.7, if the relay border is left unaltered, the perceived impedance enters the operational characteristic and could lead to malfunction.

The changed relay characteristic is not affected by the power swing and, thus, enhances the security of relay operation. To illustrate the dependability of the relay during faults, a three-phase fault is simulated in line 26–27 near bus 27. The relay at

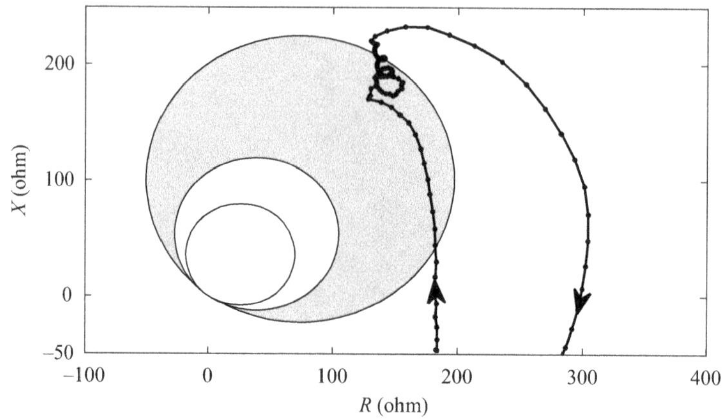

Figure 3.8 Result showing accuracy of relay at 28 to identify the fault in the far end of line 26–27

bus 28 can observe the fault in its operating characteristic as shown in Figure 3.8. This shows that the reach of the relay is not affected although the security has improved. Relays can be blocked during the power swing and then unblocked during fault situations using a variety of WAMS-based techniques [16–18]. Any data loss can have an impact on how well these techniques work. It is challenging for traditional blocking systems to identify faults when the relay is blocked under stress. When system conditions change, zone 3 features are modified for the suggested approach. Relays are operated using local data once the setting has been determined. The relay is not blocked because, as Figure 3.8 illustrates, the power swing has no effect on how it operates. In such a scenario, as seen in Figure 3.8, the relay can identify any subsequent fault using only local data. It would be challenging for the first characteristic to distinguish between an error and a power swing. The outcome demonstrates that altering the feature increases the security of relay operation under pressure without compromising dependability.

3.2.2 *Adaptive distance relay protection for series compensated line*

Series compensation is utilised to improve the power flow of a line. However, its use imposes protection challenges like voltage and current inversion, overreaching and coordination problem in between distance relays. Due to the presence of series compensation, apparent impedance seen by the relay is modulated and when the compensation is present in the fault loop, the relay overreaches. At times, series compensation may be bypassed due to maintenance purpose or by the operation of its overvoltage protective devices. For accurate operation of the relay, information regarding the compensation level in the transmission line is required.

Consider the 60% series compensation at the middle of the line MN as shown in Figure 3.9. If the relay at bus M for line MN is set without considering the compensation level in the line, then there are chances that relay will operate for fault at F_1, where F_1 comes in zone 2 of line NP. Similarly, if the relay is set considering the compensation level and the capacitor is bypassed before fault for maintenance purpose and by its overvoltage protective device during fault, the relay at M for line MN will not detect a fault at F_2 which comes in zone 2 of line MN. This is tested with a line-to-ground fault with $R_f = 1\ \Omega$ at F_2 and the impedance trajectory with capacitor bypassed condition during fault is plotted in Figure 3.10. It is observed from figure that the apparent impedance settles outside the boundary set considering series compensation and inside the boundary with compensation is not considered. This reveals that relay setting needs a change if the compensation level varies during fault. Such maloperation can be prevented if the compensation level present in the line can be communicated to the relay at M. Synchronised data available at both end of the line can be applied to compute the compensation level.

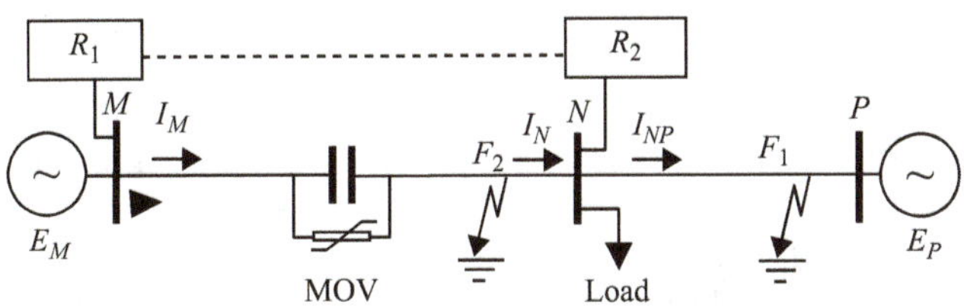

Figure 3.9 A two-source system with series compensation at the middle of the line

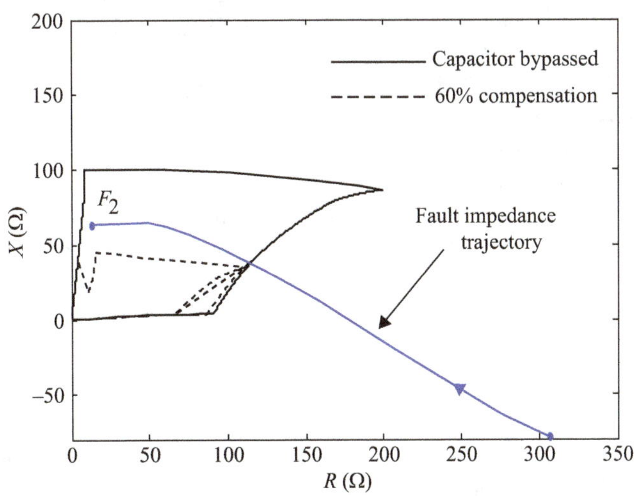

Figure 3.10 Comparison of zone 2 trip boundaries with 60% compensation at middle and capacitor bypassed

3.2.2.1 The adaptive method

A 400 kV, 50 Hz series compensated power system as shown in Figure 3.9 is considered. For this case, measurement units (distance relays with GPS facility) [20] are considered to be available at the two buses M and N. The buses are connected with a communication system which may be available for any other purpose in the system. Adaptive zone 2 and zone 3 setting of relay at bus R in the system is to be accomplished using power system condition and the level of series compensation. The positive sequence phasor data at a rate of 50 frames per second from the ends are processed at the relay end to obtain the compensation level. The impedance between two buses M and N can be calculated by

$$Z_{1\mathrm{MN}} = \frac{(\overline{V}_{\mathrm{M}})^2-(\overline{V}_{\mathrm{N}})^2}{\overline{V}_{\mathrm{M}}\overline{I}_{\mathrm{N}}+\overline{V}_{\mathrm{N}}\overline{I}_{\mathrm{M}}} \tag{3.3}$$

where $\overline{V}_{\mathrm{M}}, \overline{I}_{\mathrm{N}}$ and $\overline{V}_{\mathrm{N}}, \overline{I}_{\mathrm{N}}$ are positive sequence phasors at bus M and N for MN section, respectively. The series compensation X_{C} present in the line is obtained from the imaginary parts of the calculated impedance between bus M and N ($Z_{1\mathrm{MN}}$) and positive sequence impedance of the line ($Z_{1\mathrm{L}}$).

$$X_C = Im(Z_{1\mathrm{L}}) - Im(Z_{1\mathrm{MN}}) \tag{3.4}$$

Once compensation level is known using (3.4), zone 2/zone 3 setting are obtained at bus R. Zone 2, being the primary protection for the line beyond zone 1, cannot detect faults with a reduced reach when compensation is bypassed. Similarly zone 3 which provides backup for the next line will also face problem. Such problems will be severe when circuit breaker fails in the next line and fault is not detected by backup relay for capacitor bypassed condition. As the delay in data transfer is less than the time setting of zone 2 and zone 3, this method is feasible for zone 2 and zone 3. The flow diagram of the adaptive algorithm is given in Figure 3.11.

3.2.2.2 Apparent impedance computation and boundary generation

Series compensation may be available at the middle, at one or both ends of a line. A two-source system is considered for apparent impedance calculation as shown in Figure 3.9. The series compensation available at the middle of line section MN is considered first. Zone 2 of the distance relay located at bus M and in the similar way zone 3 can be studied. The relation for prefault voltages of both sides can be written as

$$\frac{\overline{E}_{\mathrm{P}}}{\overline{E}_{\mathrm{N}}} = he^{-j\delta} \tag{3.5}$$

where h is the voltage magnitude ratio and δ is the phase angle difference between $\overline{E}_{\mathrm{N}}$ and $\overline{E}_{\mathrm{P}}$. While deriving the characteristics for zone 2 protection for the system, the line segment between the relay and capacitor will have similar characteristics to a normal line. For capacitor portion, two issues arise during faults (i) without and

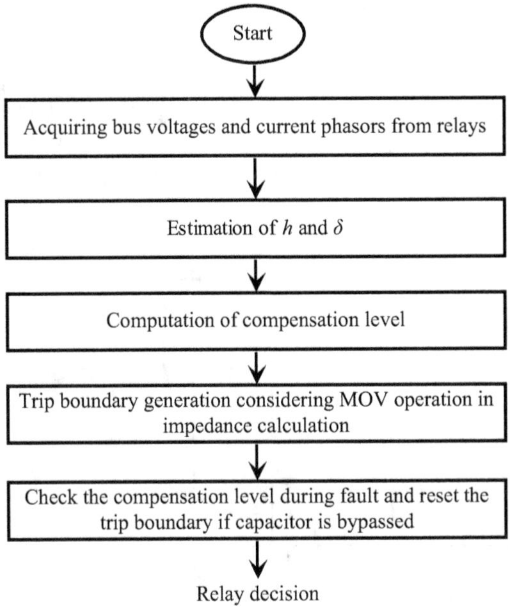

Figure 3.11 Flow diagram for the adaptive technique

(ii) with the operation of MOV. The subsequent portion of the line beyond the capacitor and 50% of the adjacent line section for zone 2 will have impedances similar to a normal line. For zone 2 protection, the seen impedance for the relay at M is obtained by analysing faults at different points on line MN and 50% of the adjacent line section NP.

For the system in Figure 3.9, consider ag-type fault at a distance d pu away from the capacitor the sequence diagram is given in Figure 3.12. Using the information on system condition calculated by (3.5) and compensation level ($-jX_C$) by (3.4), current the prefault current $\overline{I}_{\mathrm{LM}}$ is calculated using ABCD parameters approach as

$$\overline{I} = \overline{E_M}\left[\frac{\mathrm{D}}{\mathrm{B}} - \frac{he^{-\mathrm{j}\delta}}{\mathrm{B}}\right] \tag{3.6}$$

where pre-fault ABCD parameters of the transmission line are known before the fault.

For ag-type fault with fault resistance of $\mathrm{R_F}$, fault current is calculated by

$$I_{F1} = I_{F0} = I_{F0} = \frac{\overline{V}\mathrm{aF}}{Z_{\mathrm{SUM}} + 3\mathrm{R_F}} \tag{3.7}$$

where $Z_{\mathrm{SUM}} = 2\left(\frac{Z_{\mathrm{M1}}\ \ Z_{\mathrm{P1}}}{Z_{\mathrm{M1}}+Z_{\mathrm{P1}}}\right) + \left(\frac{Z_{\mathrm{M0}},\ Z_{\mathrm{P0}}}{Z_{\mathrm{M0}}+,\ Z_{\mathrm{P0}}}\right)$ and $\overline{V}\mathrm{aF}$ is the measured voltage at the relay bus during fault. $\mathrm{Z_{M1}}$, $\mathrm{Z_{P1}}$ and $\mathrm{Z_{M0}}$, $\mathrm{Z_{P0}}$ are equivalent positive and zero sequence impedances from F to M and F to P side source, respectively.

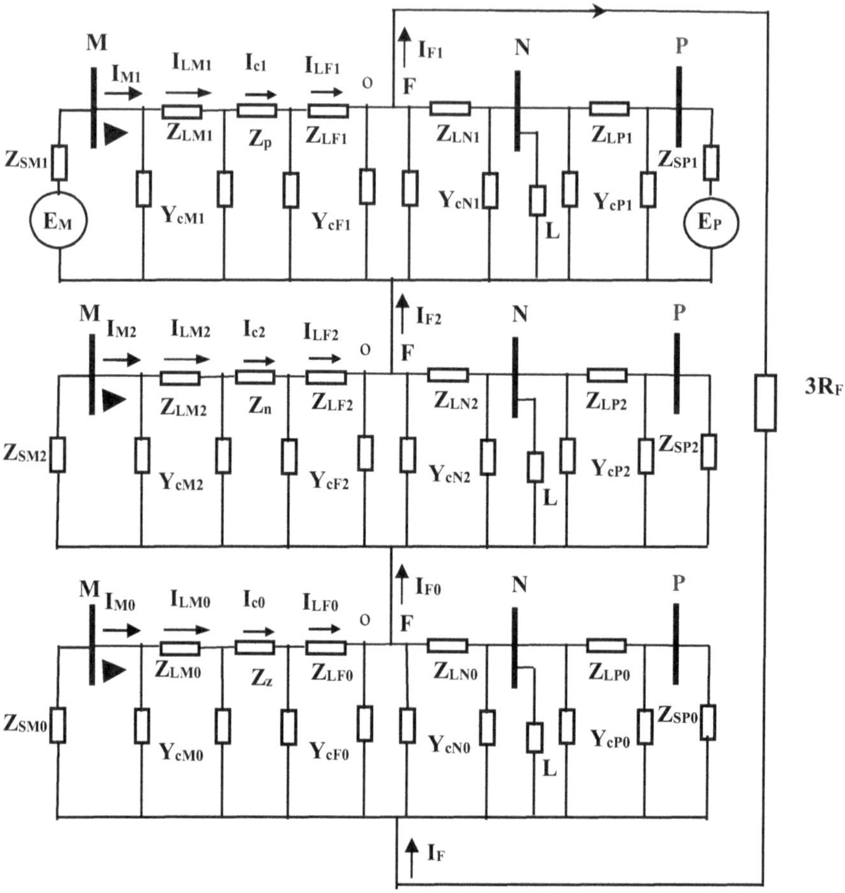

Figure 3.12 Sequence diagram for ag-type fault after the series capacitor

3.2.2.3 Impact on MOV in apparent impedance computation

The conduction of MOV is adjudged from the capacitor current level and corresponding equivalent impedance of the combination is computed [21]. The equivalent impedance offered by MOV and capacitor combination is calculated by an iterative way which converges in maximum seven iterations and the extra computation time is not an issue for setting zone 2/zone 3. However, for a MOV-capacitor combination bypassed condition, mutual coupling is not present and the adaptive setting is obtained accordingly. Using the fault current in (3.7), the capacitor current will be

$$\bar{I}\mathrm{c} = \bar{I}\mathrm{Lc} + \bar{I}\mathrm{c1} + \bar{I}\mathrm{c0} + \bar{I}\mathrm{c2} \tag{3.8}$$

$\bar{I}\mathrm{Lc}$ is the prefault current through the capacitor. If the ratio of fault current through capacitor and protective level current $I_{\mathrm{pu}} = \frac{|\bar{I}_c|}{I_{pr}}$, is more than the set value 0.98, the equivalent impedance of the MOV protected series capacitor is given by [22],

$$Z_{\mathrm{eq}} = R_{\mathrm{eq}} - jX_{\mathrm{eq}} \tag{3.9}$$

where

$$R_{\mathrm{eq}} = X_{\mathrm{C}}\left(0.0745 + 0.49e^{-0.243I\mathrm{pu}} - 35.0e^{-5.0I\mathrm{pu}} - 0.6e^{-1.4I\mathrm{pu}}\right) \tag{3.10}$$

$$X_{\mathrm{eq}} = X_{\mathrm{C}}\left(0.1010 - 0.005749I\mathrm{pu} + 2.088e^{-0.8566I\mathrm{pu}}\right) \tag{3.11}$$

The impedance matrix due to the MOV in the three phases is represented as:

$$\mathrm{Zeqabc} = \begin{bmatrix} \mathrm{Zeqa} & 0 & 0 \\ 0 & \mathrm{Zeqb} & 0 \\ 0 & 0 & \mathrm{Zeqc} \end{bmatrix} \tag{3.12}$$

Due to the unsymmetrical nature of the impedance in three phases, couplings between the sequence networks are present. The symmetrical transformation of the equivalent impedance becomes [23],

$$\mathrm{Zeq\_012} = \frac{1}{3}\begin{bmatrix} \mathrm{Zh} & \mathrm{Zm1} & \mathrm{Zm2} \\ \mathrm{Zm2} & \mathrm{Zh} & \mathrm{Zm1} \\ \mathrm{Zm1} & \mathrm{Zm2} & \mathrm{Zh} \end{bmatrix} \tag{3.13}$$

where $\mathrm{Zh} = \frac{1}{3}(\mathrm{Zeqa} + \mathrm{Zeqb} + \mathrm{Zeqc})$, $\mathrm{Zm1} = \frac{1}{3}(\mathrm{Zeqa} + \alpha\mathrm{Zeqb} + \alpha^2\mathrm{Zeqc})$, $\mathrm{Zm2} = \frac{1}{3}(\mathrm{Zeqa} + \alpha^2\mathrm{Zeqb} + \alpha\mathrm{Zeqc})$ and $\alpha = e^{-j(2\pi/3)}$

The sequence impedances considering the coupling between the series compensation portion are obtained as Z_{p}, Z_{n} and Z_z, where subscript p, n and z refer to positive, negative and zero, respectively.

$$\mathrm{Zp} = \left(\mathrm{Zh} + \frac{\bar{I}\mathrm{c2}}{\bar{I}\mathrm{c1}}\mathrm{Zm1} + \frac{\bar{I}\mathrm{c0}}{\bar{I}\mathrm{c1}}\mathrm{Zm2}\right),\ \mathrm{Zn} = \left(\mathrm{Zh} + \frac{\bar{I}\mathrm{c1}}{\bar{I}\mathrm{c2}}\mathrm{Zm2} + \frac{\bar{I}\mathrm{c0}}{\bar{I}\mathrm{c2}}\mathrm{Zm1}\right),$$
$$\mathrm{Zz} = \left(\mathrm{Zh} + \frac{\bar{I}\mathrm{c2}}{\bar{I}\mathrm{c0}}\mathrm{Zm2} + \frac{\bar{I}\mathrm{c1}}{\bar{I}\mathrm{c0}}\mathrm{Zm1}\right) \tag{3.14}$$

Since the capacitor current and the equivalent impedance offered by MOV protected capacitor are dependent, the current through the capacitor is solved iteratively using the sequence impedances. The change in capacitor current $\Delta I_{\mathrm{c}}^{k} = I_{\mathrm{c}}^{k} - I_{\mathrm{c}}^{k-1}$ is checked for the kth iteration. If $\Delta I_{\mathrm{c}}^{k}$ in two consecutive iteration becomes negligible, then the equivalent impedance offered by the combination for sequence networks $Z_{\mathrm{p}}, Z_{\mathrm{n}}$ and Z_{z} is obtained. If Ipu is less than the predefined value of 0.98, the MOV will not operate and the equivalent impedance offered by the combination will be

$$Z_p = Z_n = Z_z = -jX_C$$

The seen impedance for line-to-ground fault can be calculated from the relation

$$Z_{app} = \frac{\overline{V}_{aM}}{\overline{I}_{aM} + K_0\overline{I}_{M0}} \tag{3.15}$$

where K_0 is the zero sequence compensating factor [23], $K_0 = \frac{Z_{0L}-Z_{1L}}{Z_{1L}}$, with Z_{0L} and Z_{1L} be the zero and positive sequence impedances of the line, respectively. The K_0 value is different for the fault beyond the capacitor and depends on the MOV operation. However, in the study, we have used fixed value of K_0, for both setting and decision processes, which provides correct results and is addressed in Ref. [24].

Since shunt currents sequence network are negligible, apparent impedance can be expressed as in (3.15).

The flow diagram of the algorithm for setting the relay characteristic is shown in Figure 3.11. In the first three steps, the compensation level is calculated and in the fourth step, trip boundary with MOV operation is generated. Different ways can be applied to implement the scheme. One approach, being a protection decision, adaptive distance relay should take the synchronised data as input and do the setting calculation required in the relay. Otherwise, the system condition and level of compensation can be calculated outside the relay and the setting can be computed which should be transferred to the distance relay for proper decision.

3.2.2.4 Result for adaptive backup protection for series compensated line

Trip boundary with series compensation bypassed at the middle for a different system condition

Sometimes the capacitor may be bypassed due to the unbalance in capacitor banks or for maintenance purpose. When the capacitor bank protective devices operate, it cannot be reinserted to the system immediately. In each case, it will lead to an uncompensated line condition. A different example is simulated without series compensation in the line of the system in Figure 3.4. Synchronised phasor data collected from the two ends are provided in Table 3.1.

Table 3.1 Synchronised data for capacitor bypassed with h = 0.99, δ = −5°

Relay	Bus voltage (kV)	Line current (kA)		Calculated		
				Z_C (Ω)	h(°)	δ(°)
Bus-R	210.11−j90.9	I_R	0.045+j0.179	−j0.61	0.987	−5.06
Bus-S	217.60−j84.68	I_S	−0.126+j0.011			
		I_{ST}	−0.332+j0.068			

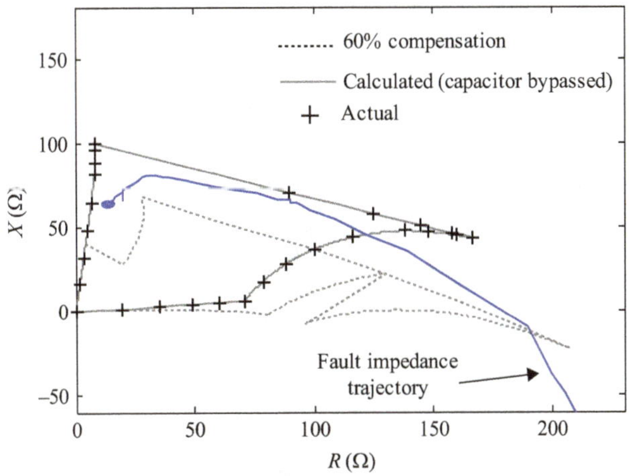

Figure 3.13 Trip boundary with capacitor bypassed at middle with h = 0.99, δ = –5°

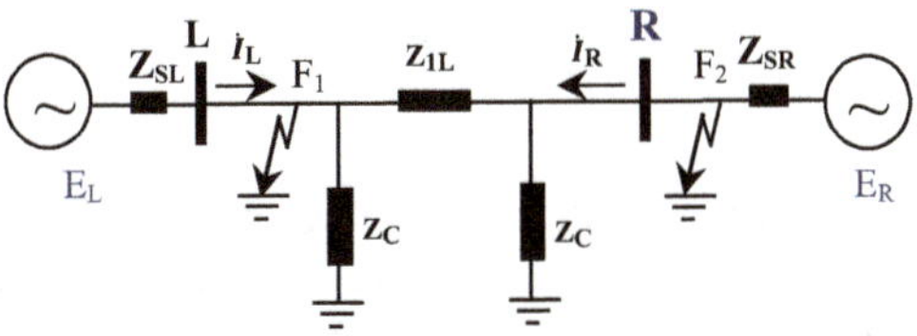

Figure 3.14 Two-source equivalent system

The calculated compensation level $Z_C = -j0.61\ \Omega$ (a low value) shows that the capacitor is bypassed. Using this information, the trip boundary is generated and plotted in Figure 3.13. The generated boundary is compared with the seen impedance obtained from the simulation using EMTDC/PSCAD for line-to-ground faults at different locations varying the fault resistance 0–200 Ω. The results also show the strength of the method in providing trip boundary correctly.

3.2.2.5 Adaptive alpha plane relay

A two-bus power system as shown in Figure 3.14 is considered to describe current-based alpha plane scheme. For any operating condition, using both end current phasors ($\dot{I}_L$ and $\dot{I}_R$), the current ratio (CR) is obtained as [25],

$$\mathrm{CR} = ke^{j\alpha} = \frac{\dot{I}_R}{\dot{I}_L} \tag{3.16}$$

where k is the magnitude ratio and α is the phase angle difference between the currents. It is to be noted that the currents for the above relation can be either using individual phase current or any sequence currents. The generalised alpha plane characteristic is provided in Figure 3.15.

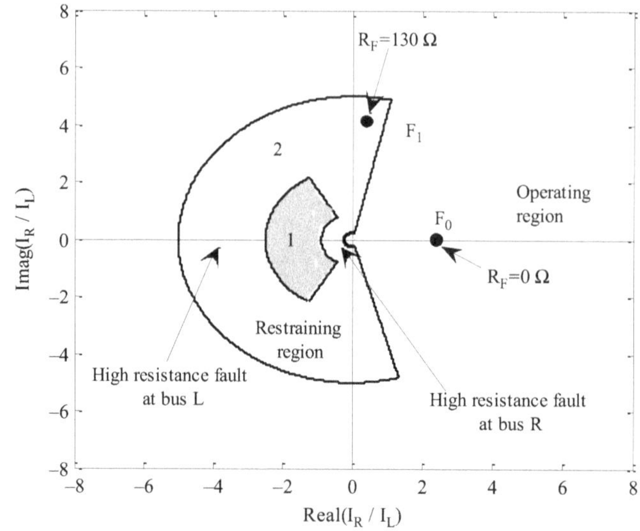

Figure 3.15 Alpha plane characteristic showing operating and restraining regions

During normal operating condition and for external fault, the CR remains within the restraining region in the alpha plane. An internal fault is ensured when CR moves into the operating region. The boundary of the restraining region is defined as: the radius of arc k (typically between 5 and 10) and the angle α (typically between 160° and 210°) [26]. Phase current based APR with $k = 5$ and $\alpha = 210°$, which bounds regions 1 and 2 has inherent immunity to relay maloperation caused by communication delay, channel asymmetry and CT saturation. It is observed from the simulation that when the fault resistance is increased for line-to-ground fault near to the relay bus, the apparent impedance enters into the restraining region which is kept large to ensure the security due to communication delay. Hence, region 2 in Figure 3.15 of the alpha plane is a conflicting region. To overcome this conflict, more discriminating functions are required. It is to be noted that region 1 ensures no internal fault, a clear restraining region.

3.2.2.6 Proposed method

For low-resistance ground fault, the phase angle between the voltage and current is nearly the transmission line impedance angle. As the fault resistance increases, this angle decreases. To improve the performance of the APR using phase currents, during high-resistance internal fault (when CR lies in region 2) an additional feature, angle between voltage and current is first selected. During a fault, the angle deviation from transmission line impedance angle (θ) is obtained at both ends

$$\begin{aligned} \phi_1 &= |\theta - \phi_L| \\ \phi_2 &= |\theta - \phi_R| \end{aligned} \tag{3.17}$$

where $\theta = \tan^{-1}\left(\frac{x}{r}\right)$ with x and r being transmission line reactance and resistance per kilometre, respectively. ϕ_L and ϕ_R are the angles between the voltage and current at L and R sides respectively. It is to be noted that ϕ_1 and ϕ_2 are high for high fault resistance cases (ϕ_L and ϕ_R being small). This feature clearly identifies a high resistance fault when CR is in region 2 and thus useful in distinguishing the condition from CT saturation, etc.

The angles ϕ_1 and ϕ_2 for both internal and external high resistance faults are similar. For any external fault or load change, the magnitude of change in current at both ends of the line is equal. This feature is used to differentiate the internal and external high resistance faults. A feature for this is defined as

$$\Delta \dot{I}^{\mathrm{a}}_{\mathrm{diff}} = \left|\Delta \dot{I}^{\mathrm{a}}_{R}\right| - \left|\Delta \dot{I}^{\mathrm{a}}_{L}\right| \tag{3.18}$$

where the change in currents

$$\Delta \dot{I}^{\mathrm{a}}_{\mathrm{R}} = \dot{I}^{\mathrm{a}}_{\mathrm{Rfault}} - \dot{I}^{\mathrm{a}}_{\mathrm{Rpre}}$$

$$\Delta \dot{I}^{\mathrm{a}}_{\mathrm{L}} = \dot{I}^{\mathrm{a}}_{\mathrm{Lfault}} - \dot{I}^{\mathrm{a}}_{\mathrm{Lpre}}$$

Similarly, it can be computed for b and c phases. During external high-resistance fault, ϕ_1 and ϕ_2 will be high but ΔI_{diff} will be small. On the other hand for internal high resistance fault, ϕ_1, ϕ_2 will be high and ΔI_{diff} will be also high. Both the features are combined to ensure the detection of internal high-resistance fault whenever the CR lies in restraining region 2. To identify internal fault when CR is in region 2, an index K_1 is defined as

$$K_1 = \begin{cases} 1 & \text{if } \Delta I_{\mathrm{diff}} > \zeta \text{ and } \phi_{\mathrm{min}} > \phi_{\mathrm{th}} \\ 0 & \text{if } \Delta I_{\mathrm{diff}} > \zeta \text{ and } \phi_{\mathrm{min}} < \phi_{\mathrm{th}} \end{cases} \tag{3.19}$$

where $\phi_{\mathrm{min}} = \min(\phi_1, \phi_2)$, ϕth and ζ are thresholds selected considering possible operating conditions. During a high-resistance fault when the CR lies in restraining region 2, K_1=1 and the proposed method concludes an internal fault. Thus, using the system condition information, the decision is derived in case of the trajectory in region 2.

3.2.2.7 Threshold selection for ΔI_{diff}

For internal fault ΔI_{diff} is very high and for an external fault it is negligible (without CT saturation) [26]. Shunt capacitance may cause difference in current magnitude at both ends but is of negligible effect on change in current. There may be difference in current during normal loading condition due to CT error but this will be small. These are considered in threshold setting to discriminate high-resistance internal fault to load change. Considering 2% error for C class protection CT at both ends and to provide a secured operation, a threshold ζ is selected as 0.08 pu. Thus by this normal condition are filtered out except CT saturation during external fault.

3.2.2.8 Threshold selection ϕth

Angle deviates for high-resistance fault and for CT saturation. The scheme finds problem during CT saturation. To differentiate CT saturation from internal high resistance fault minimum of ϕ_1, ϕ_2 is considered. Minimum change for line-to-ground fault is observed for $R_F = 0$ and it also depends on system condition and parameters are derived in the previous section. To select the threshold for $\phi_{\min}$, a line-to-ground fault with $R_F = 0$ at one end of the line is considered. The maximum deviation of ϕ at relay bus for h and δ variations is found to be 0.362 and 0.4178 rad, respectively. It depicts that variation in ϕ_1 is mostly for the changes in h and δ. With these values for heavy loading condition and from all above observations, the threshold ϕth is selected as 0.436 rad.

3.2.2.9 Steps for computation of proposed method

In this method to obtain improved performance of current-based alpha plane scheme, two features are selected depending on system condition. The different steps of the proposed method are given in the following steps

Step 1: Current ratio is calculated using (3.16).
Step 2: If the ratio lies inside region 1, the relay is blocked.
Step 3:

a) When the ratio lies inside the restraining region 2,$|\Delta\dot{I}_R|$ and $|\Delta\dot{I}_L|$for all three phases are computed, followed by calculation of ΔI_{diff}.
b) If $\Delta I_{\text{diff}} > \zeta$ then $\Phi_{\min}$ is calculated using ϕ_1 and ϕ_2.
c) Whenever $\Delta I_{\text{diff}} > \zeta$ and $\Phi_{\min} > \Phi_{\text{th}}$, K_1 is set as 1 ensuring the internal fault case.

Step 4: When the ratio is beyond region 1, it is an internal fault and K_1= 1.

3.2.2.10 Results

The performance of the proposed method is tested for an internal high-resistance fault. In the prefault condition, power flows from bus 5 to bus 4 as shown in Figure 3.16. A line-to-ground fault at 5 km from bus 4 with RF = 200 Ω, in line 5–4, is created at 1.0 s. The CR is calculated using (3.16) and plotted in Figure 3.17. From the figure, it is observed that before fault, the CR was inside restraining region 1. outside the restraining region 1 but inside the region 2. The conventional

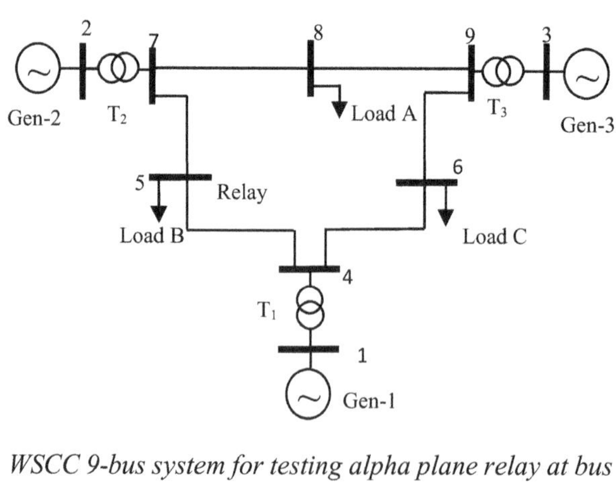

Figure 3.16 WSCC 9-bus system for testing alpha plane relay at bus 5 for line 5–4

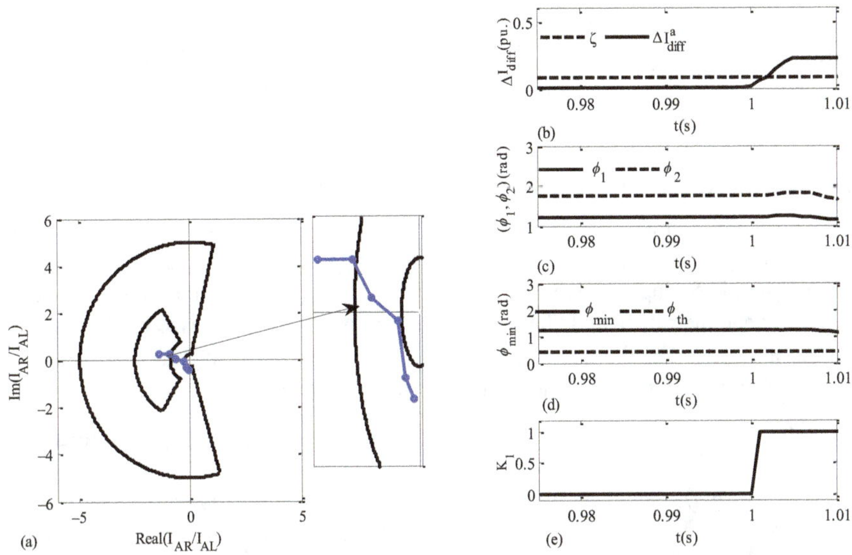

Figure 3.17 Result for internal line-to-ground fault with RF = 200 Ω. (a) alpha plane characteristic, (b) ΔI_{diff}, (c) ϕ_1 and ϕ_2 in rad, (d) ϕ_{min} in rad, (e) index K_1.

APR will send a block signal for this internal fault. During this period, current phasors obtained from both ends are applied to compute ΔI_{diff} using (3.18) and plotted in Figure 3.17(a). With $\Delta I_{diff} > \zeta$, angle deviations between voltage and current is calculated for each end and plotted in Figure 3.17(b). The minimum of both the angles is obtained and plotted in Figure 3.17(c). The index K_1 is obtained using (3.19) and plotted in Figure 3.17(d). With K_1=1 for the case, an internal fault is confirmed which is correct. The proposed scheme provides a correct decision where the conventional alpha plane fails during high-resistance fault.

For external fault with CT saturation or channel asymmetry or communication delay, the CR remains in an extended portion, region 2 as marked in Figure 3.17 and is outside region 1. However, to overcome this conflict, additional features selected will be helpful. It is mandatory that both the CTs will not saturate at the same instant. Even if one CT will see a change in phase angle, the other CT phase angle change will be less. Hence, the feature in (3.17) prevents the maloperation during external fault CT saturation cases and improves the security.

3.3 Online identification of protection element failure

Fault in a power system is cleared by different protection schemes comprising of different relays, circuit breakers and accessories. Collective action of all these elements in a protection scheme is necessary to isolate a fault. Failure of an element in such a protection arrangement may lead to unwanted line trip risking further outages in the network. This work proposes a WAMS data-based technique for online identification of protection component failure of transmission line. Voltage current phasors along with relay decision and circuit breaker status signals are used to identify the faulted line in the system followed by the identification of the protection component failure, if any, in the selected section.

3.3.1 Faulted line identification

Initiation of fault at any point in a network results in huge current flow. This causes severe drop of voltage at buses close to the fault point. Also due to the inductive nature of fault, the phase angle between voltage and current phasors at these buses deviates. Hence, change of voltage and phase angle at buses in a network is used to identify fault point in a system. For this purpose, two indices are applied as in Ref. [28] which are obtained using synchronised voltage and current phasors are as defined below:

a) *v_abs (n)*: absolute value of positive sequence voltage deviation during fault for bus-n.
b) *vi_phase (nm)*: deviation of phase angle between the positive sequence voltage of a bus and current in a line connected to the bus during fault where n and m correspond to buses connected by the line.

The bus with maximum deviation in positive sequence voltage phasor is selected. Next, the change in phase angle between the voltage and current phasors for the lines connected to the selected bus is observed. The line with highest deviation in the phase angle value is identified as the faulted line.

3.3.2 Component failure indices

Correct operation of various components such as relay, circuit breaker and accessories is imperative for error-free functioning of a protection scheme. Failures of any such component can affect the power system stability. A simple description of a protection scheme is shown in Figure 3.18.

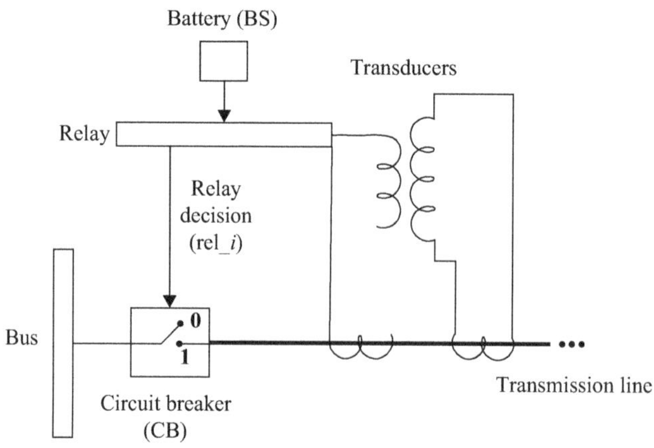

Figure 3.18 Elements of the protection arrangement

It shows the basic elements involved in the detection and removal of faults. Various signals on status of monitored devices such as circuit breaker, protective relay decision signal can be obtained in real time through WAMS [27].

The algorithm proposes four indices that are based on the relay decisions and circuit breaker status signals obtained for a system. They are defined as:

(I) Relay operation signal (*rel_i*): $i = 1, 2, 3$ for zone 1, zone 2 and zone 3, respectively. A value of 1 for operation and 0 for non-operation is assigned. If zone 1 has operated, then *rel*_1 = 1 and *rel*_2 and *rel*_3 are 0.

(II) Circuit breaker status signal (CB): 1 if breaker is open and 0 if closed.

(III) Battery status (BS): Index value is 0 if battery has defect and 1 if functioning properly. A defective battery indicates that the relay operation is not active [28].

(IV) Timer signal (*tim_sig_i*): Value depends on rel_i. For zone 1, its value is *tim_sig*_1 = 1 always and *tim_sig*_2, *tim_sig*_3 = 0. For zone 2 operation, the index value is 0 till 15 cycles and for zone 3 operation till 50 cycles after the inception of fault and becomes 1 thereafter. The time delay values are selected based on the standard backup protection operation time.

Relay status (RS): 0 for relays connected to the faulted line. For backup relays, it is 2 for the first 15 cycles (300 ms) after fault initiation and 0 thereafter. For reverse operating relays, it has a value of 2.

3.3.3 *Signal communication*

Protection operation is accomplished at high speed; therefore, an online assessment and correction of these schemes require high-speed data communication. Today's wide area communication topologies such as synchronous optical network are capable of delivering messages from one area of a power system to multiple nodes in few milliseconds. IEEE synchrophasor standard [29] uses a binary format for data package transfer through communication channels. In such systems, a configuration frame is a machine-readable binary data set containing information and processing parameters for a synchrophasor data stream. The data fields in a configuration frame are used for communicating information. In this work, along with synchronised phasor data, protection element operation signals are obtained from WAMS. For this purpose, a 4-bit binary signal on the status of the elements is used. This 4-bit data can be transmitted through one of the data fields of a configuration frame. The binary data structure for the scheme is shown in Figure 3.19. The first two bits are for BS and CB signals, respectively, while bits 3 and 4 are information on relay operation signals.

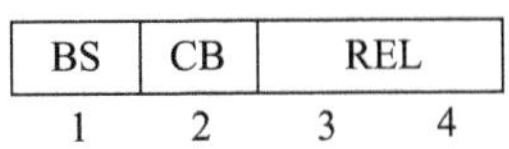

Figure 3.19 4-bit signal information obtained from WAMS data for the method

3.3.4 The method

Identification of protection maloperation for distance relay components during a fault condition in a power system is being proposed in Ref. [29]. The steps for the method are shown in Figure 3.20. In the first step, the algorithm identifies the line affected by fault situation in the system. For this, synchronised positive sequence voltage and current data are used. Depending on the fault location, system topology and zones of operation, relays in the system are assigned weights (*rel_w*) [30]. In the second part of the algorithm, indices described are computed using data obtained from WAMS to identify the failed component, if any, for a protection operation.

After fault affected line is detected, the relays connected to the faulted line and its adjacent lines are weighted according to their zone of operation for the concerned fault situation. Relays connected to the faulted line are assigned each a weight of 1. For adjacent lines, relays located at far end from faulted line are each weighted 0.5 and relays at near end are assigned weights of −3 each. The algorithm uses WAMS data as shown in Figure 3.17 to identify if there is any protection component maloperation. This information is obtained through WAMS from relays that are weighted after identifying the line affected by fault. The process has two levels of computation.

For identification of maloperating protection elements, two *protection component failure* indices are used as defined below:

(I) Malfunction index *(mal_fn)*: determine whether a failure of any protection scheme operation has taken place and the class of maloperation, i.e. whether a zone 1 or zone 2 or zone 3 maloperation has occurred.
(II) Failure index *(fail_in)*: pinpoints the component for which the protection scheme has maloperated.

The details for the computation process for the identification are provided below:

Stage 1: To check any protection maloperation

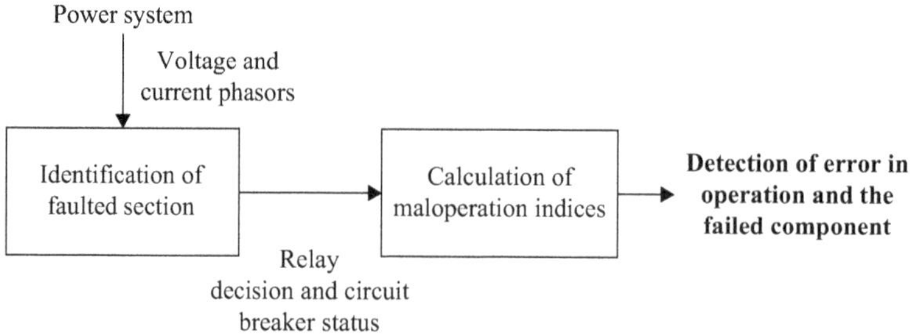

Figure 3.20 Flow diagram of the method

This level detects whether maloperation of any protection scheme in the network has taken place or not. For this, the mal_fn index is computed using the following relation:

$$mal\_fn = \sum_{i=1}^{m} rel_{w_i} \times w_i \tag{3.20}$$

where m is the buses connected to the faulted line and its adjacent lines rel_{w_i} is the weight associated with each relays.

Zone 1 relays are weighted with a value of 1. Thus, if circuit breakers corresponding to relays connected to faulted line operate, then *mal_fn* value becomes 2, which corresponds to a correct protection operation.

Stage 2: To identify failed component

After the identification of class of maloperation and computation of the failure index, the next step is to pinpoint the protection component failure. Using the indices defined in Section 3.2, failure index (*fail_in*) is computed as:

$$fail\_in = BS + rel\_i + CB \times 0.5 + tim\_sig\_i \times 1.5 + RS \tag{3.21}$$

where i = 1, 2 and 3 for zone 1, zone 2 and zone 3, respectively.

Table 3.2 lists the range of values of the two indices (*mal_fn* and *fail_in*) and the remarks therefrom. Some of the simulated cases are presented in the following section for a test system.

3.3.5 Case study: relay underreaching

A single line to ground fault is simulated in the line connecting bus 26 and 28 at 0.1 s. The relay setting at bus 28 is such that it is underreaching for the particular situation as a result of which fault could not be cleared at bus 28 for system in Figure 3.6. Circuit breaker opens at bus 26 successfully but fault current continues from bus 28 because of relay malfunction. Synchronised measurements are available from all nearby buses close to the fault. Deviations in synchronised voltage-phasor measurements of selected buses are shown in Figure 3.21. It is observed that buses 26 and 28 have the maximum voltage deviation. The plots for deviation of

Table 3.2 Indices and corresponding inference

Malfunction index	$0 < mal\_fn \leq 1$ (zone 1)			$1 < mal\_fn < 2$ or $mal\_fn > 2$ (zone 2 / zone 3)					$Mal\_fn \leq 0$
Failure Index	$fail\_in \leq 0$	$fail\_in = 1$	$fail\_in = 3.5$	$fail\_in = 1$	$fail\_in = 1.5$	$fail\_in = 3.5$	$fail\_in = 4.5$ or $fail\_in = 2.5$	$fail\_in = 6$	$fail\_in = 6$
Inference	Transducer error	Relay under reach	Circuit breaker error	Relay under reach	Transducer error	Circuit breaker error	Timer failure	Relay overreach	Unwanted reverse operation
$fail\_in = 4$: correct operation									

phase angle between voltage and current for different lines connected to buses 26 and 28 are shown in Figure 3.22, respectively. It is observed that for measurements at both buses 26 and 28, phase angle deviation for line 26–28 is maximum. From this, the technique concludes that the fault is in line 26–28, which is correct.

Depending on the fault location, relays in the system are assigned weights. Relays connected to the faulted line are weighted 1. Relays that provide backup for this fault condition are weighted 0.5. The relays for which the fault is in the reverse

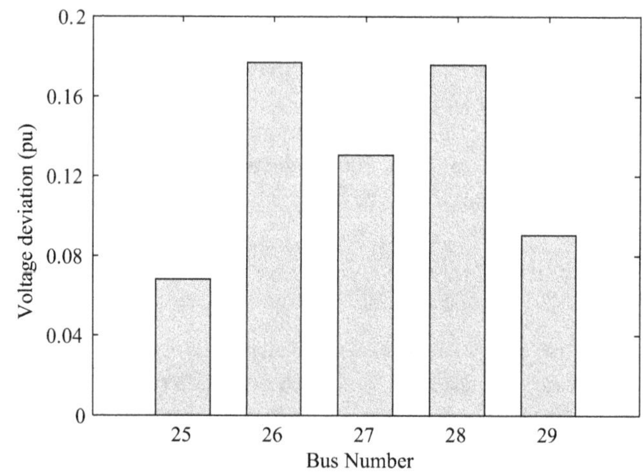

Figure 3.21 Bus with significant voltage deviation for fault in line 26–28

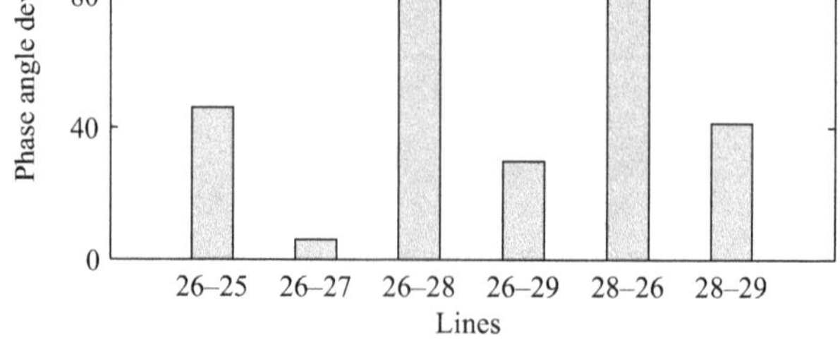

Figure 3.22 Phase angle deviation for lines connected to bus 26 for fault in line 26–28

direction are weighted −3. For the case, relay weights are:

$$\text{rel\_w}(26-28(26))=1,\ \text{rel\_w}(28-26(28))=1,\ \text{rel\_w}(25-26(25))=0.5,$$
$$\text{rel\_w}(27-26(27))=0.5,\ \text{rel\_w}(29-26(29))=0.5,\ \text{rel\_w}(29-28(29))=0.5,$$
$$\text{rel\_w}(26-25(26))=-3,\ \text{rel\_w}(26-27(26))=-3,\ \text{rel\_w}(26-29(26))=-3,$$
$$\text{rel\_w}(28-29(28))=-3.$$

After faulted line identification and assigning relay weights, the bitwise information for protection schemes of weighted relays is obtained. The information obtained for protection schemes for fault in line 26–28 is shown in Table 3.3 (information for reverse operating relays is not shown).

The next step is to assess the protection operation and identify failure, if any, in the protection scheme:

(I) Depending on the faulted line identification, relays at bus 26 and 28 for line 26–28 have been given a weight of 1 each. As CB circuit breaker status at bus 26 is 1 and for all other breakers in the system, *CB* is 0. Calculation of malfunction index using (16) gives:

$$mal\_fn = rel_w\,(26-28(26)) \times CB(26-28(26)) = 1 \times 1 = 1.$$

Value of *mal_fn* confirms a zone 1 maloperation, i.e. one of the zone 1 protection components has failed.

(II) Calculating the failure index for protection operation at bus 26 using (3.2), we get:

$$\begin{aligned} fail\_in(26-28(26)) &= \text{BS} + rel\ 1 + \text{CB} \times 0.5 + tim\ sig\ 1 \times 1.5 + \text{RS} \\ &= 1 + 1 + 1 \times 0.5 + 1 \times 1.5 + 0 = 4 \end{aligned}$$

Value of *BS*, *CB* and *Rel_i* is 1, obtained from Table 3.3. Timer sig value is also 1 and *RS* is 0 as the relay is connected to the faulted line. A value of 4 for *fail_in* shows a correct operation at bus 26 for fault in line 26–28.

(III) Calculating the failure index for protection operation at bus 28 using (3.17), we get:

$$\begin{aligned} fail\_in(26-28(28)) &= \text{BS} + rel\ 1 + \text{CB} \times 0.5 + tim\ sig\ 1 \times 1.5 + \text{RS} \\ &= 1 + 0 + 0 \times 0.5 + 0 \times 1.5 + 0 = 1 \end{aligned}$$

Table 3.3 Information received for fault in line 26–28

Bus	**Line**	**Information**		
		BS	**CB**	**Rel *i***
26	26–28	1	1	01
28	28–26	1	0	10
25	25–26	1	0	11
27	27–26	1	0	10
29	29–26	1	0	11
29	29–28	1	0	11

A value of 4 for *fail_in* shows that there is no maloperation at bus 27 for fault in line 26–28.

The protection failure indices for the system clearly conclude that the relay at bus 28 has underreached. As a result of which no tripping operation took place in bus 28 for a fault in the line connecting bus 26 and 28. Protection scheme at bus 26 for the line operates properly. The algorithm is able to identify the protection component failure correctly.

With fast communication availability for WAMS data, the detection can be done in about eight cycles (0.16 s) after fault inception. The backup relay for this case is at bus 29 connected to line 29–28. This relay sees the fault in its zone 3 region. So the timer is set to operate after 1 s. After detection of the discrepancy in the relay operation at bus 28, the relay at bus 29 can be redirected to clear the fault, thus speeding up the backup process.

3.4 Conclusions

Maloperation of power system protection schemes can result in the escalation of a small disturbance to a severe situation in a power system. Events such as power swing, load encroachment and failure of components involved in the protection schemes lead to incorrect operation. This causes unwanted tripping or delay in clearing of faults from the system. This work attempts to use wide area measurement data to adapt the relay setting to improve the security of power system. The method for supporting relay decisions during stressed conditions and prevention of maloperation due to infeed current is presented. Indices obtained from WAMS data in a power system are used for identifying malfunction of any element in a protection scheme. Further, such informative data improves series compensated by applying adaptive settings for distance relays. Using data at both ends, the adaptive APR is also achieved. The WAMS-based techniques enhance protection schemes by making them more reliable and speed up protection for power networks.

Abbreviations

APR	alpha plane relay
CDR	current differential relay
DFT	discrete Fourier transform
GPS	global positioning system
OST	out-of-step trip
PDC	phasor data concentrator
PMU	phasor measurement unit
PSB	power swing blocking
RAS	remedial action scheme
SPS	special protection scheme
WAMS	wide area measurement system

References

[1] A. G. Phadke and S. H. Horowitz, *Relaying, Power System*. Taunton, UK: Research Studies Press, 2008.

[2] J. L. Blackburn, *Protective Relaying Principles and Applications*. New York: Marcel Dekker, 1987.

[3] A. G. Phadke and J. S. Thorp, "Expose hidden failures to prevent cascading outages [in power systems]," *IEEE Comput. Appl. Power*, vol. 9, no. 3, pp. 20–23, 1996.

[4] S. Tamronglak, S. H. Horowitz, A. G. Phadke, and J. S. Thorp, "Anatomy of power system blackouts: Preventive relaying strategies," *IEEE Trans. Power Deliv.*, vol. 11, no. 2, pp. 708–715, 1996.

[5] G. Andersson, P. Donalek, R. Farmer, *et al.*, "Causes of the 2003 major grid blackouts in North America and Europe, and recommended means to improve system dynamic performance," *IEEE Trans. Power Syst.*, vol. 20, no. 4, pp. 1922–1928, 2005.

[6] D. Junce and C. Zexiang, "Mixed measurements state estimation based on wide-area measurement system and analysis," in 2005 *IEEE/PES Transmission & Distribution Conference & Exposition: Asia and Pacific*, 2005, pp. 1–5.

[7] M. Gol and A. Abur, "A fast decoupled state estimator for systems measured by PMUs," *IEEE Trans. Power Syst.*, vol. 30, no. 5, pp. 2766–2771, 2015.

[8] L. Hu, Z. Wang, I. Rahman, and X. Liu, "A constrained optimization approach to dynamic state estimation for power systems including PMU and missing measurements," *IEEE Trans. Control Syst. Technol.*, vol. 24, no. 2, pp. 710–730, 2015.

[9] A. R. Salgado, C. R. F. Esquivel, and J. G. Calderon Guizar, "SCADA and PMU measurements for improving power system state estimation," *IEEE Lat. Am. Trans.*, vol. 13, no. 7, pp. 2245–2251, 2015.

[10] M. M. Eissa, M. E. Masoud, and M. M. M. Elanwar, "A novel back up wide area protection technique for power transmission grids using phasor measurement unit," *IEEE Trans. Power Deliv.*, vol. 25, no. 1, pp. 270–278, 2010.

[11] M. G. Adamiak, A. P. Apostolov, M. M. Begovic, *et al.*, "Wide area protection—Technology and infrastructures," *IEEE Trans. Power Deliv.*, vol. 21, no. 2, pp. 601–609, 2006.

[12] J. De La Ree, V. Centeno, J. S. Thorp, and A. G. Phadke, "Synchronized phasor measurement applications in power systems," *IEEE Trans. Smart Grid*, vol. 1, no. 1, pp. 20–27, 2010.

[13] H. J. Altuve, J. B. Mooney, and G. E. Alexander, "Advances in series-compensated line protection," in 2009 *62nd Annual Conference for Protective Relay Engineers*, 2009, pp. 263–275.

[14] D. Erwin, M. Anderson, R. Pineda, D. A. Tziouvaras, and R. Turner, "PG&E 500 kV series-compensated transmission line relay replacement: Design requirements and RTDS® testing," *in 2011 64th Annual Conference for Protective Relay Engineers*, 2011, pp. 191–202.

[15] I. P. S. R. Committee, *Protection Aspects of Multi Terminal Lines.* IEEE Publ. 79 TH0056-2-PWR, 1979.

[16] NERC, "A technical document prepared by the systems protection and control task force of the NERC planning committee—The complexityof protecting three terminal transmission line," 2006. Available online: http://2006.nerc.com/comm/RSTC/SPCWG/PRC-019-2 Technical Reference Document.pdf.

[17] R. Hasan, R. Bobba, and H. Khurana, "Analyzing NASPInet data flows," in *2009 IEEE/PES Power Systems Conference and Exposition*, 2009, pp. 1–6.

[18] A. G. Phadke and J. S. Thorp, "Communication needs for wide area measurement applications," in *2010 5th International Conference on Critical Infrastructure (CRIS)*, 2010, pp. 1–7.

[19] P. Kundu and A. K. Pradhan, "Enhanced protection security using the system integrity protection scheme (SIPS)". *IEEE Trans. Power Deliv.*, vol. 31, no. 1, pp. 228–235, 2015.

[20] A. K. Pradhan and S. Sarangi, "Synchronised data-based adaptive backup protection for series compensated line," *IET Gener. Transm. Distrib.*, vol. 8, no. 12, pp. 1979–1986, 2014.

[21] D. L. Goldsworthy, "A linearised model for MOV-protected series capacitors," *IEEE Trans. Power Syst.*, vol. 2, no. 4, pp. 953–957, 1987.

[22] M. Coursol, C. T. Nguyen, R. Lord, and X.-D. Do, "Modeling MOV-protected series capacitors for short-circuit studies," *IEEE Trans. Power Deliv.*, vol. 8, no. 1, pp. 448–453, 1993.

[23] J. Mahseredjian, P. J. Lagace, S. Lefebvre, and A. Chartrand, , "Superposition technique for MOV-protected series capacitors in short circuit calculations," *IEEE Trans. Power Deliv.*, vol. 10, no. 3, pp. 1394–1400, 1995.

[24] P. Kundu and A. K. Pradhan, "Synchrophasor-assisted zone 3 operation," *IEEE Trans. Power Deliv.*, vol. 29, no. 2, pp. 660–667, 2014.

[25] G. Benmouyal, "The trajectories of line current differential faults in the α-plane," in *32nd Annual Wester Protective Relay Conference*, Spokane, WA, 25–27 October 2005.

[26] G. Benmouyal and J. B. Mooney, "Advanced sequence elements for line current differential protection," 2007. Available online: https://www.selinc.com/WorkArea/Download.

[27] S.C. Sun and R.E. Ray "A current differential relay systems using fiber optics communications," *IEEE Trans. Power Appar. Syst.*, vol. 102, no. 2, pp. 410–419, 1983.

[28] A. K. Pradhan and P. Kundu, "Online identification of protection element failure using wide area measurements," *IET Gener. Transm. Distrib.*, vol. 9, no. 2, pp. 115–123, 2015.

[29] M. Gilany, A. Al-Kandari, and J. Madouh, "A new strategy for determining fault zones in distance relays," *IEEE Trans. Power Deliv.*, vol. 23, no. 4, pp. 1857–1863, 2008.

[30] D. Novosel, D. G. Bartok, and G. Henneberg, "IEEE PSRC report on performance of relaying during wide region stressed conditions," *IEEE Trans. Power Deliv.*, vol. 25, no.1, pp. 3–16, 2010.

Chapter 4
Synchrophasor-assisted visualization and protection of power systems

Sukumar Brahma[1]

Traditional visualization of power system is performed through supervisory control and data acquisition (SCADA) systems and state estimator algorithm. Though decades of development and implementation efforts coupled with modern computing resources have resulted in a robust and dependable visualization tool, the measurements fed to the SCADA system are scalar values and are not synchronized at a precise time-stamp. Typically, a new set of measurements is available every 2–4 s, which is sufficient for steady-state analysis. With recent advances in sensing, it is possible to get measurements of better quality at a faster rate. Recent generation of phasor measurement units (PMUs) measures and transmits synchronized phasor values of voltages and currents at 120 fps. The purpose of this chapter is not to show how to assimilate these measurements with the current SCADA system, but to explore the possibility of creating another visualization layer independent of SCADA that can provide more insight into real-time dynamic events taking place in power systems. A concept of supervisory protection is also explored to increase security of protection schemes by detecting misoperations. Results are presented from published research papers to facilitate critical evaluation of the feasibility of these concepts.

Traditional visualization with SCADA

A block diagram showing SCADA processes is shown in Figure 4.1. SCADA acquires data related to system state, network topology, and control settings from remote terminal units (RTUs) at substations. These consist of breaker statuses, transformer taps and phase shifts, scaler values of voltages and currents, real and reactive power, and settings of control devices. Using these data where measurements outnumber the system states, and stochastically modeling errors in measurements, a state estimator algorithm solves an overdefined set of equations to create the best estimate of the state of the power system—phasor values of voltages

[1]The Holcombe Department of Electrical and Computer Engineering, Clemson University, USA

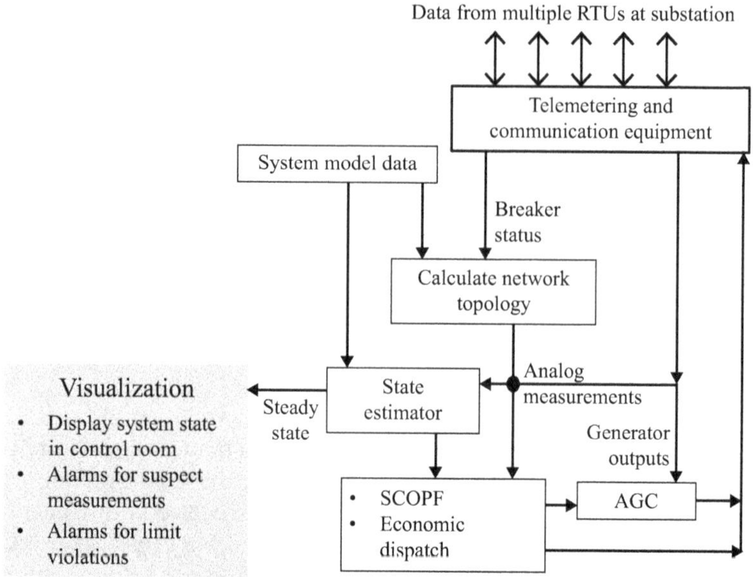

Figure 4.1 SCADA processes

at all buses and power flow through all lines are estimated and displayed on large screens in control centers for visualization. Probable bad data are also identified. Alarms for bad data and overloads are generated. The estimated state is then used for further analysis to run a security constrained optimal power flow (SCOPF) and economic dispatch to finally generate new settings for automatic generation control (AGC). With recent computing resources, a SCOPF can take up to 15 min. If SCADA system is lost, power system operations would be seriously, and probably fatally, affected, as was the case during the August 13 blackout in the Northeastern United States.

State monitoring with synchrophasors

Synchrophasors are phasor measurements that are synchronized using an accurate time reference. PMUs measure voltages and currents in power systems as synchrophasors using global positioning system (GPS) to provide time synchronization. The number of PMUs has grown at a rapid rate in North America, increasing from 200 research-grade PMUs in 2009 to 1,700 production-grade PMUs in 2014. Typically, several PMUs transmit their measurements to a central server called phasor data concentrator (PDC) through communication channels. Compared to

*Reprinted with permission from IEEE Power System Relaying Committee.
†Reprinted with permission from Schweitzer Engineering Laboratories, Inc.

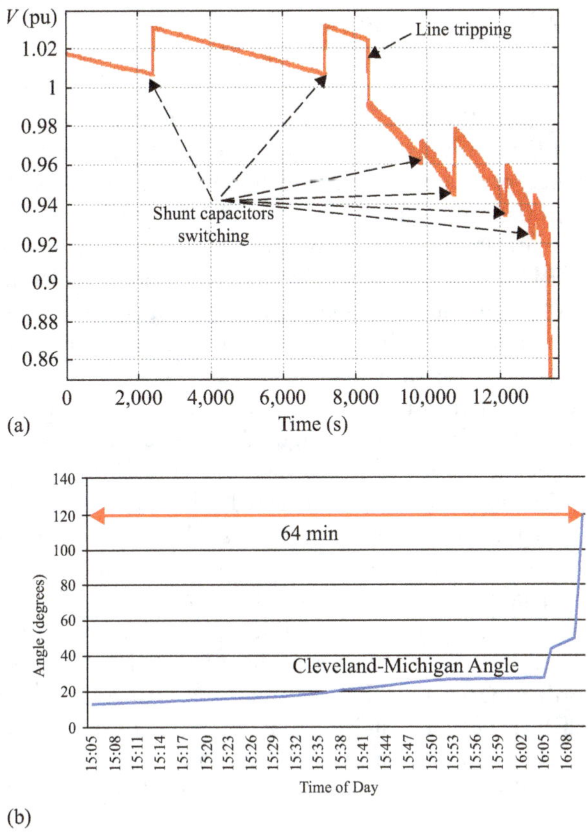

Figure 4.2 *Dynamic events showing (a) voltage instability and (b) angle instability*

SCADA systems reporting loosely synchronized *scalar* values of grid state once every 2–4 s, PMU data, though sparser, capture the physics of the grid with much higher precision and speed. These measurements therefore open a window of opportunity for finer visualization of power systems by detecting and identifying *dynamic* events in real time. Figure 4.2 shows a voltage instability event in (a)[*] and an angle instability event in (b)[†]. If the system state were monitored with PMUs, providing 60 measurements every second, these dynamic events could be vividly visualized. It takes approximately 5 min for precipitation of the final phase of voltage instability, and 3 min for angle instability, which would have been observed through 300 and 180 PMU measurements, respectively. Such high fidelity visualization and direct measurement of angles can provide fast and accurate insight into dynamic events and facilitate timely response from operators to avert system-wide instability.

4.1 Disturbance monitoring and identification using synchrophasors

Current generation of PMUs measure three-phase voltages and currents at 120 fps; this rate is likely to increase in near future. PMUs also measure frequency and rate of change of frequency that are important parameters for detecting a disturbance. Such data rates are unprecedented for electric utilities. As of now, these data are simply archived and only used for postmortem analysis. PMUs do have the capability of detecting a disturbance and creating separate files storing disturbance data to facilitate such analysis. North American Reliability Council (NERC), the regulating entity in North America, has published document PRC 002-2 that mandates disturbance monitoring and reporting requirements. This document mandates either of the following to detect a disturbance: (i) frequency <59.55 Hz or >61 Hz, (ii) rate of change of frequency $df/dt > 0.124$ Hz/s, (iii) undervoltage trigger set no lower than 85% of the normal operating voltage for a duration of 5 s. When a disturbance is detected using these criteria, a disturbance file is created that includes at least two cycles of pre-disturbance data as mandated by NERC. NERC mandates at least 30 cycles worth of data including the pre-disturbance data to be stored in the disturbance file. Utilities have been known to adopt more stringent criteria for detection, and longer disturbance files than the mandated norms. For example, disturbance files created by a set of PMUs for a north American utility are 3-min long. It has 55 s (3,300 cycles at 60 Hz) of pre-disturbance data, and 125 s of data (7,500 cycles) after disturbance was detected. The trigger criteria selected by this utility were voltage change of $\pm 10\%$ change in the nominal voltage, frequency change of ± 0.05 Hz, and $df/dt = \pm 5$ Hz/s.

Figure 4.3 shows field recordings of PMU data for different types of disturbances obtained from four PMUs owned by this utility. Voltage and frequency values are plotted in per unit based on 345 kV and 60 Hz, respectively. These are from older generation of PMUs and show positive sequence voltages and frequency measured at 30 fps. The plots show truncated data files covering 0.5 s before and 1.5 s after a disturbance is detected. It can be seen that most of the dynamics are captured within half a second, or 30 cycles (for 60 Hz system) after the disturbance, but voltages for line trip and capacitor switching events do take longer to settle down. Fault and generation loss events in the figure are confirmed through corresponding entries in the utility log; the rest of the events are educated guesses made by comparing the disturbance patterns with simulated data [1].

It can be seen that each type of disturbance leaves a certain signature on voltage and frequency waveforms. The other important observation is that data recorded by all four PMUs show similar trends. This is because PMUs are capturing the same physics during the dynamic event. The PMU electrically closest[‡] to

[‡]A metric for electrical closeness could be the bus impedance matrix element corresponding to the bus with PMU and the fault point.

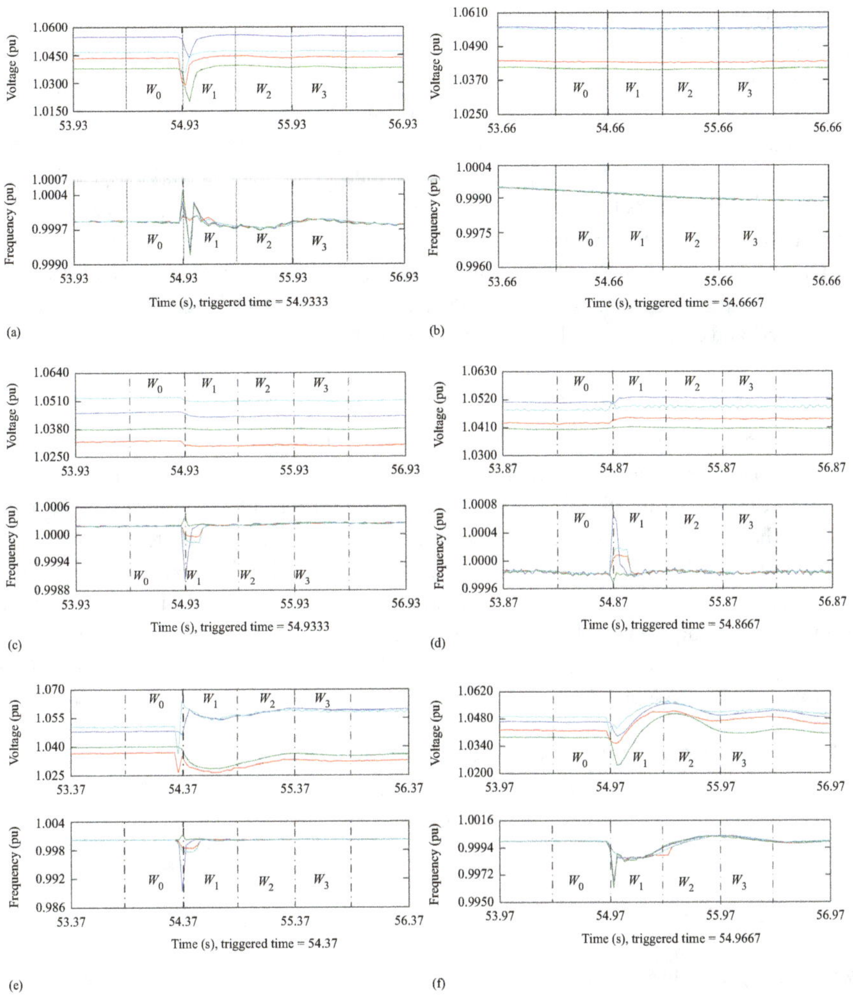

Figure 4.3 Plots of field-recorded PMU data for: (a) fault, (b) generation loss, (c) large load switched on, (d) large loaded switch off, (e) line trip, and (f) series capacitor switched on

a disturbance would show the "strongest" signature, or largest deviations. Measurement noise is also present on the waveforms. A recent study of measurement noise in field PMU data from transmission, primary distribution, and secondary distribution networks show that the noise follows a Gaussian distribution with zero mean and can be approximated with a signal to noise ratio (SNR) of 45 db [2].

Based on the observations made so far, it is reasonable to ask the question—"Instead of simply storing the disturbance files for postmortem use, can we classify them in real time to identify the associated disturbance, thus creating visualization tool for dynamic disturbances?" The answer to this question is explored in this section. This essentially is a classification problem, where classes are the types of disturbance. For any classification problem, the first stage is feature extraction, where data are typically transformed to a domain where representative features are extracted while reducing the volume/dimensionality of original time-series data. Classifiers would use these features to identify the type of the associated disturbance. The science and engineering of feature extraction and classification forms an area in itself. Exhaustive discussion of this area is not within the scope of this chapter. Instead, this section will bring out the difficulties of using field data for credible classification and how these difficulties were overcome in the study that is presented. Since the visualization is required in real time, both speed and accuracy of classification are important.

4.1.1 Feature extraction

Though many feature extraction methods exist in literature, there is no qualitative approach to determine which method will work the best for a given dataset. This section reviews some classical feature extraction methods and a relatively new method that has been found promising for PMU data. Data from different PMUs are highly correlated due to the same underlying physics. This fact can be exploited to first identify the PMU dataset that is most affected by a disturbance, and then simply use that dataset for feature extraction. One such method to identify the most affected dataset is described in Ref. [3]. All of the feature extraction methods summarized here assume such dataset to be available.

4.1.1.1 Discrete Fourier transform

Discrete Fourier transform (DFT) converts data in time domain to features in frequency domain using complex exponentials as the basis.

$$X(k) = \frac{1}{N}\sum_{n=0}^{N-1} x(n)e^{-j2\pi kn/N} \tag{4.1}$$

where $X(k)$ are the DFT coefficients.

DFT coefficients form the representative features for the original time series. DFT works well with stationary signals, but is not designed to capture temporal variation of a time series.

4.1.1.2 Discrete wavelet transform

The discrete wavelet transform (DWT) decomposes a time series into multiple levels of resolution using basis functions derived from a mother wavelet. DWT coefficients form the representative features for the original time series. DWT, unlike DFT, offers a choice of large number of basis functions called mother wavelets. However, its performance is highly dependent on the choice of the mother wavelet, which required trial and error approach. The right choice typically results in a sparse representation, meaning the least number of representative DWT coefficients. Due to data visualization at different resolution levels, temporal variations of a time series are captured by this method.

4.1.1.3 Principal component analysis

This technique represents the coordinates of data in $\Re^D$ dimension as D principal axes, differentiated by their variance. Principal component analysis (PCA) represents the coordinates of data in $\Re^D$ dimension as linear combination of D new orthogonal basis vectors or principal components (PCs). PCs capture the variance of the original correlated data in decreasing order, allowing the user to discard the PCs having low variance without losing critical information, thus allowing dimensionality reduction. PCA has been shown to be capable of capturing hidden dynamics underlying a complex dataset through the PCs.

4.1.1.4 Shapelets

DFT, DWT, and PCA employ domain conversion to create features. Shapelets have been recently introduced by Ye and Keogh as simply "snippets" or sub-sequences of the time-series data [4]. The essence of this method is to discover the most representative subset of a time series as its feature. Shapelets provide easy visualization and interpretation since they are not transformed to another domain. However, they are computationally expensive. The method to extract shapelets essentially requires scanning the entire time series for a sub-sequence of certain length that is most representative of the class associated with the time series. This would mean for a time series of length n, the number of subsequences to be tested is $(n) \times (n+1)/2$, a daunting number for a large time series. After their initial introduction, shapelets discovery has been vigorously researched [5–8]. Efficient algorithms for shapelet discovery are proposed [6,9] because the exponential nature of shapelet candidates prohibits the use of a brute-force algorithm. These algorithms may not extract the same shapelets, but they can still achieve good classification accuracies.

In order to reduce the computation burden of discovering shapelets, domain knowledge can be used. In the case of PMU data, as seen in Figure 4.3, critical changes occur for a short time after the disturbance. So the search can be restricted to a subset of the waveform close to the triggering point of disturbance. Such shapelets are termed *Dshapelet* (*DS* in short) due to the influence of the application domain.

4.1.2 Classification

Popular classification methods work on data with known class labels. Classification of PMU-generated disturbance files technically does not fall into this category. Author has access to field data from a North American utility. The data are from four PMUs in 345-kV network reported at 30 fps reporting positive sequence quantities. Although a total of 1,013 disturbance events were recorded by these PMUs over a 4-year period, the corresponding utility logs are available only for 84 events: 23 generation loss, 58 faults, and 3 line-trips. Thus, there is a large section of data potentially belonging to unknown types of disturbance events; in other words, the number of classes and their labels (disturbance types) are unknown. Before using traditional classifiers over the extracted features from PMU data, the number of classes within the dataset needs to be established. This type of problem lends itself to unsupervised clustering, where partition-based, hierarchical, and density-based approaches exist. A summary of these approaches can be found in Ref. [1]. This paper explains the reasoning behind choosing an agglomerative hierarchical clustering approach that can identify natural clusters and outliers without parametric input from users, and is computationally more efficient compared to divisive clustering approaches. However, domain expertise is critical in identifying the clusters that are formed by this method. The limited field data can also help identifying some clusters. The study reported in Ref. [1] explains this process and reveals that the clusters or disturbance types hidden in the unlabeled disturbance files were (1) faults, (2) generation loss, (3) series capacitor switched on, (4) series capacitor switched off, (5) load switched on, (6) load switched off, (7) synchronous motor switched on. Once the class labels are known, traditional classifiers can be used. Two classifiers—k-nearest neighbor (kNN) and support vector machines (SVM) that performed best in our study [10]—are summarized in this section.

4.1.2.1 kNN classifier

kNN algorithm is an instance-based lazy classification method. Different from most classification algorithms, which consist of a training stage to train a model by utilizing the training data and a testing (or classification) stage to classify testing data using the learned model, kNN does not have a training stage. Instead, it treats all the training data as a model. In the testing stage, this method finds the k most similar nearest neighbors of a testing instance and predicts the testing instance's class label to be the one that is shared by the majority of its kNNs. To use kNN algorithm, a proximity measure to calculate the similarity/distance between data points (e.g., Euclidian distance) is required. k is typically a small positive integer. For the special case of $k = 1$, the test data are assigned to the class of the training data that is closest to the test data. The advantage of this classification method is its easy interpretation and its flexibility of producing arbitrarily shaped decision boundaries. The disadvantage of this method is that its testing stage is expensive when there is large amount of training data.

4.1.2.2 SVM classifier

SVM performs classification based on a few support vectors derived from the training dataset. These support vectors correspond to the position vectors of the training data located near the classification boundary (a hyperplane) within a specified margin—the larger the margin, the better the performance of the classifier. The class to which a test data point belongs is determined by projecting its position vector over the stored support vectors. This basic SVM principle is applicable for classifying linearly separable data. For data with nonlinear decision boundary, an appropriate nonlinear mapping is used to transform the data to a linearly separable subspace. This mapping is achieved by introducing kernel functions for SVM. Appropriate choice of kernel function and its associated parameters can accomplish highly nonlinear classification tasks by storing only a few support vectors instead of storing the entire training data. This generalization of the training data results in faster testing response, but this approach has high memory and computational requirement during the training stage, where support vectors are created. The SVM is applicable for a binary classification problem. The multi-class problems are decomposed into several binary classification problems using one-versus-all or one-versus-one approach to accommodate the use of SVM. Other alternatives also exist. In recent years, SVM have gained much interest in the computer science community due to their complex computations and other application communities due to their accurate prediction. The disadvantage of SVM is that it is hard to interpret the misclassified results, especially when the decision boundary is non-linear. The non-intuitive kernel functions make it difficult to trace a misclassified result back to the data and analyze directly the reason(s) for misclassification.

4.1.3 Case study: disturbance identification using PMU-generated disturbance files

In this case study, we assume that the disturbance types (class labels) are known through unsupervised clustering described in Section 4.1.2. Since there is no universally fitting set of features and classifiers, suitability of features and classifier for every classification problem needs some trials. In this study, results from using the features and classifiers described in Sections 4.1.1 and 4.1.2 on real as well as simulated PMU data are described. The purpose is to find a combination of feature extraction methods and classifier that is best suited to disturbance data generated by PMUs. The full study is reported in Ref. [11].

4.1.3.1 Data

For credibility of any classifier, use of field data and availability of ground truths are two important metrics. As mentioned in Section 4.1.2, author has access to field data for a total of 1,013 disturbance events recorded by four PMUs over a 4-year period, but the utility logs are only available for 23 generation loss events and 58 faults—we neglect the line-trip event for this study. This poses two problems: (1) the data are too few to test a classifier, and (2) ground truths are available for only two events. A slightly non-ideal solution had to be adopted to circumvent this difficulty. General Electric's PSLF dynamic simulation tool has the utility's system simulated in great detail. This software was used to simulate data for more types and number of disturbances as shown in Table 4.1. Two-second long event files

Table 4.1 Events simulated to generate disturbance files

Class	Events
Fault	1,260
Generation loss	558
Load switching off	348
Load switching on	354
Reactive power switched off	497
Reactive power switched on	442
Synchronous motor switched off	36
Total events	3,495

were generated, including 0.5 s of pre-disturbance data. The reporting rate was kept the same as the field data—30 fps. Thus, each event file, for both field-recorded (truncated as shown in Figure 4.3) and simulated events, had 60 data-points. Field data were preprocessed for bad/missing data. This process is described in detail in Ref. [12].

4.1.3.2 Selection of features and classifiers

For feature extraction, all coefficients were selected for DFT and DWT. Least asymmetric wavelet was chosen as the mother wavelet for DWT; however, other mother wavelets like Daubechies and Coiflet also gave comparable results. For the PCA-based features, the largest 20 principal components out of a possible 60 were selected based on trials.

The original disturbance files generated by PMUs are 3-min long, having 5,400 data points for each variable. For an exhaustive search for shapelets, it would take $5{,}400 \times 5{,}401/2 = 14.5827 \times 10^6$ subsequences to process. However, it is clear from Figure 4.3 that the critical information lies around the disturbance trigger. Due to this reason, a subset of the disturbance file was formed and used as mentioned in subsection *Data* of Section 4.1.3.1 to extract *DS*. This substantially reduced the search space. Further, it was observed that the critical information resided around the largest deviations or peaks. Therefore, the first extreme point was selected as the anchor point for extracting *DS*. The algorithm first finds the global extreme points—global maximum and global minimum—from the chosen subset, and chooses the first extreme point as the anchor point. Next, the algorithm extracts one subsequence that is centered at the anchor extreme point and has length $2L+1$. If the anchor point is at index n_1, then the extracted subsequence (or *DS*) is $V_{n_1-L}, \cdots, V_{n_1-1}, V_{n_1}, V_{n_1+1}, \cdots, V_{n_1+L}$. For instance, if the anchor extreme point is 20, which can be the global maximum point, for length $L = 4$, the *DS* that we extract is the subsequence from samples 16 to 24. $L = 4$ was chosen in this study after experimenting with different values. Again, domain knowledge was used to determine the maximum value of L to consider by visual inspection of event waveforms.

Because the variation patterns in disturbance waveforms are different for different types of disturbances, a new feature—slope sequence of *DS*—denoted as (*SSD*), was also created as described below to capture the nature of change in *DS* data using slope.

For a data point $x(n)$ within *DS*, its K-step slope value is calculated as

$$\lambda(n) = \frac{x(n+K) - x(n-K)}{2K}. \tag{4.2}$$

For $K = 1$, given a *DS* with N sampled points, *SSD* is the sequence $\lambda(2), \lambda(3), \cdots, \lambda(N-1)$. To find *SSD* in this study, the *DS* were generated with $L = 5$, and then k was chosen as 1. The reason for choosing a larger L is that *SSD* reduces the length of the feature vector *DS*.

kNN and SVM were used as classifiers. For the kNN classifier, $k = 1$ gave good results, and a radial basis function with Gaussian kernel provided the best results with SVM. Classification was performed using features from (1) just voltage data, (2) just frequency data, and (3) both voltage and frequency data. For the third case, the features derived in the first two cases were concatenated.

4.1.3.3 Results

Due to the constraint that ground truths and field data are available only for two classes, i.e., fault and generation loss, the robustness of the features was tested through different choices of training and testing data as follows:

1. use simulated data corresponding to generation loss and fault for training a two-class classifier, and use field data for testing (results in Table 4.2),
2. use simulated data corresponding to all types of disturbances for training a seven-class classifier, and use field data (two classes) for testing (results in Table 4.3),
3. use simulated data corresponding to generation loss and fault for training a two-class classifier *and* testing, using tenfold cross-validation (results in Table 4.4),
4. use simulated data corresponding to all types of disturbances for training a seven-class classifier *and* testing, using tenfold cross-validation (results in Table 4.5).

Table 4.2 Training (1,818 simulated events with faults and generation loss) and testing (81 actual events)

	Voltage		**Voltage and frequency**	
	1NN	**SVM**	**1NN**	**SVM**
Raw data	77.7	93.8	64.2	71.6
DFT	88.8	72.8	80.2	71.6
DWT	80.2	81.4	62.9	71.6
PCA	71.6	71.6	71.6	71.6
DS	87.6	**100**	69.1	71.6
SSD	95.1	97.5	88.9	**100**

Note: Bold values indicate the best-performing combinations of features and classifiers.

Table 4.3 Training (3,495 simulated events from seven disturbance types), testing (81 actual events with FLT and GL)

	Voltage		**Voltage and frequency**	
	1NN	**SVM**	**1NN**	**SVM**
Raw data	49.4	81.5	61.7	71.6
DFT	55.6	71.6	72.8	71.6
DWT	50.6	75.3	60.5	71.6
PCA	71.6	71.6	71.6	71.6
DS	56.9	74.1	61.7	71.6
SSD	77.8	82.7	72.8	**100**

Note: Bold values indicate the best-performing combinations of features and classifiers.

Table 4.4 Training (90% of 1,818 simulated events with FLT and GL), testing (10% of 1,818 simulated events with FLT and GL); tenfold cross-validation

	Voltage		**Voltage and frequency**	
	1NN	**SVM**	**1NN**	**SVM**
Raw data	99.1	92.7	98.9	89.5
DFT	98.5	87.3	98.3	83.9
DWT	99.1	90.4	98.9	88.1
PCA	69.8	69.3	69.3	69.3
DS	**99.4**	97.9	**99.7**	96.0
SSD	**99.9**	99.3	**99.6**	**99.3**

Note: Bold values indicate the best-performing combinations of features and classifiers.

Table 4.5 Training (90% of all 3,495 simulated events with seven disturbance types), testing (10% of all 3,495 simulated events with seven classes); tenfold cross-validation

	Voltage		**Voltage and frequency**	
	1NN	**SVM**	**1NN**	**SVM**
Raw data	80.6	75.2	86.9	77.8
DFT	74.8	66.5	81.8	64.3
DWT	80.5	74.1	86.9	74.9
PCA	36.7	36.1	36.1	36.1
DS	78.2	69.9	86.3	79.6
SSD	79.8	71.7	**92.5**	88.5

Note: Bold values indicate the best-performing combinations of features and classifiers.

Tables show performance with raw data to establish a baseline. Since the last two cases involve only simulation data, tenfold cross-validation was used. In this approach dataset is partitioned into ten folds, where each fold consists of equal (as far as possible) number of disturbance files for every disturbance type. Configuration x/y means that x% of the disturbance files are used as training data, and the remaining $100 - x\% = y\%$ used as testing data. For each training/testing configuration, *ten* different tests are performed to guarantee that each fold is used at least once as training data and used at least once as testing data. In this study, 90/10 configuration is used. Best results are boldfaced in tables.

Results in the tables show that concatenated features give better results. A combination of *SSD* and SVM performs with near perfect accuracy to classify faults and generation loss (Tables 4.2–4.4). For the experiment described through Table 4.5, where all seven classes are being classified, the accuracy drops, although still respectable. An interesting observation is that raw data and feature extraction methods that choose subsets of raw data and their derivatives (*DS*, *SSD*) perform better than the classical methods performing change of domain. This can be explained by observing the waveforms in Figure 4.3, and noting that the changes in voltage and frequency waveforms are *visually different* for faults and generation loss. In other words, the characteristic features are already present in raw data. Finally, it is observed that simulation data used for training has no noise, but the field data docs contain noise; however, *SSD* is robust against this noise, unlike some traditional methods that fare poorly when used with field data. For the results described in Table 4.5 where accuracy is lower, it was observed that most misclassifications were between load/reactive power switching and generation loss, which is due to similar physics underpinning these events, resulting in similar disturbance patterns. Increase in PMU deployment can probably alleviate this problem, because the PMU nearer to the currently misclassified event may produce stronger signatures, and hence clearer features. In addition, a physics-based approach can be employed to properly classify these events [13].

4.1.3.4 Physics based models

Energy functions are Lyapunov-like structure-preserving functions that have been used for power systems and are well documented [14]. Although they capture the energy of the entire system, they are formed of components that capture energy in specific physical components of power systems. Changes in energy stored in these *components* [15] can be used to identify events that cannot be discriminated with machine learning-based classification. As documented in Ref. [15], changes in the potential energy of the energy function due to active and reactive parts of the load are captured in W_{22} and W_{23}, respectively. W_{25} captures the change in potential energy of the energy function due to the change in magnetic energy stored in all lines, and W_{21} provides the change in mechanical input from all generators. The approach is documented in Ref. [13] and describes how these energy function components are calculated. Plots shown in Figure 4.4 show how these components of energy function track various disturbances created on the IEEE 39-bus test system. The plots are drawn using only partial measurements from the system, as it is assumed only part of the system buses are monitored by PMUs. It is clear that components W_{21} and W_{22} can easily discriminate between generation loss and load loss. Although W_{25} and W_{23} show similar patterns, the amount of change can easily distinguish between reactive power loss and fault. This shows that physics-inspired models can augment the event identification performed by machine learning-based classifiers.

4.2 Use of synchrophasors for power system protection

4.2.1 Real time constraints of power system protection

Power system protection is designed to protect the components of power system from damage and save the system from instabilities. The precipitating reason for potential damage or instability is typically a fault. Faults generate currents several times higher than rated values and must be cleared (by opening appropriate circuit breakers) within a few cycles to save both equipment and stability of power systems. Since this function is crucial, it is bolstered by main and backup functions. Main protection

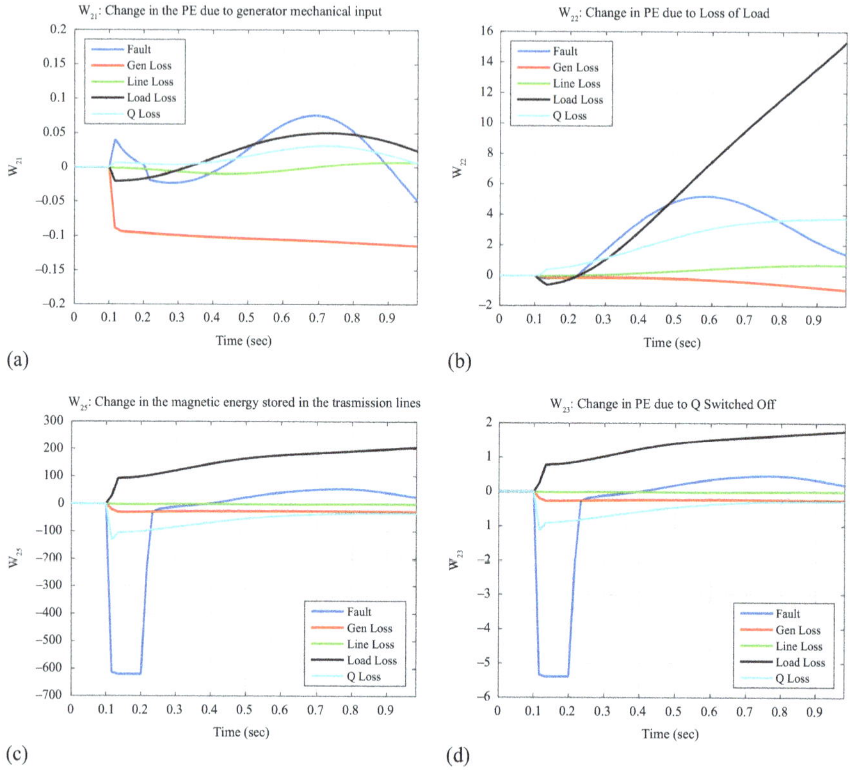

Figure 4.4 Energy function traces with partial measurement for disturbances created on the IEEE 39-bus test system

clears the fault in typically 100 ms or less; if it fails to operate, backup protection operates and clears the fault typically in about half a second, although at the cost of selectivity—meaning some healthy part of the system will also be unnecessarily isolated. Designed and refined through more than half a century of experience, insight, and technological advances, protection function is the fastest operating component of a power system that has kept the system exceptionally safe, in light of the fact that power grid is the largest nonlinear dynamic machine in the world, a large part of which is exposed to extreme weather conditions.

4.2.2 Suitability of PMU measurements for protection

In order to evaluate the possibility of using synchrophasor measurements for power system protection, it is important to assess the accuracy and real-time availability of these measurements with respect to the needs for protective relaying—dependability, security, speed, and selectivity. With current generation of PMUs, there is one synchrophasor measurement available every cycle; protective relays provide typically up to 64 phasor estimates per cycle. According to the guidelines in IEEE C37.118.2-2011-a2014 standard, the *M* Type PMUs consider a seven-cycle sliding window for phasor calculation, whereas *P* type PMUs consider a two-cycle sliding window. Relays consider a one-cycle window. The phasor estimates of a voltage waveform using these three window lengths are shown in Figure 4.5. The voltage waveform was generated using time-domain simulation of a transmission line, and fault was applied at 0.05 s into the simulation. Waveform was processed by a low-pass filter to avoid aliasing. Clearly, relays provide a stable measurement of the fault voltage the fastest—in one cycle after the inception of fault.

It should also be known that most PMUs used for dynamic analysis and state estimation are fed through metering current transformers (CTs) for accurate current measurement during normal operating conditions. These CTs will saturate heavily under faults and cannot be used for any protection applications. If PMU measurements are to be used for protective relaying, PMUs must be fed from protection CTs, which means the metering applications will suffer from larger errors. Moreover, standards do not specify any accuracy limits for waveforms affected by faults. This means PMU estimates for faults, especially for currents with dc offset, may not be accurate enough. It is therefore safer to wait for a few cycles until fault transients damp out when using PMU measurements for protective applications.

Considering these aspects of synchrophasor measurements, it is clear that their use for protection would primarily impact the speed of protection schemes. Most protection schemes in power systems use local measurements, where synchronization with a global time stamp is not required. Since capabilities of a numerical relay exceed those of a PMU for protection applications, it does not make sense to use PMUs for local protection schemes. In cases where such synchronization is desired, it is currently not advisable to use PMU measurements for main protection schemes that typically require the fault cleared in 3–5 cycles. The capabilities of synchrophasors would therefore be better suited to backup or wide area-based protection schemes, and for offline functions like fault location. If, however, PMU manufacturers start providing relay-grade phasor estimates at 120 or 240 fps (for a 60 Hz system) from *P* type PMUs, it may become feasible to use PMUs for important primary protection applications like differential protection, where synchronized phasors can be used in relay logic. However, these applications will have to be designed differently, and a direct communication path for PMU data to be processed by the protection logic may be necessary instead of routing them through a PDC.

4.2.3 Potential improvements in existing protection applications

Currently, there are few protection schemes in practice that rely solely on synchrophasors. Due to the direct availability of angles, there are some documented installations of improved out-of-step (OOS) protection schemes implemented using synchrophasors. These schemes essentially use direct measurement across critical tie-lines to detect power swings. If a swing is detected, a prediction tool is used to predict the trajectory of the swing to determine if the swing is likely to be stable or not. A scheme in Florida,

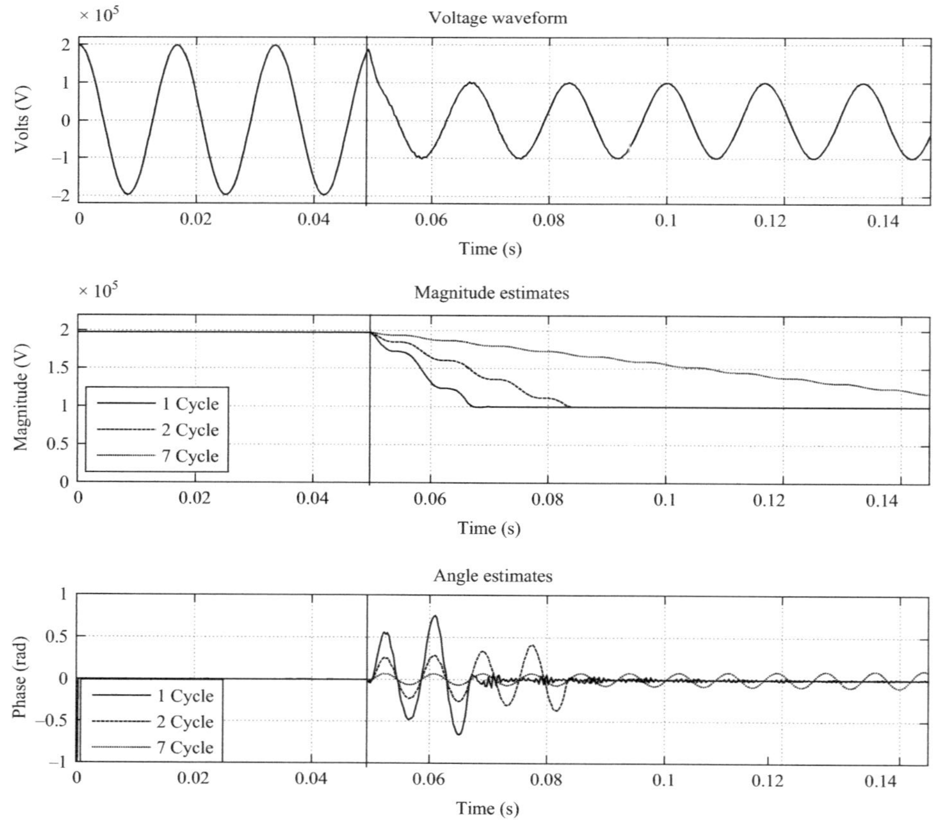

Figure 4.5 Phasor measurement using DFT with different window lengths

USA, and another in Tokyo, Japan are described in Ref. [16]. The scheme in Florida uses a single machine infinite bus (SMIB) model and equal area criteria to predict stability of swing after observing it for up to 250 ms, whereas the scheme in Tokyo uses a time-series prediction tool to predict the swing trajectory 200 ms in future.

A secure backup distance protection for transmission lines can be cited as an example of the potential use of PMUs for backup protection. Load encroachment is a problem that often crops up with the backup protection feature of distance relays. Most blackout logs indicate this problem as a contributing factor to the progression of instability. In this phenomenon, backup protection provided by a distance relay through Zone 3 confuses extreme overload with fault and trips unnecessarily, taking out crucial power corridors at a time when they are most needed. If PMU measurements in the area are used to take over this backup function, this problem can be averted. From multiple data streams arriving at a PDC, a more informed decision can be made if there is a fault or not, and backup protection can be provided much faster than the traditional 500 ms time, even with communication latency.

One of the more ambitious applications would be to improve the performance of a system integrity protection scheme (SIPS). Such schemes acquire measurements from across a wide area of power systems. Currently, settings of such schemes are based on pre-calculated offline studies for foreseeable contingencies. Responses to such contingencies are also predetermined and implemented in design of SIPS. However, there can always be an unforeseen contingency that can destabilize the system. In view of the advances in computing, it may become possible to perform stability analysis in real time using synchronized information of the system state at key locations and generate responses to mitigate a particular contingency as it develops. For example, inter-area oscillations, which represent swinging of one coherent group of generators in one area swinging against another group in an adjoining area, can potentially destabilize the system if they are not damped. Usually, such contingencies are envisaged at the planning stage by simulating many cases of line or generator tripping. Remedial actions to damp such oscillations are predetermined and built into SIPS logic. However, inaccuracy of dynamic system model and inability of exhaustively predicting all dangerous contingency is a drawback. Synchrophasors arriving at a PDC can be treated by a modal analysis tool (prony analysis for example) and damping ratios can be calculated and used for prediction of instability, and subsequent implementation of remedial actions.

However, if PMU measurements are to be used for wide area protection schemes (or other schemes where communication is required), delays introduced by communication channels also need to be considered. Such delays are not capped by standards. The following broad guidelines can be considered based on the IEEE C37.118.2-2011 standard.

- Data takes 3–5 ms to travel 500 miles; this is predictable for a given network.
- Buffering, de-multiplex/multiplex points, forwarding, and routing can add tens to hundreds of milliseconds; this may not be predictable
- Error detection/correction can add hundreds of milliseconds.
- Realistically achievable delay would range between 50 and 130 ms.

4.2.4 Supervisory protection—Potential use of synchrophasors to supervise local protection

There is one area in protection that has eluded protection engineers until now—detection of relay misoperations. If a relay fails to operate for a fault in its primary zone, there is always a backup relay to cover that fault, although at the cost of selectivity and speed. However, if a relay misoperates, meaning it operates without any fault, it cannot be detected in real time. Misoperations can be caused by faulty relay components, also known as hidden faults, incorrect settings, or misinterpretation of system conditions, for example load encroachment in distance relays. In the past, misoperations have been known to trigger blackouts anywhere between a few seconds to a few minutes to even hours [10]. This section explores the possibility of the concept of *supervisory protection* using synchrophasors.

As seen in Section 4.1, it is possible to identify dynamic events using PMU measurements. In the case study described in Section 4.1.3, all fault cases—field-recorded or simulated—were identified correctly using a data segment of 1.5 s after inception of fault. It has been shown that fault identification can be performed much faster in less than 15 ms using dedicated methods that do not require feature extraction or classification [3]. The block diagram of the proposed supervisory protection is shown in Figure 4.6. The scheme takes as input the circuit breaker status from SCADA. If a circuit breaker opens, and fault is *not* sensed by the supervisory layer, it is flagged as misoperation. In this case, either the breaker could be closed back and the associated relay temporarily disabled, or it could be left to the system operator as a warning in real time. In case a fault is detected, the relay action is corroborated.

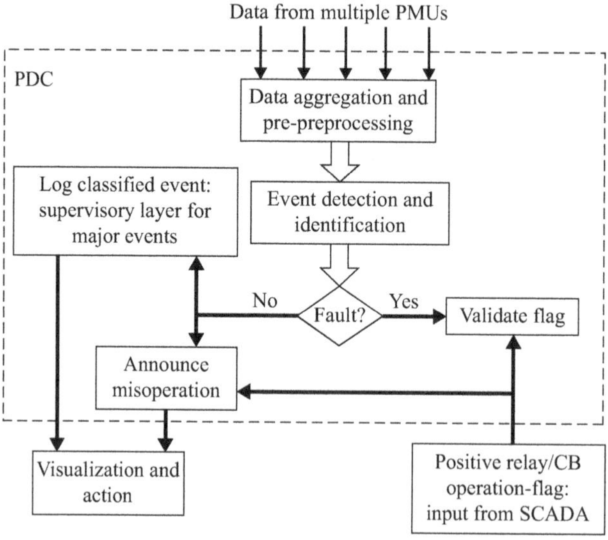

Figure 4.6 PMU-assisted visualization of dynamic events and supervisory protection

4.2.5 Case study: synchrophasor assisted out-of-step protection for generators

Generators usually have a OOS detecting relay (device 78). OOS conditions occur when a generator (or a coherent group of generators) loses synchronism with the rest of the system. This operation needs to be secure, and the most secure scheme used in practice is the single-blinder scheme with trip-on-the-way-out, as shown in Figure 4.7. It is essentially an offset Mho element that maps the angle of the generator rotor with respect to an equivalent source on to an impedance plane. If the impedance enters blinder *B1* and leaves blinder *B2*, OOS is declared. Sometimes, the boundary of the Mho circle is also considered instead of blinder *B*2 to declare OOS. This mapping however is made with simplifying assumptions—generator controls are neglected, and the rest of the system is presented as a Thevenin equivalent. If the rotor angle could be *directly* known even in a dynamic state, the OOS detection can be improved. In this case study, a scheme that uses PMU measurements at the generator bus to estimate the rotor angle is presented.

The scheme is shown in Figure 4.8. It is assumed that the voltage at the generator bus is measured with a PMU and the measurements are available to a dynamic state estimator (DSE) tool. Since the voltage magnitude V and angle θ are *both* available, the state estimator estimates the internal parameters of the generator—in this case the rotor angle δ. The angular difference $\tilde{\delta} = \delta - \theta$ is a direct indicator of swing dynamics. There are no simplifying assumptions needed in this model and generator controls are fully taken into account. When $\tilde{\delta}$ crosses a certain threshold, OOS can be declared.

In this study, the threshold is taken as 120°, which is consistent with general consensus. Particle filter (PF) is used as a DSE. Details of this filter can be found in Ref. [17]. Compared with other widely used DSE algorithms (extended Kalman filter and unscented Kalman filter), the PF is not restricted by model assumption (e.g., probability distribution of measurement noise is Gaussian) and yields superior

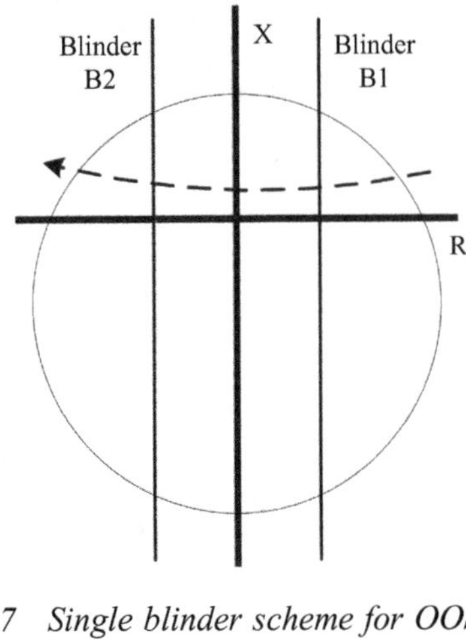

Figure 4.7 Single blinder scheme for OOS detection

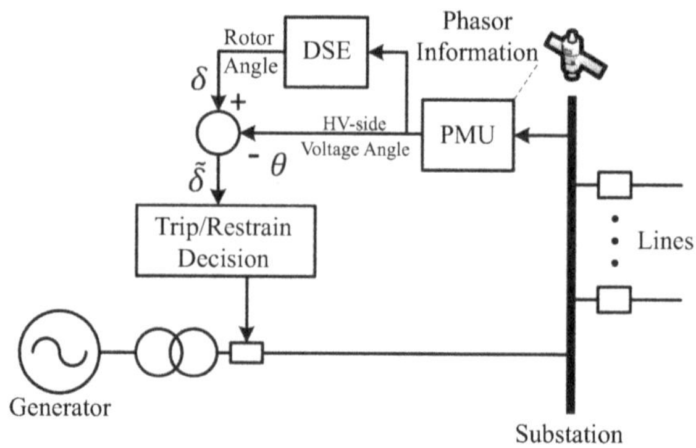

Figure 4.8 Block diagram showing direct estimation of rotor angle to detect OOS

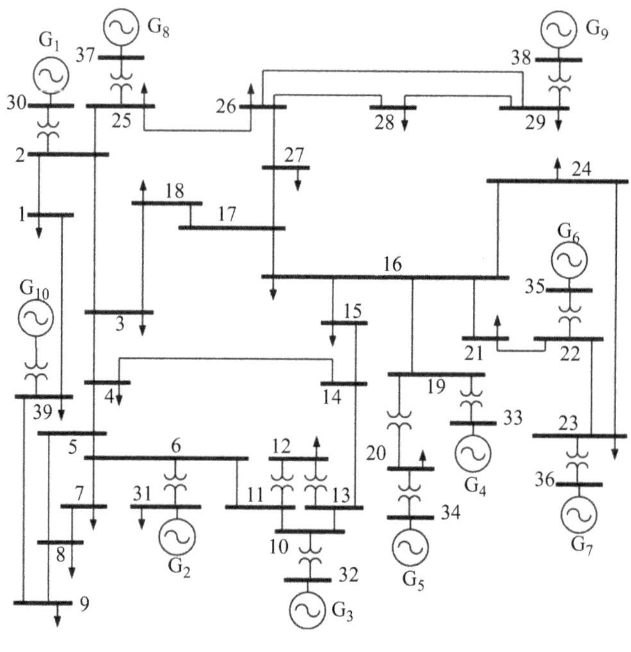

Figure 4.9 New England ten-generator 39-bus system

results on nonlinear/non-Gaussian systems at the expense of increased computational effort. PF is also robust against measurement noise. The proposed scheme is tested on the New England ten-generator system [14] shown in Figure 4.9 simulated on SIMULINK® platform. For all test cases, the measurement reporting rate of the PMU is set to a conservative value of 30 fps.

A three-phase fault was created on line 25–26 at $t = 6$ sec. The faulted line is removed by opening the circuit breaker at both ends of the line. The curves of angular difference ($\tilde{\delta}$) for different clearing times are shown in Figure 4.10. The clearing times are unusually large simply for the purpose of creating swing conditions. Clearing time of 14 cycles results in an unstable swing experienced by *generator 8*, and the worst stable swing is created by choosing to trip the breaker one cycle earlier.

A 3% white Gaussian noise is added to PMU measurements. The generators are modeled accounting for subtransient dynamics (sixth-order model) [18]. The prime mover dynamics (steam-turbine governor) and the excitation system (IEEE DC1A) models are considered for each generator except for *generator #10* that has constant excitation input. The number of particles were selected to be 80. Figures 4.11 and 4.12 show the performance of the conventional single-blinder trip-on-the-way-out scheme and the DSE-based scheme described in this section. It can be observed that DSE is generating usable estimates of the rotor angle, but unlike the conventional relays, its formulation does not make any simplifying assumptions about system topology or controls. This study is described in more detail in Ref. [19]. It is also possible to introduce swing analysis and prediction processes tailored for real-time implementation. This means all the calculations for estimation and prediction should be completed during the time interval between consecutive phasor measurements. Students are encouraged to reproduce the study with different DSE algorithms.

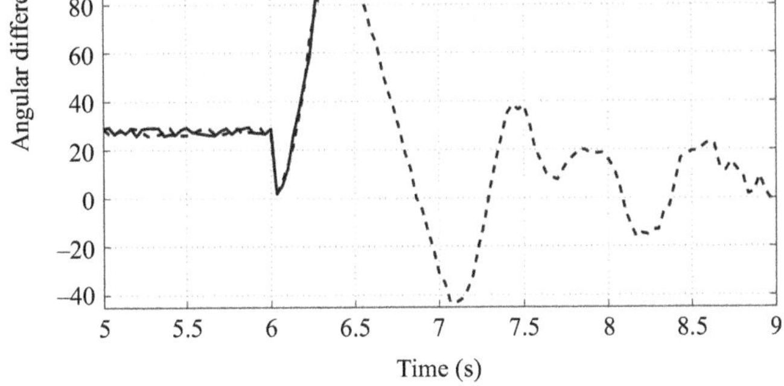

Figure 4.10 Angular difference curves for different clearing times

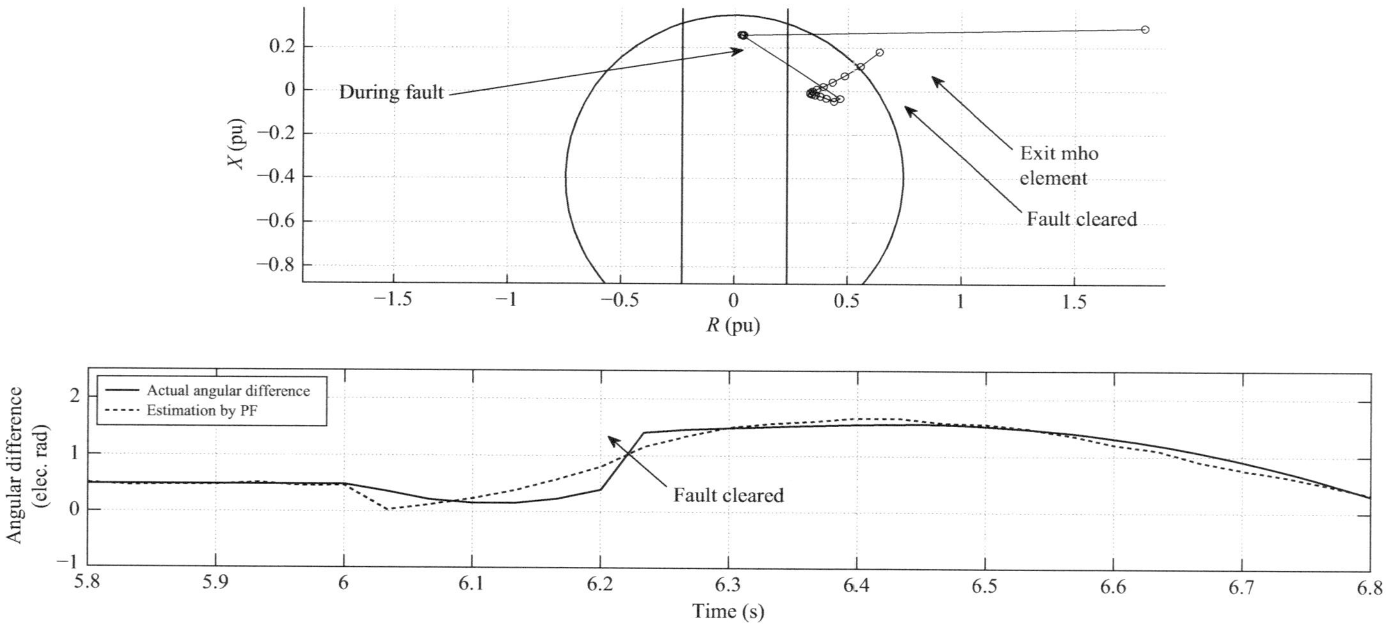

Figure 4.11 Impedance locus for the worst stable swing and estimated angular difference for G_8

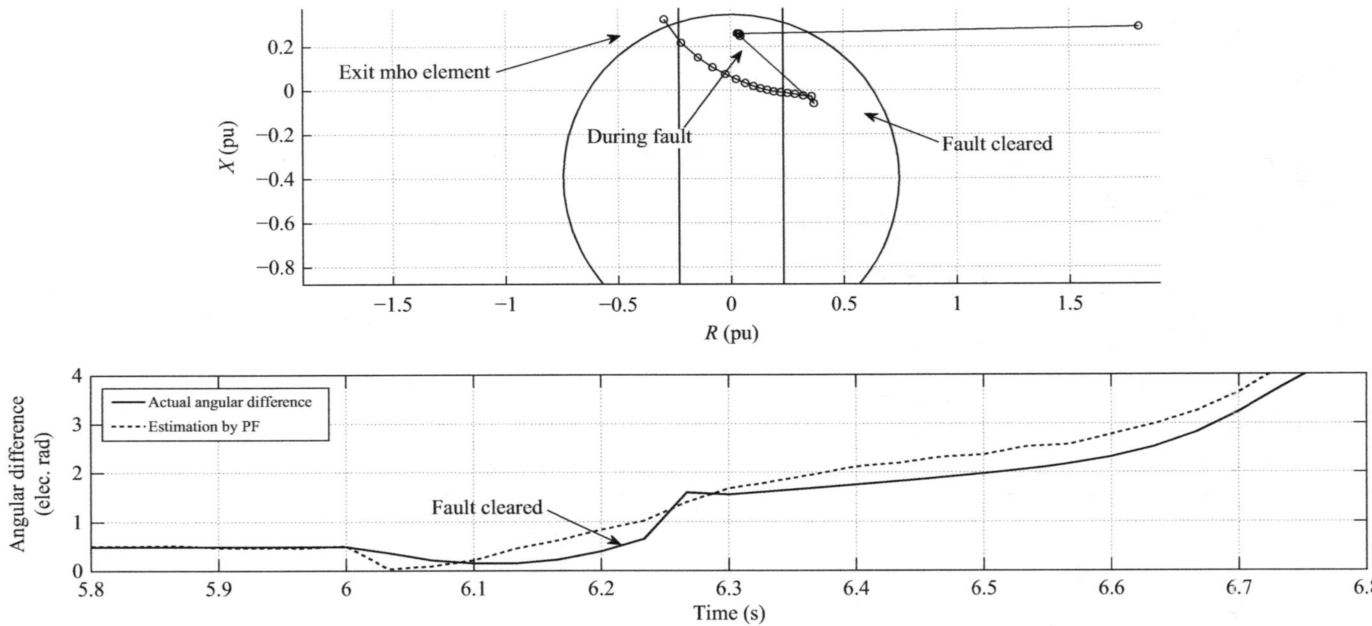

Figure 4.12 Impedance locus for the unstable swing and estimated angular difference for G_8

4.3 Summary

This chapter briefly describes the traditional visualization of power systems using SCADA and postulates the feasibility of adding another layer of situational awareness using PMU measurements. This layer would provide real-time detection and identification of major dynamic events taking place in power systems. Pattern creation and classification models that underpin this layer are described, and a case study is presented to illustrate the performance of some representative models on real and simulated disturbance data. The case study also highlights the difficulties with using field data, like not knowing the classes, having very few ground truths, and describes how these difficulties were overcome.

The chapter also assesses the feasibility of using PMU measurements for protection functions. Discussion includes the real-time constraints imposed by protection, measurements generated by different types of PMUs, communication latencies in PMU data, finally identifying potential facets of power system protection that can benefit from PMU measurements. A new concept of supervisory protection to detect relay misoperations is explained. A case study describing synchrophasor-assisted generator OOS protection is presented.

References

[1] Dahal OP, Brahma SM, and Cao H. Comprehensive clustering of disturbance events recorded by phasor measurement units. *IEEE Transactions on Power Delivery*. 2014;29(3):1390–1397.

[2] Brown M, Biswal M, Brahma S, Ranade S, and Cao H. Characterizing and quantifying noise in PMU data. In: *2016 IEEE Power and Energy Society General Meeting*; 2016. pp. 1–6.

[3] Biswal M, Brahma S, and Cao H. Supervisory protection and automated event diagnosis using PMU data. *IEEE Transactions on Power Delivery*. 2016;31(4):1855–1863.

[4] Ye L and Keogh EJ. Time series shapelets: a new primitive for data mining. In: *Proceedings of the ACM SIGKDD International Conference on Knowledge Discovery and Data Mining*; 2009. pp. 947–956.

[5] Grabocka J, Schilling N, Wistuba M, and Schmidt-Thieme L. Learning time-series shapelets. In: *SIGKDD*; 2014. pp. 392–401.

[6] Keogh EJ and Rakthanmanon T. Fast shapelets: a scalable algorithm for discovering time series shapelets. In: *SDM*; 2013. pp. 668–676.

[7] Chang K, Deka B, Hwu WW, and Roth D. Efficient pattern-based time series classification on GPU. In: *ICDM*; 2012. pp. 131–140.

[8] Lines J, Davis LM, Hills J, and Bagnall A. A shapelet transform for time series classification. In: *SIGKDD*; 2012. pp. 289–297.

[9] Mueen A, Keogh EJ, and Young N. Logical-shapelets: an expressive primitive for time series classification. In: *SIGKDD*; 2011. pp. 1154–1162.

[10] Dahal OP, Cao H, Brahma S, and Kavasseri R. Evaluating performance of classifiers for supervisory protection using disturbance data from phasor

measurement units. In: *2014 IEEE PES Innovative Smart Grid Technologies Conference Europe (ISGT-Europe)*; 2014. pp. 1–6.

[11] Biswal M, Hao Y, Chen P, Brahma S, Cao H, and DeLeon P. Signal features for classification of power system disturbances using PMU data. In: *Power Systems Computation Conference (PSCC), 2016*; 2016. pp. 1–7.

[12] Dahal OP and Brahma SM. Preliminary work to classify the disturbance events recorded by phasor measurement units. In: *2012 IEEE Power and Energy Society General Meeting*; 2012. pp. 1–8.

[13] Brahma S, Kavasseri R, Cao H, Chaudhuri NR, Alexopoulos T, and Cui Y. Real-time identification of dynamic events in power systems using PMU data, and potential applications—models, promises, and challenges. *IEEE Transactions on Power Delivery*. 2017;32(1):294–301.

[14] Pai MA. *Energy Function Analysis for Power System Stability*. Norwell, MA: Kluwer Academic Publishers; 1989.

[15] Kavasseri RG, Cui Y, and Brahma SM. A new approach for event detection based on energy functions. In: *2014 IEEE PES General Meeting|Conference Exposition*; 2014. pp. 1–5.

[16] O'Brien J, Deronja A, Apostolov, A, *et al.* Use of synchrophasor measurements in protective relaying applications. In: *Power System Relaying Committee Report of Working Group C-14 of the System Protection Subcommittee*; 2013. Available from: https://pes-psrc.org/kb/report/103.pdf [Accessed 11 Feb 2025].

[17] Cui Y and Kavasseri R. A particle filter for dynamic state estimation in multi-machine systems with detailed models. *IEEE Transactions on Power Systems*. 2015;30(6):3377–3385.

[18] Kundur P. *Power System Stability and Control*. New York: McGraw-Hill Professional; 1994.

[19] Cui Y, Kavasseri R, and Brahma S. Dynamic state estimation assisted posturing for generator out-of-step protection. In: *2016 IEEE Power and Energy Society General Meeting*; 2016. pp. 1–8.

Chapter 5

Using PMU measurements for enhanced power grid monitoring and protection

Yuan Liao[1] *and Xiangqing Jiao*[1]

5.1 Introduction

Power grid monitoring and protection play a pivotal role in ensuring safe and reliable system operation. Compared with conventional measurements from supervisory control and data acquisition (SCADA) system, phasor measurement units (PMUs) provide time-synchronised and direct measurements of phasors at a higher sampling frequency [1,2]. With increasing deployment of PMUs in power systems, monitoring and protection of power systems may be greatly improved. PMU measurements are capable of capturing system dynamics and thus can be used for dynamic model validation and calibration. PMU measurements can also be used to increase the efficiency of state estimation. IEEE Standard [1] addresses the definition of a synchronised phasor, time synchronisation, application of timetags, method to verify measurement compliance with the standard and message formats for communication with a PMU. Standard [2] defines functional and performance requirements of a phasor data concentrator and provides guidance on concentrator testing.

This chapter discusses the opportunities that PMUs bring to enhance system monitoring and protection. Three topics are included: monitoring of power system operation, monitoring of power system model and wide-area fault detection and location based on PMU measurements. Particularly, latest developments in ZIP load modelling, Kalman filtering-based line parameter estimation and travelling wave (TW) and graph theory-based fault detection and location have been added to this revised chapter.

5.2 Monitoring of power system operation: State estimation

5.2.1 Background

State estimation serves to obtain the best estimate of the internal states of power grids based on redundant measurements. The states of a power system refer to voltage magnitudes and phase angles at every bus. State estimation is one of the most

[1]Department of Electrical and Computer Engineering, University of Kentucky, USA

important functions in energy management systems (EMS), because it provides inputs for advanced analyses and control applications. Without the knowledge of the system's real-time operation condition, applications such as automatic generation control, optimal power flow, economic dispatch, etc. cannot be performed [1]. Due to measurement redundancy, state estimation is able to provide an optimal estimate of the grid's states. This information is then passed on to supervisory control system, which in turn takes control actions. State estimator also serves as a filter between the raw measurements received and the application functions. Measurement noises and gross data are detected and filtered out for more reliable and dependable estimation results.

Conventionally, state estimation takes measurements from devices like intelligent electronic devices (IEDs) and remote terminal units (RTUs) in EMS/SCADA systems. For the areas where measurements are unavailable, pseudo-measurements can be generated based on approximation methods such as historical records, and short-term load forecast, etc. To reduce the impact of inaccuracy, the weights assigned to pseudo-measurements are relatively low.

The state estimator consists of five main functions: topology processor, observability analysis, bad-data processor, parameter and structure error processor and state estimation solution. The topology processor establishes a topology model of the system, based on the initial system model and real-time measurements including the status of circuit breakers and switches, transformer taps and any other discrete information related to system model. Observability analysis determines if the available measurements are sufficient to derive a feasible state estimation solution, based on techniques such as analysis of the property of the measurement Jacobian matrix. If the Jacobian matrix has full column rank, the system is considered to be fully observable. Bad-data processor detects and eliminates any gross errors in the measurements. Parameter and structure error processor is to estimate various network parameters and detect structural errors in the network configuration. Finally, state estimation solution provides an optimal estimate of the power grid's states.

5.2.2 State estimation solution

The inputs of state estimator include system model and measurements, and the major output of state estimator is system state. This section introduces the basic state estimation solution and detection of errors in measurements, network topology and parameters is not included.

Most common measurements for state estimation include line power flows, bus power injections, bus voltage magnitudes and line current flow magnitudes [3,4]. Assume there are m measurements available, based on which a $m \times 1$ measurement vector z can be formed. Measurements inevitably include errors, which can be described by the following equation:

$$z = \begin{bmatrix} z_1 \\ z_2 \\ \vdots \\ z_m \end{bmatrix} = \boldsymbol{h}(\boldsymbol{x}) + \boldsymbol{e} = \begin{bmatrix} h_1(x_1, x_2, \ldots, x_{2n-1}) \\ h_2(x_1, x_2, \ldots, x_{2n-1}) \\ \vdots \\ h_m(x_1, x_2, \ldots, x_{2n-1}) \end{bmatrix} + \begin{bmatrix} e_1 \\ e_2 \\ \vdots \\ e_m \end{bmatrix} \tag{5.1}$$

where z_i is the ith measurement ($i = 1, \ldots, m$), $\boldsymbol{h}$ is the nonlinear measurement function vector which relates the state vector $\boldsymbol{x}$ with the measurement vector $\boldsymbol{z}$, $\boldsymbol{h}$ represents the expected values of measurements $\boldsymbol{z}$, $\boldsymbol{e}$ is a zero-mean Gaussian measurement error vector and the state vector $\boldsymbol{x}$ equals $[x_1, x_2, \ldots, x_{2n-1}]^T$, with $[.]^T$ providing the transpose of a matrix/vector. The first $n-1$ elements of $\boldsymbol{x}$ are voltage angles, and the latter n elements are voltage magnitudes.

To get optimal estimate of state variables, two estimation algorithms, maximum likelihood estimation (MLE) and weighted least square (WLS) estimation, are widely used [1,4,5]. MLE, by maximising the likelihood function, determines the most likely states. The WLS method, by minimising the sum of square of errors, finds the model that deviates the least from the measurements. Given that the measurements errors are commonly assumed to have a Gaussian distribution, the two algorithms will finally yield the same objective function and they are considered to be equivalent.

- Maximum likelihood estimation: The likelihood function in the MLE method is a measure of the probability of obtaining the particular set of measurements. It is represented by the joint probability density function (p.d.f.) for all measurements. Since each measurement is independent of the rest and has the same Gaussian p.d.f., the joint p.d.f. is the product of individual p.d.f.s:

$$\mathscr{L} = f(\boldsymbol{z}) = f(z_1)f(z_2)\cdots f(z_m) \tag{5.2}$$

where L is the likelihood function, $f(.)$ is the p.d.f. function, $\boldsymbol{z}$ is the measurement vector and z_i is the ith measurement. Substitute the Gaussian p.d.f. into (5.2) will lead to the following equation:

$$\mathscr{L} = \prod_{i=1}^{m} \frac{1}{\sqrt{2\pi}\sigma_i} e^{-\frac{1}{2}\left(\frac{z_i-\mu_i}{\sigma_i}\right)^2}, \quad i = 1, 2, \ldots, m \tag{5.3}$$

where $\prod(.)$ returns the product of its arguments, μ_i and σ_i are the expected value and the standard deviation of the ith measurement z_i, respectively. In practice, it is often more convenient to use the logarithm of the likelihood function which is called the log-likelihood function:

$$\begin{aligned} \ln\mathscr{L} &= \ln f(\boldsymbol{z}) = \sum_{i=1}^{m} \ln f(z_i) \\ &= -\frac{1}{2}\sum_{i=1}^{m} \frac{(z_i-\mu_i)^2}{\sigma_i^2} - \sum_{i=1}^{m} \ln \sigma_i - \frac{m}{2}\ln(2\pi) \end{aligned} \tag{5.4}$$

To maximise $\ln\mathscr{L}$, the following objective function is to be minimised:

$$\begin{aligned} J_1(\boldsymbol{x}) &= \sum_{i=1}^{m} \frac{(z_i-\mu_i)^2}{\sigma_i^2} = \sum_{i=1}^{m} (z_i-\mu_i)^2\sigma_i^{-2} \\ &= \sum_{i=1}^{m} [z_i - h_i(\boldsymbol{x})]^2\sigma_i^{-2} \end{aligned} \tag{5.5}$$

- Weighted least square estimation: Based on the measurement accuracy, each measurement is assigned with a weighting factor w_i, $i = 1, \ldots, m$. The more accurate an observation is, the more it should be weighted. The objective is to minimise the sum of mismatch between each measurement and its expected value. And the objective function of WLS estimation can be expressed as

$$J_2(\boldsymbol{x}) = \sum_{i=1}^{m} w_i [z_i - h_i(\boldsymbol{x})]^2 \tag{5.6}$$

where w_i is the weighting factor associated with the ith measurement. The standard deviation of each measurement error distribution reflects the corresponding measurement's accuracy. Hence, the weighting matrix $\boldsymbol{W}$ is written as the inverse of measurement error covariance matrix. Since the measurements are independent, the covariance matrix of measurement error is

$$\boldsymbol{R} = E\left[\boldsymbol{e}\boldsymbol{e}^T\right] = diag\left\{\sigma_1^2, \sigma_2^2, \ldots, \sigma_m^2\right\} \tag{5.7}$$

where $E[.]$ returns the expected value of its argument and $diag\{.\}$ returns a square diagonal matrix with the elements of the input vector on its diagonal. Therefore, $w_i = \sigma_i^{-2}$, and thus the objective function of WLS method is the same as that of MLE method.

Given the objective function, the state estimation problem can be solved by finding

$$\boldsymbol{x} = arg \quad \min \sum_{i=1}^{m} [z_i - h_i(\boldsymbol{x})]^2 \sigma_i^{-2} = arg \quad \min[\boldsymbol{z} - \boldsymbol{h}(\boldsymbol{x})]^T \boldsymbol{R}^{-1} [\boldsymbol{z} - \boldsymbol{h}(\boldsymbol{x})] \tag{5.8}$$

Due to nonlinearity of $\boldsymbol{h}$, $\boldsymbol{x}$ cannot be directly calculated. Thus, iterative algorithm is adopted to solve $\boldsymbol{x}$.

Possible problems of state estimation algorithm include computational burden, excessive requirement of storage memory and numerical instability. Various approaches have been proposed in the past decades to address these issues such as Cholesky decomposition of gain matrix, forward/back substitution, QR factorisation of Jacobian matrix, decoupled formulation, blocked formulation, etc.

State estimation shares some similarities with power flow, but they are different. The essence of solving state estimation and power flow is the same: to find the state vector based on measurements. The measurements for state estimation, however, include more types and quantities than that of power flow. Measurements for power flow only include real/reactive power injections at PQ buses and real power injections and voltage magnitudes at PV buses. Measurements for state estimation are redundant with respect to the unknown state variables. More types of measurements are available in state estimation problem: bus power injections, line power flows, line current flow magnitudes as well as bus voltage magnitudes.

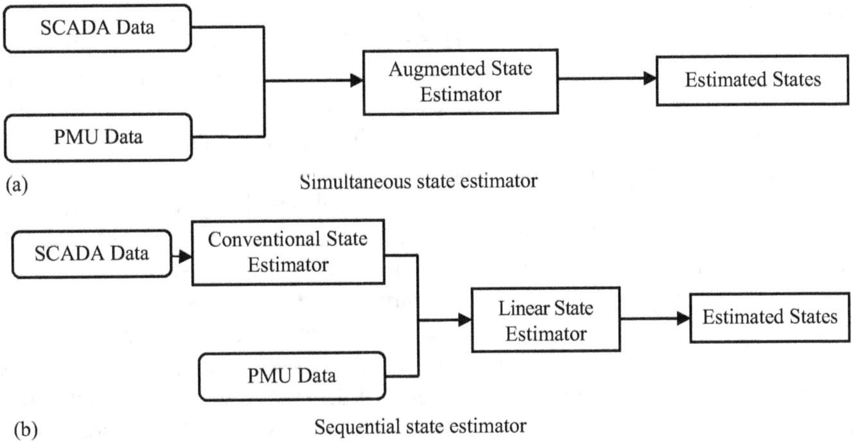

Figure 5.1 Inclusion of SCADA and PMU data in state estimation

5.2.3 Use of PMUs for state estimation

Computational efficiency and numerical stability can be achieved by improving state estimation algorithms. However, revising computational techniques cannot solve other problems associated with measurements such as time skew between measurements, etc. Nevertheless, the use of PMU data may mitigate some of the problems.

If all the measurements are PMU measurements, the state estimator will be linear regardless whether PMUs are installed at all buses or at some buses as long as the system is observable.

If the measurements include both SCADA and PMU data, there are two formulation approaches: simultaneous state estimation formulation and sequential state estimation formulation as shown in Figure 5.1. Simultaneous state estimator mixes phasor measurements with SCADA measurements and uses the traditional nonlinear iterative process to estimate states. Sequential state estimator is a two-step estimator that uses the outcome of traditional state estimation as the input of linear state estimator. PMU measurements and the initially estimated voltage phasors with their respective error covariance matrices are considered to be a set of measurement for the linear estimator. Note that the error of the initial estimate is not zero-mean Gaussian distribution. Its error covariance needs to be appropriately calculated.

5.3 Monitoring of power system model

5.3.1 Introduction

To simulate a power system in software environment, the mathematical model of the power system including its components needs to be developed. The mathematical model is represented by a set of mathematical equations that characterise the behaviour of the power system and its components under specified operating conditions. These equations can be implemented in software modules to develop a software simulation model of the power system. The simulation model is the basis for most power system studies including power flow, state estimation, stability analysis, system planning, etc.

Only with appropriate models that can truly replicate the system behaviour can the model-based analysis, control and planning be conducted with confidence. If there is a significant mismatch between prediction and actual system performance, unexpected power system instability, disturbances and large-scale outages may occur. Such inconsistency triggered serious power outages in history. An example is the 1996 Western Interconnection blackout [8]. The system model miscalculated the maximum transfer capabilities, and expected that the system would still remain in a stable condition under a sequence of disturbances. However, serious violation of system stability occurred, which in turn caused system oscillation, breakup and the blackout. If the transfer limits could have been correctly calculated based on accurate models, such blackout might have been prevented. Another notable example is the 2003 US–Canada blackout that affected nearly 50 million people. It is revealed that the models used in the EMS software during that blackout underestimated reactive power load [6]. The model failed to accurately predict the severe instability. Therefore, the US–Canada Power System Outage Task Force recommended utilities to 'improve quality of system modelling data and data exchange practices' [9].

Moreover, future power systems will keep evolving. It is important to keep the system model accurate and updated, and the accuracy of the models needs to be regularly monitored to ensure the reliability of model-based analyses.

One approach is to use the PMUs to monitor the system models. In other words, the data collected by PMUs can be used to estimate model parameters and validate system models. For example, if PMUs are installed at two terminals of a transmission line, line parameters can then be estimated online based on PMU measurements. Generator model validation and tuning can also be realised using disturbances data collected by PMUs. Model validation refers to a procedure where simulation is performed based on the model to obtain simulated response to a disturbance, which is compared with measured response to a similar disturbance. If a reasonable matching is obtained, the model is said to be acceptable.

PMU data has a few distinguished advantages over conventional measurements collected by SCADA. First, the sampling rate of PMU is much higher than that of SCADA measurements, which facilitates observation of dynamic system phenomena and construction of dynamic models. Second, the PMU measurements are time-synchronised. PMUs installed at different locations of the system can provide measurements that correspond to the same event, which can be readily used to validate the system model. Without PMUs, it is difficult to correlate measurements captured at different locations and use them for model validation or any other analysis.

In the aspect of model validation, PMUs make online model monitoring possible. Previously, offline testing is the most common method to collect data for model validation and parameter estimation. For example, in order to obtain the model for a power plant, the plant needs to be taken offline so that various testing can be conducted. Short-circuit testing has also been used in load modelling [10]. Online model monitoring has a number of advantages over traditional testing. First, online monitoring provides prompt observation of models. Compared with conventional testing, PMU can use real-time responses to estimate parameters or validate model. For some advanced applications that are sensitive to model parameters, real-time parameter estimation is critical. For example, most fault location algorithms require the knowledge of transmission line parameters. However, the line parameters can change significantly due to line sag, conductor temperature, etc. [11]. With the help of PMU-based online parameter estimation, the fault location result can be more accurate. Besides, since the mismatch between simulated and actual outputs can be identified promptly, severe accidents like outage and low-frequency oscillation may be avoided. Frequent model validation will boost system operators' confidence in the power system model. Second, online model monitoring is much more cost-effective. Compared with taking power system components offline to conduct testing, online monitoring allows the power system assets to continue operating without causing any additional costs. By using natural disturbance data, inconvenience for customers due to short-circuit testing and the demand of manpower to conduct testing can be omitted. Third, the measurement-based online monitoring reflects the actual component behaviour. The component's behaviour under off-line testing may be different with its online performance, considering impacts of other components in the system. Compared with testing of individual components, online monitoring gives a closer look at the components' online dynamic characteristics.

Furthermore, both steady-state and dynamic models can be validated by PMU measurements. Power system models can also be classified into steady-state model and dynamic model. Each of the components in the system can be represented by steady-state models. The combination of individual component models is a complete system model that represents the entire system's steady-state behaviour. Such model is known as power flow model, and it generally includes only positive-sequence quantities. PMU measurements during normal operation can be used to validate steady-state models. Dynamics property of the components can be represented by dynamic models. The response of a dynamic model depends on the historical inputs as well as on the present inputs. A dynamic model structure is usually expressed in a form of ordinary differential equations, or difference equations, or partial differential equations, or transfer functions, etc. PMUs can record not only steady-state behaviour but also dynamics response of the system. Hence, PMU measurements makes dynamic model validation possible.

To demonstrate how to perform parameter estimation and model validation, the following part of this section presents the method of online transmission line parameter estimation and dynamic load model validation using PMU data.

5.3.2 Online transmission line parameter estimation

Transmission lines are essential components for delivering power from generation units to distribution substations. Transmission line parameters, including series resistance, series reactance and shunt susceptance, are important for analyses like protective relay setting, fault location and power flow studies. Precise line parameters can ensure the reliability and accuracy of these analyses. The parameters are usually calculated based on design data including conductor geometric parameters, conductor type, etc. However, actual values of these factors may not be the same as assumed, which can result in inaccurate estimation of line parameters. Moreover, the line conductor temperature varies depending on the current flow and weather conditions. Line length also changes due to line sag. Therefore, the actual line parameters are likely to differ from the calculated values. So, it is desirable to design effective methods for online estimation of transmission line parameters for improved accuracy.

The following section presents two line-parameter estimation methods. One is based on the classical analytical method that utilises the relationship of voltages and currents at different terminals of a line. The second method harnesses Kalman filter-based estimation method considering statistical properties of measurements.

5.3.2.1 Line parameter estimation based on a classical analytical method

This section presents an online optimal estimation method utilising PMU measurements [11]. This method is capable of estimating positive-sequence line parameters and filtering potential bad measurements. The studied transmission line is a transposed line between terminals P and Q. Distributed parameter line model is adopted to fully consider the shunt capacitance of the line. Assume PMUs are installed at both terminals, on which the steady-state voltage and current phasors can be measured in real time at different time instants during normal operation. The following notations are adopted: $V_{Pi}^{(1)}$ and $V_{Qi}^{(1)}$ are the ith phasor measurement of positive-sequence voltage at P and Q, respectively; $I_{Pi}^{(1)}$ and $I_{Qi}^{(1)}$ are the ith phasor measurement of positive-sequence current at P and Q, respectively; $Z_c^{(1)}$ is the positive-sequence characteristic impedance of the line; $\gamma^{(1)}$ is the positive-sequence propagation constant of the line; l is the length of the line in miles or kilometres.

Based on the synchronised measurements obtained by PMUs at P and Q, the following two equations can be derived:

$$V_{Pi}^{(1)} - Z_c^{(1)}\sinh\left(\gamma^{(1)}l\right) + \sinh\left(\gamma^{(1)}l\right)\tanh\left(\gamma^{(1)}l/2\right)V_{Pi}^{(1)} - V_{Qi}^{(1)} = 0 \quad (5.9)$$

$$I_{Pi}^{(1)} - \tanh\left(\gamma^{(1)}l/2\right)\frac{V_{Pi}^{(1)}}{Z_c^{(1)}} + I_{Qi}^{(1)} - \tanh\left(\gamma^{(1)}l/2\right)\frac{V_{Qi}^{(1)}}{Z_c^{(1)}} = 0 \quad (5.10)$$

$$Z_c = \sqrt{z^{(1)}/y^{(1)}} \quad (5.11)$$

$$\gamma = \sqrt{z^{(1)}y^{(1)}} \quad (5.12)$$

where $z^{(1)}, y^{(1)}$ are the positive-sequence series impedance and shunt admittance of the line per mile or kilometre, respectively. $z^{(1)}$'s real part $r^{(1)}$ and imaginary part $a^{(1)}$, along with $y^{(1)}$'s imaginary part $b^{(1)}$ are the variables to be determined. For each set of the synchrophasor measurement, two complex equations can be obtained, which can be separated into four real equations. By making use of multiple measurement sets, possible measurement errors can be eliminated based on least square optimal estimation. Statistical approaches can be adopted to detect, identify and remove any possible bad measurements.

Assume N sets of measurements are available. The measurements are

$$\boldsymbol{M} = \left[V_{P1}, I_{P1}, V_{Q1}, I_{Q1}, \dots, V_{PN}, I_{PN}, V_{QN}, I_{QN}\right] \tag{5.13}$$

Introduce $\boldsymbol{S}$ and $\boldsymbol{F}(\boldsymbol{X})$ as the measurement vector and function vector, respectively. They are related by:

$$\boldsymbol{S} = \boldsymbol{F}(\boldsymbol{X}) + \boldsymbol{\mu} \tag{5.14}$$

where, $\boldsymbol{\mu}$ is characterised by the following equation:

$$\boldsymbol{R} = E\left(\boldsymbol{\mu}\boldsymbol{\mu}^T\right) = diag\left(\sigma_1^2, \dots, \sigma_N^2\right) \tag{5.15}$$

Elements of $\boldsymbol{R}$ can be specified according to the accuracy of the metres, a smaller value of which indicates a more accurate measurement. The formation of measurement vector and function vector can be found in Ref. [12].

The optimal estimate of $\boldsymbol{X}$ is obtained by minimising the cost function defined as

$$J = [\boldsymbol{S} - \boldsymbol{F}(\boldsymbol{X})]^T \boldsymbol{R}^{-1} [\boldsymbol{S} - \boldsymbol{F}(\boldsymbol{X})] \tag{5.16}$$

The chi-square test-based bad data identification method can be used. The expected value of the cost function is to be computed first, which is equal to the number of degrees of freedom k. Then, the estimated value of the cost function C_J is calculated. If $C_J \geq \chi^2_{k,\alpha}$, then the presence of bad data is suspected with probability of $1 - \alpha$, where the value of $\chi^2_{k,\alpha}$ can be calculated for a specific k and α based on chi-square distribution. If bad measurement is detected, the measurement corresponding to the largest standardised error will be identified as the bad measurement.

Such online estimation methodology is also applicable to some other transmission network topologies [11], and can also be adopted in fault location algorithms. Following is an example of locating fault that occurred on a series compensated transmission line. Consider the transmission line between terminals P and Q to be a transposed line. The series compensation device is installed at the point of C. A fault may occur either to the left side or to the right side of the compensator.

Assume there are N sets of measurements during normal operation, and one set of measurements during the fault. Then the obtained measurement is

$$\boldsymbol{M} = \left[V_{P1}, I_{P1}, V_{Q1}, I_{Q1}, \dots, V_{PN}, I_{PN}, V_{QN}, I_{QN}, V_{fP}, I_{fP}, V_{fQ}, I_{fQ}\right] \tag{5.17}$$

where $V_{Pi}, I_{Pi}, V_{Qi}, I_{Qi}$ are the ith set of normal operation measurement, $V_{fP}, I_{fP}, V_{fQ}, I_{fQ}$ are the during-fault measurement.

The unknown variable vector is defined as

$$X = [x_1, x_2, ..., x_{8N}, x_{8N+1}, ..., x_{8N+8}, x_{8N+9}, x_{8N+10}, \ldots, x_{8N+18}]^T \tag{5.18}$$

where $x_{8i-7}, ..., x_{8i}$ $(i = 1, \ldots, N)$ are the variables required to represent the 4N prefault complex measurements; $x_{8N+1}, ..., x_{8N+8}$ are the variables required to represent the during-fault complex measurements; x_{8N+9} is the fault location in per unit; $x_{8N+10}, ..., x_{8N+18}$ are the variables required to represent the series impedance, series reactance and shunt suspectance in positive-, negative- and zero-sequence.

The derivation of measurement vector $\boldsymbol{S}$ and function vector $\boldsymbol{F}(\boldsymbol{X})$ is described as follows.

Based on the equivalent π circuit during normal operation shown in Figure 5.2, the following three equations can be establish for sections PC and CQ:

$$I_{0cL}^{(1)k} = -\frac{1}{Z_C^{(1)}}\sinh\left(\gamma^{(1)}l_{PC}\right)V_{0P}^{(1)k} + \cosh\left(\gamma^{(1)}l_{PC}\right)I_{0P}^{(1)k} \quad k = 1, ..., N \tag{5.19}$$

$$I_{0cR}^{(1)k} = -\frac{1}{Z_C^{(1)}}\sinh\left(\gamma^{(1)}l_{CQ}\right)V_{0Q}^{(1)k} + \cosh\left(\gamma^{(1)}l_{CQ}\right)I_{0Q}^{(1)k} \quad k = 1, ..., N \tag{5.20}$$

$$I_{0cL}^{(1)k} = -I_{0cR}^{(1)k} \quad k = 1, ..., N \tag{5.21}$$

where k is the index of set of prefault measurements; l is the distance between P and Q; l_C is the distance between P and C; $V_{0P}^{(1)k}$, $V_{0Q}^{(1)k}$ are the kth positive-sequence voltage phasor measurement at P and Q preceding fault, respectively; $I_{0P}^{(1)k}$, $I_{0Q}^{(1)k}$ are the kth positive-sequence current phasor measurement at P and Q preceding fault, respectively; $Z_c^{(i)}$ and $\gamma^{(i)}$ are the ith sequence characteristic impedance and propagation constant of the line, respectively, $i = 1$ in (5.19)–(5.21).

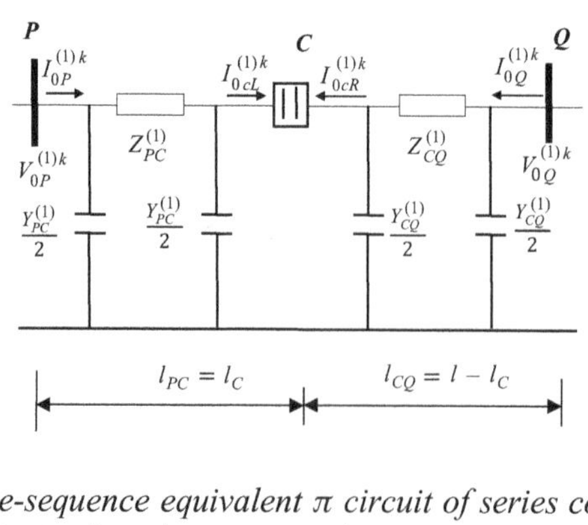

Figure 5.2 Positive-sequence equivalent π circuit of series compensated transmission line during normal operation

Substitute (5.19) and (5.20) into (5.21) leads to

$$f_k(X) = \left[-\frac{1}{Z_C^{(1)}}\sinh\left(\gamma^{(1)}l_{PC}\right)V_{0P}^{(1)k} + \cosh\left(\gamma^{(1)}l_{PC}\right)I_{0P}^{(1)k}\right] + \left[-\frac{1}{Z_C^{(1)}}\sinh\left(\gamma^{(1)}l_{CQ}\right)V_{0Q}^{(1)k} + \cosh\left(\gamma^{(1)}l_{CQ}\right)I_{0Q}^{(1)k}\right] = 0, k = 1, ..., N \tag{5.22}$$

The first N functions in the function vector can be developed based on (5.22). The other functions are to be constructed based on the equivalent circuit during fault and the fault type.

If the fault is an AG fault, it holds that

$$V_f^{(1)} + V_f^{(2)} + V_f^{(0)} = 3R_f I_f^{(1)} \tag{5.23}$$

where $V_f^{(i)}$ and $I_f^{(i)}$ are the ith sequence fault voltage and current, respectively, $i = 0, 1, 2$; R_f is the fault resistance. Since R_f is a real number, $\left(V_f^{(1)} + V_f^{(2)} + V_f^{(0)}\right)$ and $I_f^{(1)}$ are in phase, which can be represented as follows:

$$f_{N+1}(X) = Im\left\{\frac{V_f^{(1)} + V_f^{(2)} + V_f^{(0)}}{I_f^{(1)}}\right\} = 0 \tag{5.24}$$

If the fault is a BC fault, it holds that

$$V_f^{(1)} - V_f^{(2)} = 3R_f I_f^{(1)} \tag{5.25}$$

Thus, the function can be written as

$$f_{N+1}(X) = Im\left\{\frac{V_f^{(1)} - V_f^{(2)}}{I_f^{(1)}}\right\} = 0 \tag{5.26}$$

If the fault is a BCG fault, it holds that

$$V_f^{(0)} - V_f^{(1)} = 3R_f I_f^{(0)} \tag{5.27}$$

Therefore, the function can be written as

$$f_{N+1}(X) = Im\left\{\frac{V_f^{(0)} - V_f^{(1)}}{I_f^{(0)}}\right\} = 0 \tag{5.28}$$

If the fault is an ABC fault, it holds that

$$V_f^{(1)} = R_f I_f^{(1)} \tag{5.29}$$

Hence, the function can be expressed by

$$f_{N+1}(X) = Im\left\{\frac{V_f^{(1)}}{I_f^{(1)}}\right\} = 0 \tag{5.30}$$

The $V_f^{(i)}$ and $I_f^{(i)}$ in (5.23), (5.25), (5.27) and (5.29) can be expressed as functions of the unknown fault location and line parameters based on analysis of the during-fault circuit. It notes that the series compensation device, which is installed at a fixed place along the transmission line, divides the line into two sections. Since on which side the fault occurs is unknown to us, it is necessary to develop two subroutines addressing possible fault on either side.

- Subroutine one: fault occurs to the left side of the compensator.

The equivalent circuit of *i*th sequence transmission line during the fault is shown in Figure 5.3. Based on section *PF* in Figure 5.3, the voltage at the fault point $V_f^{(i)}$ and current on the right-hand side of the fault point $I_{fR}^{(i)}$ can be calculated by the following equations:

$$V_f^{(i)} = \cosh\left(\gamma^{(i)} l_{PF}\right) V_P^{(i)} - Z_C^{(i)} \sinh\left(\gamma^{(i)} l_{PF}\right) I_P^{(i)} \tag{5.31}$$

$$I_{fR}^{(i)} = I_{fL}^{(i)} - I_f^{(i)} \tag{5.32}$$

where $V_P^{(i)}, V_Q^{(i)}$ are the *i*th sequence voltage phasors at *P* and *Q* during fault, respectively, $i = 0, 1, 2$; $I_P^{(i)}, I_Q^{(i)}$ are the *i*th sequence voltage phasors at *P* and *Q* during fault, respectively, $i = 0, 1, 2$ and the current on the left-hand side of the faulted point $I_{fL}^{(i)}$ can be obtained by the following equation:

$$I_{fL}^{(i)} = -\frac{1}{Z_C^{(i)}} \sinh\left(\gamma^{(i)} l_{PF}\right) V_P^{(i)} + \cosh\left(\gamma^{(i)} l_{PF}\right) I_P^{(i)} \tag{5.33}$$

Based on section *CQ* in Figure 5.3, the current on the left-hand side of the compensator can be calculated by the following equation:

$$I_{cL}^{(i)} = I_{cR}^{(i)} \tag{5.34}$$

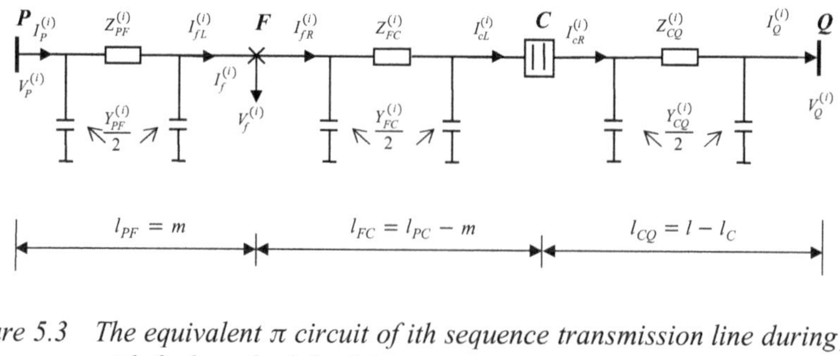

Figure 5.3 The equivalent π circuit of ith sequence transmission line during fault with fault to the left of the compensator

where

$$I_{cR}^{(i)} = \frac{1}{Z_C^{(i)}} \sinh\left(\gamma^{(i)} l_{CQ}\right) V_Q^{(i)} + \cosh\left(\gamma^{(i)} l_{CQ}\right) I_Q^{(i)} \tag{5.35}$$

For section FC in Figure 5.3, it holds that

$$I_{cL}^{(i)} = -\frac{1}{Z_C^{(i)}} \sinh\left(\gamma^{(i)} l_{FC}\right) V_f^{(i)} + \cosh\left(\gamma^{(i)} l_{FC}\right) I_{fR}^{(i)} \tag{5.36}$$

By substituting (5.31), (5.32) and (5.34) into (5.36), the ith sequence fault current can be calculated by using the measurements at both ends.

$$I_f^{(i)} = \frac{\left[\frac{1}{Z_C^{(i)}} \sinh\left(\gamma^{(i)} l_{CQ}\right) V_Q^{(i)} + \cosh\left(\gamma^{(i)} l_{CQ}\right) I_Q^{(i)}\right] + \frac{1}{Z_C^{(i)}} \sinh\left(\gamma^{(i)} l_{FC}\right) V_f^{(i)}}{\cosh\left(\gamma^{(i)} l_{FC}\right)} + \frac{1}{Z_C^{(i)}} \sinh\left(\gamma^{(i)} l_{PF}\right) V_P^{(i)} - \cosh\left(\gamma^{(i)} l_{PF}\right) I_P^{(i)} \tag{5.37}$$

- Subroutine two: fault occurs to the right side the compensator.

Based on section FQ in Figure 5.4, the voltage at the fault point $V_f^{(i)}$ and the current on the left-hand side of the faulted point $I_{fL}^{(i)}$ can be calculated by the following equations:

$$V_f^{(i)} = \cosh\left(\gamma^{(i)} l_{FQ}\right) V_Q^{(i)} + Z_C^{(i)} \sinh\left(\gamma^{(i)} l_{FQ}\right) I_Q^{(i)} \tag{5.38}$$

$$I_{fL}^{(i)} = I_{fR}^{(i)} + I_f^{(i)} \tag{5.39}$$

where

$$I_{fR}^{(i)} = \frac{1}{Z_C^{(i)}} \sinh\left(\gamma^{(i)} l_{FQ}\right) V_Q^{(i)} + \cosh\left(\gamma^{(i)} l_{FQ}\right) I_Q^{(i)} \tag{5.40}$$

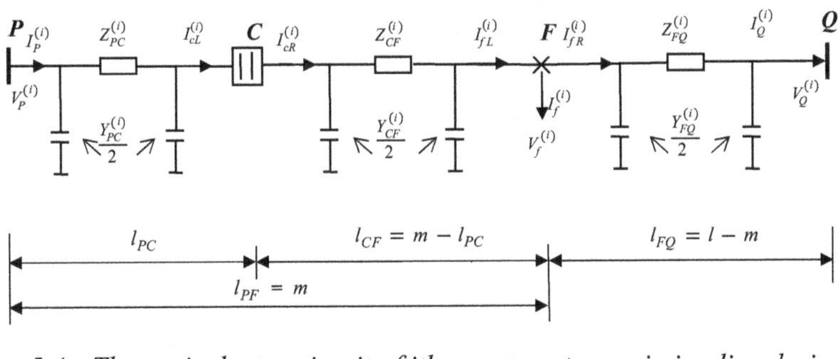

Figure 5.4 The equivalent π circuit of ith sequence transmission line during fault with fault to the right of the compensator

Based on section *PC* in Figure 5.4, the current on the right-hand side of the series compensator can be calculated by the following equation:

$$I_{cR}^{(i)} = I_{cL}^{(i)} \tag{5.41}$$

where

$$I_{cL}^{(i)} = -\frac{1}{Z_C^{(i)}}\sinh\left(\gamma^{(i)} l_{PC}\right) V_P^{(i)} + \cosh\left(\gamma^{(i)} l_{PC}\right) I_P^{(i)} \tag{5.42}$$

For section *CF* in Figure 5.4, it holds that

$$I_{cR}^{(i)} = \frac{1}{Z_C^{(i)}}\sinh\left(\gamma^{(i)} l_{CF}\right) V_f^{(i)} + \cosh\left(\gamma^{(i)} l_{CF}\right) I_{fL}^{(i)} \tag{5.43}$$

By substituting (5.38), (5.39) and (5.41) into (5.43), the ith sequence fault current can be calculated.

$$I_f^{(i)} = \frac{\left[-\frac{1}{Z_C^{(i)}}\sinh\left(\gamma^{(i)} l_{PC}\right) V_P^{(i)} + \cosh\left(\gamma^{(i)} l_{PC}\right) I_P^{(l)}\right] - \frac{1}{Z_C^{(i)}}\sinh\left(\gamma^{(i)} l_{CF}\right) V_f^{(i)}}{\cosh\left(\gamma^{(i)} l_{CF}\right)} - \frac{1}{Z_C^{(i)}}\sinh\left(\gamma^{(i)} l_{FQ}\right) V_Q^{(i)} - \cosh\left(\gamma^{(i)} l_{FQ}\right) I_Q^{(i)} \tag{5.44}$$

Therefore, the variables used in function vector $\boldsymbol{F}(\boldsymbol{X})$ are calculated, and $\boldsymbol{F}(\boldsymbol{X})$ is formulated as:

$$\boldsymbol{F}_{2i-1}(\boldsymbol{X}) = \mathrm{Re}(f_i(\boldsymbol{X})), i = 1, \ldots, N+1 \tag{5.45}$$

$$\boldsymbol{F}_{2i}(\boldsymbol{X}) = Im(f_i(\boldsymbol{X})), i = 1, \ldots, N+1 \tag{5.46}$$

$$\boldsymbol{F}_{(2N+2)+(2i-1)}(\boldsymbol{X}) = x_{2i-1}, i = 1, \ldots, 4N+4 \tag{5.47}$$

$$\boldsymbol{F}_{(2N+2)+2i}(\boldsymbol{X}) = x_{2i}, i = 1, \ldots, 4N+4 \tag{5.48}$$

where Re(.) and *Im*(.) yield the real and imaginary part of their arguments, respectively.

The measurement vector $\boldsymbol{S}$ is formulated as:

$$\boldsymbol{S}_i = 0, i = 1, \ldots, 2N+2 \tag{5.49}$$

$$\boldsymbol{S}_{(2N+2)+(2i-1)} = \mathrm{Re}(\boldsymbol{M}_i) i = 1, \ldots, 4N+4 \tag{5.50}$$

$$\boldsymbol{S}_{(2N+2)+2i} = Im(\boldsymbol{M}_i) i = 1, \ldots, 4N+4 \tag{5.51}$$

where $\boldsymbol{M}$ is defined in (5.17).

The optimal estimate of $\boldsymbol{X}$ can be obtained by minimising the cost function as defined in (5.16).

5.3.2.2 Line parameter estimation based on Kalman filtering technique

The Kalman filtering approach consists of two steps: prediction and update. In the prediction step, the priori estimate of the states at time k is predicted based on the states at time $k-1$. In the update step, the posteriori estimate of the states is obtained based on the observation matrix and observation vector.

For line parameter estimation application, it is assumed that at consecutive sampling moments, the line parameters remain the same and thus the states remain the same. The states, or state vector, will be formulated as a function of line parameters, so derivation of the states will readily lead to line parameters. When line parameters vary over time due to ambient conditions and current flow, the method will be able to dynamically track the value of line parameters [13].

The following equations describe the prediction phase:

$$\widehat{x}_k^- = \widehat{x}_{k-1} \tag{5.52}$$

$$P_k^- = P_{k-1} + Q \tag{5.53}$$

where the subscript k is the instant, the superscript $-$ means priori, $\widehat{x}$ is the estimate of the state vector, P is the error covariance and Q is the process noise covariance. Then, the Kalman gain K, the posteriori estimate $\widehat{x}$ and the posteriori error covariance P at instant k are obtained as follows:

$$K_k = P_k^- H_k^T \left(H_k P_k^- H_k^T + R\right)^{-1} \tag{5.54}$$

$$\widehat{x}_k = \widehat{x}_k^- + K_k\left(w_k - H_k \widehat{x}_k^-\right) \tag{5.55}$$

$$P_k = (I - K_k H_k)P_k^- \tag{5.56}$$

where R is the observation noise covariance, I is the identity matrix, H is the observation matrix and w is the observation vector.

To apply the two steps for line parameter estimation, the relationship between the states and the voltage and current measurements at line terminals is derived, and relevant vectors and matrices that are used in the Kalman filtering steps are identified. Transmission lines have all sorts of configurations, e.g. lines without compensation, lines with series compensation, lines with taps, etc. The principle for solving for line parameters for different line configurations is the same. The following section will illustrate the principle using the example of a mid-compensated line with a series compensation capacitor bank being installed at the middle of the line. See Figure 5.2, where the equivalent series impedance and shunt admittance to the left side of the capacitor and those to the right side of the capacitor are equal. The following equations can be obtained:

$$\begin{bmatrix} I_{0P}^{(1)k} \\ I_{0Q}^{(1)k} \\ 0 \\ 0 \end{bmatrix} = M \begin{bmatrix} V_{0P}^{(1)k} \\ V_{0Q}^{(1)k} \\ V_{CL} \\ V_{CR} \end{bmatrix} \tag{5.57}$$

where $M = \begin{bmatrix} a & 0 & b & 0 \\ 0 & a & 0 & b \\ b & 0 & a+Y_{cap} & -Y_{cap} \\ 0 & b & -Y_{cap} & a+Y_{cap} \end{bmatrix}$, $a = \frac{Y_{PC}^{(1)}}{2} + \frac{1}{Z_{PC}^{(1)}}$ and $b = -\frac{1}{Z_{PC}^{(1)}}$. V_{CL} and V_{CR} are the voltage to the left and right side of the capacitor. Y_{cap} is the capacitor admittance.

Using node elimination technique gives

$$\begin{bmatrix} I_{0P}^{(1)k} \\ I_{0Q}^{(1)k} \end{bmatrix} = \begin{bmatrix} A & \mathrm{B} \\ \mathrm{B} & A \end{bmatrix} \begin{bmatrix} V_{0P}^{(1)k} \\ V_{0Q}^{(1)k} \end{bmatrix} \tag{5.58}$$

where $A = a - \frac{b^2}{a} + \frac{Y_{cap}b^2}{a^2+2aY_{cap}}$ and $B = -\frac{b^2 Y_{cap}}{a^2+2aY_{cap}}$. The equation is rewritten as

$$\begin{bmatrix} I_{0P}^{(1)k} \\ I_{0Q}^{(1)k} \end{bmatrix} = \begin{bmatrix} V_{0P}^{(1)k} & V_{0Q}^{(1)k} \\ V_{0Q}^{(1)k} & V_{0P}^{(1)k} \end{bmatrix} \begin{bmatrix} A \\ B \end{bmatrix} \tag{5.59}$$

The matrices to be used in the Kalman filtering techniques are obtained as follows: the state vector is $x = \begin{bmatrix} A \\ B \end{bmatrix}$, the observation vector $w = \begin{bmatrix} I_{0P}^{(1)k} \\ I_{0Q}^{(1)k} \end{bmatrix}$ and the observation matrix is $H = \begin{bmatrix} V_{0P}^{(1)k} & V_{0Q}^{(1)k} \\ V_{0Q}^{(1)k} & V_{0P}^{(1)k} \end{bmatrix}$.

The Kalman filter method will yield the state vector x, from which a and b can be obtained. The per unit length series impedance and shunt admittance of the line can be further calculated.

This section presents the method for estimating the positive sequence line parameters. Zero-sequence line parameters can be obtained similarly using available zero-sequence measurements, which may be available under unbalanced operating conditions.

5.3.3 *Load model monitoring*

Load model monitoring is periodic validation of the load model used in simulation software such as the EMS software using both steady-state and disturbance data.

A load model is a set of equations that represent the relationship between voltage and/or frequency at a load bus and the real and reactive power consumed by load. Voltage dependency is the main concern, while frequency dependency is usually ignored. In this chapter, only voltage dependency is discussed.

Load model can be categorised as static and dynamic load models. A static load model is not dependent on time, and the model expresses a load's real and reactive powers at any time instant as functions of the bus voltage magnitude and/or frequency at the same instant. An example static load model is:

$$P(t) = c[V(t)]^2 \tag{5.60}$$

where $P(t)$ is the power response at the time instant t, c is a constant coefficient and $V(t)$ is the voltage magnitude at time instant t.

However, a dynamic load model expresses the relation, as functions of current and past voltage and/or frequency. An example of dynamic load model is:

$$P(t) = c[V(t)]^2 + d[V(t-1)]^2 \tag{5.61}$$

where $P(t)$ is the power response at the time instant t, c and d are constant coefficients, $V(t)$ is the voltage magnitude at time instant t and $V(t-1)$ is the voltage magnitude at time instant $t-1$.

When the load response reaches steady state, the steady-state power demand is a function of the steady-state voltage. This function defines the steady-state load characteristics known as voltage-dependent load models in load flow studies. A sudden voltage change would cause an instantaneous power demand change. This change defines the dynamics/transient characteristic of the load. Most of the loads in a power system respond very quickly to the change in voltage. The steady state of the load response is achieved rapidly. Such cases can be analysed by using static models. However, when it comes to stability analysis, static load model is no longer adequate since load representation has critical effects in voltage stability.

Accurate load models are vital for many operation and analysis applications such as power system stability analysis. The fact that power system instability is dynamic requires dynamics representation of the loads. If the load representation is not sufficiently accurate, significant mismatch will occur between the simulation results and the actual response of the load, which will affect the system stability assessment. In the past, power system instability usually occurs in weak systems. However, with increasing loads in the system, this problem has begun to appear in highly developed networks. Furthermore, power systems have been operated closer to their limits nowadays. As power system is more heavily stressed, reliable stability assessment is in greater demand than before, which in turn requires reliable load models.

For most utilities, load models have already been produced in the past. The accuracy of these load models, however, is not monitored based on actual measured power system data. In other words, these models are not regularly checked and validated through operation procedures. The studies based on these invalidated models may be unreliable. Besides, a lot of the existing models approximate loads as static models that neglect loads' dynamic properties. Usually, in the step of model conversion in the dynamic studies using software such as PSS/E, the real power of the load is represented as constant current load, and the reactive power of the load is represented as constant admittance load [6,7]. With growing penetration level of power electronics devices and motors, many dynamic phenomena cannot be replicated by existing models, such as air-conditioner stalling. Given that the dynamics characteristics of system load have a major impact on system stability, static load models are increasingly considered as inadequate for representing loads in dynamics studies. So, it is necessary to monitor the load models and calibrate them if needed.

PMU measurements enables on-line monitoring of load models. The load response captured during steady-state operation can be used to validate model's steady-state performance, and the load response captured during disturbances can be used to validate the dynamic property of the existing models. If serious mismatch is observed, then the model needs to be re-constructed.

5.3.3.1 Online load model validation

Online model verification and validation can be accomplished in the following steps:

1. Data collection: install PMU at load aggregation point, and obtain measurement data from PMUs.

   If field measurement devices have not been installed yet, monitoring devices should be first selected and installed. At medium and high voltage levels, the power system loads have to be aggregated in order to obtain manageable models that are appropriate for simulation and analysis. As a matter of fact, in the context of bulk power system studies, the term 'load' refers to an aggregated load of a feeder, substation or major delivery point on a system; and transmission and sub-transmission substations are usually the location of choice for load aggregation [14]. To achieve the most accurate estimation/validation result, the PMUs should be placed at the load aggregation points. In this way, the behaviour of other system components can be filtered. If PMUs are not installed at aggregation point, the collected data would reflect the combined performance of aggregated load and other asset that is between the load and the PMU. To use these data for validation, the models for the other components have to be included in the analysis.

   The quantities to be recorded are voltage and current phasors, based on which the real and reactive power can be calculated. The quantities should be recorded in two manners: continuous and exception-triggered. The continuous recording is used to validate the steady-state load model. The exception-triggered recording is used to validate dynamics load model; the data should cover the exceptional events such as voltage deviations and current deviations, and should be recorded in a much higher sampling rate.
2. Event identification: for validation of dynamics property, the dynamic disturbances that are suitable for replication should be identified; for validation of steady-state responses, the steady-state normal operation data can be used.

   The extraction of steady-state data is straightforward: select a certain instant during steady state, and record the measurements.

   Disturbances such as system faults and loss of generation usually cause large voltage variations, and can be used for dynamic model validation. In the context of online model monitoring, natural disturbance data rather than testing-based data are used. The use of natural disturbance data is advantageous from a technical and economical point of view.

   In addition to the disturbance data, normal operation data could be used in load modelling. First, the steady-state data can be used to validate model's steady-state response. Second, daily normal operation involves voltage variations as well, and these voltage variations can be used to validate model's dynamic response. Examples of voltage variations are switching on/off capacitors related to reactive power compensation, voltage step change due to tap changer operations, etc. The voltage variations during normal operation are relatively small, but may occur more frequently than large transients. These voltage variations may result in changing power consumed by the load. Such data can be employed to validate the dynamics property of the load. These data can also help to determine the load model structure.

3. Event replication and results comparison: use measured voltage as stimulus, and obtain the response based on its model. Then compare the simulated response with the actual response collected by PMUs and determine the model's validity.

Suppose that the voltage and current at the bus connecting the component to the external system are obtained. To validate the model of the component, the voltage measurements are used to calculate the current of the load. Then the calculated currents are compared with the measured currents to determine whether the load model is acceptable, as shown in Figure 5.5.

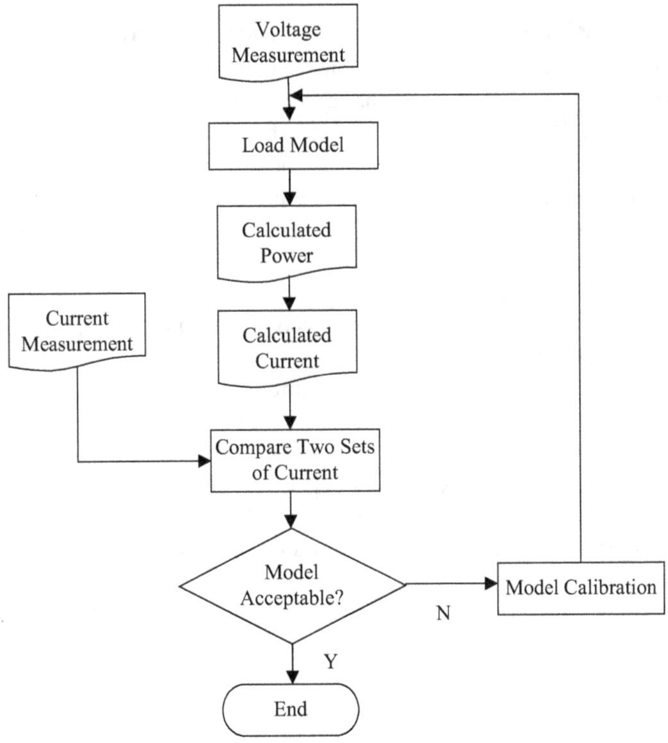

Figure 5.5 Flowchart of event replication and results comparison in load model validation

Model validity is confirmed if the simulated response to a disturbance reasonably matches the measured response to a similar disturbance. The predicted outputs are compared with the measured data (active and reactive power). If there are significant differences between the load's actual behaviour and the model's prediction, calibration of model is required. The differences are to be used to tune the model. More appropriate model structures will be selected, and parameters will be adjusted in order to obtain the best match between the simulated and measured responses. The revised model should be revalidated.

1. Generalisation capability test: repeat validation procedure for various disturbances over a range of different conditions.

Generalisation capability of a model is the model's ability to reflect the load's behaviour under unseen conditions. A load model built based on one set of measurements may not be able to fit the load under other conditions. Therefore, the model's generalisation capability needs to be examined as well. Assume measurements under different conditions are available, and these measurements are regarded as testing data. Use these testing data to the model as stimulus, and obtain the model's validity under these conditions. Based on the model's performance, a rough estimation of generalisation capability can be made.

5.3.3.2 Discussion on load model validation procedure

Although a lot of work has been done on developing load model validation procedure, there are still improvements that can be made.

First, both 'verification' and 'validation' should be included in the procedure. Verification addresses whether the model implements the assumption correctly, and validation examines whether the assumption which has been made is reasonable. In the context of load modelling, the assumption refers to the load structure and model parameters. Therefore, verification means to check if the model is implemented correctly based on the assumed load type and model parameters. Then the validation procedure checks if the assumed type and parameters are correct.

Second, obtaining a confidence level about the model validity based on statistical analysis is more appropriate than a simple binary pass/fail conclusion. The accuracy and uncertainty of a model can be quantified by computing an index called 'accuracy index (AI)' and the respective 95% confidence intervals (CIs). Significance test accepts or rejects a model based on the 95% CI. Such statistical validation result will enable both the model developer and the model user to have quantitative support to decide whether the model is adequate for its intended purpose of use.

Assume in a recorded event, there are N time instants. At each time instant, an actual output value and a simulated output value is recorded. The accuracy index that measures the closeness between the actual and simulated outputs can be calculated by the following equation:

$$Accuracy\ Index\ (AI) = \left[\frac{X_{im}(\theta)}{X_{is}(\theta - 1)}\right]^2, i = 1, \ldots, N \tag{5.62}$$

where X_{im} is the ith measured output, X_{is} is the ith simulated output and N is the total number of time instants. Thus, the overall accuracy for this event can be quantified by

$$\sqrt{\frac{1}{N}\sum_{i=1}^{N}\left[\frac{X_{im}(\theta)}{X_{is}(\theta-1)}\right]^2} \tag{5.63}$$

The uncertainty level of accuracy is to be expressed by an interval within which the true values of the accuracy index are expected to lie with a specified level of confidence. To calculate the CI, first, the mean value of accuracy index $\bar{AI}$, and the standard deviation for the mean value σ should be determined.

Then, the margin of error (MOE) is found by

$$MOE = Critical\ Value \times \sigma \tag{5.64}$$

where the critical value for 95% confidence level is 1.96 [15]. Other confidence levels can be chosen as well, and corresponding critical values should be calculated accordingly.

Finally, the CI is calculated as

$$CI = \bar{AI} \pm MOE \tag{5.65}$$

If the simulation model exactly represents the characteristics of the system or component under study, the simulated outputs will be the same as measured outputs, i.e. the difference between the simulated and measured output is zero. CI gives a plausible range of such difference. All values in the CI are plausible values, whereas values outside the interval are rejected as plausible values for the difference. Therefore, if the CI does not include zero, it can be concluded that the measured outputs are significantly different from the simulated outputs. And the load model is considered as an invalid model.

In addition to adopting 'verification & validation' and statistical judgement, the types of invalidity can be explored in order to help model calibration. There are a few invalidity types that could account for the mismatch between simulated response and the measured response:

- Invalidity Type I: external conditions in the simulation case are erroneous.
- Invalidity Type II: loads are not aggregated appropriately.
- Invalidity Type III: load model parameters are not accurate.
- Invalidity Type IV: the load structure and model parameters are correct, but the model does not implement the assumptions correctly.
- Invalidity Type V: the load model is not applicable to this particular case.

Based on different invalidity types, load model should be calibrated accordingly.

A few reasons could result in Invalidity Type I, such as bad models of other power system elements in the system, inaccurate estimation of power components from measurements, etc. Invalidity Type I should be examined primarily if the model is considered not accurate. This type of invalidity should be excluded primarily.

If the load aggregation site is not appropriate, or the load model structure is not representative of the aggregated loads, the loads are said to be not aggregated well, which results in Invalidity Type II.

Invalidity Type III is straightforward: inaccurate model parameters would almost certainly result in inaccurate responses. From a practical point of view, the parameter estimation problem can be regarded as a curve-fitting problem. Nonlinear least squares-based optimisation algorithms can be used for parameter estimation.

Invalidity Type IV can be checked through the verification process. The computer code that implements the model requires thorough examination.

Invalidity Type V addresses the model's generalisation capability, i.e. the model's ability to predict load's actual response of unseen events. The load characteristics are constantly changing over time. In extreme situations, such as those near the voltage collapse point, the load's characteristic is dramatically different from that during normal operation. Hence, it is acceptable if the model does not work for some particular situations. A dedicated model representing such kind of situation can be built and used whenever such situation occurs.

If a model needs calibration, the load model structure and parameters should be re-estimated.

5.3.3.3 ZIP and exponential load modelling

Two common static load models for steady-state applications are ZIP model, i.e. constant impedance, constant current and constant power model and the exponential load model [16]. The ZIP load model is represented by

$$P = P_0\left[a_p\left(\frac{V}{V_0}\right)^2 + b_p\left(\frac{V}{V_0}\right) + c_p\right] \tag{5.66}$$

$$a_p + b_p + c_p = 1 \tag{5.67}$$

where P are P_0 are the real power consumption of the load at voltage V and the base power at base voltage V_0. a_p, b_p and c_p are real power load model coefficients, namely load model parameters, for constant impedance, constant current and constant power, respectively. The load model for reactive power is similar.

One popular application of ZIP load model is to study the CVR (conservation voltage reduction) factor for distribution systems. When the load model is known, CVR factor can be easily obtained as

$$f_{cvr} = \frac{dP/P_0}{dV/V_0} = 2a_p\frac{V}{V_0} + b_p \tag{5.68}$$

So, a ZIP load has a CVR factor ranging from 0 to about 2.0 around base voltage operating conditions.

The ZIP parameters can be obtained using the least squares approach or optimisation approach, where the optimisation approach can readily consider the parameter ranges. The optimisation method is represented as

$$\min_{P_0,a_p,b_p,c_p} \sum_{k=1,2,\ldots,N} \left(P_0 \left[a_p \left(\frac{V_k}{V_0} \right)^2 + b_p \left(\frac{V_k}{V_0} \right) + c_p \right] - P_k \right)^2 \tag{5.69}$$

subject to all coefficients being between 0 to 1 with a sum of 1 and base power being non-negative. In the equation, the subscript k is the time moment when the measurements are taken.

The exponential load model is expressed as

$$P = P_0 \left(\frac{V}{V_0} \right)^{\alpha} \tag{5.70}$$

where P are P_0 the real power at voltage V and the base power at base voltage V_0. α is the load model parameter and can be estimated as

$$\alpha = \ln\left(\frac{P_2}{P_1} \right) / \ln\left(\frac{V_2}{V_1} \right) \tag{5.71}$$

where the subscript of power and voltage represents measurements at different time moments.

The CVR factor based on the exponential load model is obtained as

$$f_{cvr} = \frac{dP/P_0}{dV/V_0} = \alpha \left(\frac{V}{V_0} \right)^{\alpha-1} \tag{5.72}$$

f_{cvr} is approximately α around base voltage operating conditions.

Accurately estimating the load model parameters requires proper equipment measuring devices that capture voltage and power at sufficiently small time resolutions to make sure the base power P_0 during the sampling window remains the same while there are sufficient voltage variations at different sampling moments in the sampling window. Otherwise, the obtained results may not be meaningful or reliable.

In real-world power systems, load changes over time even under a fixed voltage say the base voltage. The key to accurate load parameter estimation is to distinguish between the natural load variation over time and the load variation in response to voltage changes. Usually, staged tests such as load tap changer operations or disturbance events with sufficient voltage variations may provide needed data to reliably estimate the load parameters [16]. The instrument that records the measurements need to have a sufficiently small sampling resolution, which may be a fraction of second, seconds or minutes.

Load parameters may change over time since load composition changes over time. So, load parameters may be obtained at different moments say at the peak load moment. Average load parameters can then be obtained. If individual load models are obtained in a power distribution grid, the aggregate load model at a feeder header can be analytically obtained by using the load model aggregation method as described in Ref. [17].

5.3.4 System-level model monitoring

If sufficient synchronised measurements are available in a power grid, the validation of individual component models can be extended to system-level model monitoring. To validate the system model's steady-state performance, power flow simulation results are compared with measured voltage phasors and current phasors.

To validate the model's dynamics performance, simulated responses to a recorded event are compared with the measured responses. Before dynamics simulation can be run, the initial condition of the system model needs to be set. The initial condition should match the observed system condition immediately prior to the event. This task can be done by developing the power flow case. The power dispatch of generators, voltage set-point of generators, network topology, transformer tap position, status of switched reactive compensators and load profiles should be set.

5.4 Wide-area fault location based on PMU measurements

5.4.1 Introduction

Electric power is delivered from generation stations to distribution substations through transmission lines. Transmission lines occasionally suffer from faults that are generally created by random and unpredictable events such as lightning, tree touch, animal contact, insulation breakdown due to animal waste, severe weather, etc. When a fault occurs on a line, it is very important for the utilities to identify the fault location as quickly as possible to speed up maintenance work and improve the system reliability. If the fault location cannot be pinpointed quickly, system restoration cannot be performed and the line will be kept out of service, which will result in considerable economic losses. Therefore, accurate and fast transmission-line fault location plays a critical role in quick power system restoration.

Various algorithms for pinpointing faults have been proposed in the past. Algorithms for both single and parallel transmission lines are proposed to help utilities to perform quick restoration. Most of the existing methods require measurements at either one or both terminals of the line. However, such measurements may not always be available. A method that is able to accurately pinpoint the fault solely based on remote measurements is highly desirable. This session presents an accurate fault location method based on wide-area sparse measurements [18]. The wide-area measurements are provided by PMUs, and these measurements may be captured from buses far away from the faulted transmission line. Even if the measurements are available from a single remote bus, the fault can still be reliably located. If measurements from multiple buses are available, optimal estimation theory is exploited to make the most of available measurements for enhanced estimation, and possible bad measurements can be filtered. The method is applied to both transposed and untransposed transmission lines as long as the line parameters are known.

5.4.2 Fault location algorithm

The method is based on the bus impedance matrix concept [18] and is described as follows. To represent the fault, a fictitious bus r at the assumed fault location is added. The transfer impedances between the fictitious bus and the buses with measurement devices can be expressed as functions of the fault location. The driving point impedances of the faulted nodes as well as the transfer impedances between the faulted nodes can also be represented. Based on short-circuit analysis theory, the during-fault voltage and current at the fictitious bus can be derived in terms of the according transfer impedance and driving point impedance. Consequently, they can be written as functions of the fault location. Hence, the reactive power consumed at the fictitious bus can be revealed in terms of fault location. By utilising the fact that the fault resistance only consumes real power, the fault location will be found. Assume the voltage measurements at bus k is available. It can be shown that the reactive power consumed by the fault resistance can be computed as

$$Q = \mathrm{Imag}\left\{-\boldsymbol{E}_r\left[\left(\boldsymbol{Z}_{kr}{}^T\boldsymbol{Z}_{kr}\right)^{-1}\left(\boldsymbol{Z}_{kr}{}^T\boldsymbol{\Delta E}_k\right)\right]^*\right\} \tag{5.73}$$

where Imag{.} returns the imaginary part of its argument, and $(.)^*$ denotes the complex conjugate, $\boldsymbol{E}_r$ is the during-fault voltage at the fault bus, $\boldsymbol{Z}_{kr}$ contains the transfer impedances between the bus k and the fault bus, $\boldsymbol{\Delta E}_k$ is the superimposed voltage at the fault bus. It notes that $\boldsymbol{E}_r$ and $\boldsymbol{Z}_{kr}$ are functions of the fault location. Since the fault resistance only consumes real power (5.58) should equal to zero. The fault location can then be solved by the Newton–Raphson technique. It is noted that the algorithm is independent of fault resistance.

If the faulted line is a double-circuit line, a fictitious bus containing six nodes is to be added. Suppose the line is between buses p and q. The shunt admittance and the series impedance of the line between p and the fault location are $\boldsymbol{y}_1$ and z_1, respectively. The shunt admittance and the series impedance of the line between q and the fault location are $\boldsymbol{y}_2$ and z_2, respectively. The transfer impedances between nodes on bus k and the faulted nodes can be derived as

$$\boldsymbol{Z}_{k_ir} = \left(\frac{y_1}{2}+\frac{y_2}{2}+z_1^{-1}+z_2^{-1}\right)^{-1}\left(z_1^{-1}\left[\boldsymbol{Z}_{k_ip},\boldsymbol{Z}_{k_ip}\right]^T+z_2^{-1}\left[\boldsymbol{Z}_{k_iq},\boldsymbol{Z}_{k_iq}\right]^T\right), i=1,2,\ldots,6 \tag{5.74}$$

where $\boldsymbol{Z}_{k_ip}$ contains the transfer impedances between the ith node on bus k and the nodes on bus p, and $\boldsymbol{Z}_{k_iq}$ contains the transfer impedances between the ith node on bus k and the nodes on bus q.

The transfer impedances and the driving point impedances related to the faulted nodes are expressed as

$$\boldsymbol{Z}_{rr_i} = \left(\frac{y_1}{2} + \frac{y_2}{2} + z_1^{-1} + z_2^{-1}\right)^{-1}\left(z_1^{-1}\left[\boldsymbol{Z}_{pr_i}, \boldsymbol{Z}_{pr_i}\right]^T + z_2^{-1}\left[\boldsymbol{Z}_{qr_i}, \boldsymbol{Z}_{qr_i}\right]^T + \boldsymbol{v}_i\right),\ i = 1, 2, \ldots, 6 \tag{5.75}$$

where $\boldsymbol{Z}_{pr_i}$ contains the transfer impedances between the bus p and the ith node on bus r, $\boldsymbol{Z}_{qr_i}$ contains the transfer impedances between the bus q and the ith node on bus r and $\boldsymbol{v}$ is a 6×6 identity matrix whose ith column is denoted as $\boldsymbol{v}_i$.

Compared to [18], this double-circuit line fault location method obviates the necessity of specifying which circuit of the two circuits is faulted.

In practical power systems, not all fault conditions remain unchanged during fault. An evolving fault refers to a fault where the faulted phase or phases change over time. Locating evolving faults is challenging due to the change in fault type after the fault inception. The presented fault location algorithm in this chapter, however, is immune to fault type. Therefore, to use the proposed algorithm to locate evolving faults, the need of detection of changing fault types is omitted. By applying a moving window to read the phasors from the beginning to the end of the fault data, continuous estimation of fault location can be achieved, regardless of the changing fault type. A window refers to the number of sample points analysed at a certain time instant. The running of the window is triggered by the fault occurrence and terminated at the end of the fault. A locus of estimation can be drawn as the window moves. Since the transition of fault conditions, such as fault resistance and fault type, does not affect the arguments in (5.58), the need of detection and classification of evolving fault is no longer a necessity.

5.4.3 Graph theory and travelling wave-based fault detection and location method

Power distribution grids are traditionally protected by overcurrent protection elements, which are based on sensing the increase of short circuit currents due to faults. Future power systems will have high penetration of inverter-based distributed energy resources (DER) such as renewables, storages and charging stations. In contrast to traditional radial systems, active distribution systems have characteristics of varied fundamental frequency fault current contributions and lack of zero and negative sequence fault currents, which are caused by inverter control methods and practices. So, overcurrent protection will no longer be effective. Moreover, intermittent arcing faults generate little fault currents and traditional overcurrent relay will not function at all. However, both temporary and permanent faults usually generate high-frequency transients that are affected little by the output of DERs. So, harnessing the fault-generated high-frequency transients may be a promising way to design new, effective protection schemes for inverter-based DER dominated power grids. In addition, IEDs installed in the system capture an increasing amount of measurements. Data-driven and machine learning-based methods provide a potential way of utilising these data. This section presents transients, data-driven and machine learning-based novel protection algorithms that are suitable for power grids with any type of DERs. PMUs is one type of synchronised IEDs.

Strategically placed PMUs continuously collect measurements from the system and communicate extracted features into a central location for detecting and locating faults. When a fault occurs, a TW is generated and travels along both directions along the feeder, which is governed by telegraph equation: $\frac{\partial u}{\partial x} = -\left(Ri + L\frac{\partial i}{\partial t}\right)$ and $\frac{\partial i}{\partial x} = -\left(Gu + C\frac{\partial u}{\partial t}\right)$, where u, i, denote voltage and current along the line at location x; R, L, G, C represents per unit length line resistance, inductance, conductance and capacitance. The TW will reflect and refract along the network. Then, each IED will record a series of TW that may include the initial wave and reflected and refracted waves that will arrive at IEDs at different times with different magnitudes and specific polarity. By analysing the information, the faulted feeder segment can be decided, and then the relays that are designed to isolate the faulted feeder can trip the associated breakers to isolate the fault.

Based on graph theory, a network description matrix $[N]$ is defined based on the connection of nodes and direction of feeder sections. Its element N_{kl}, the kth row and lth column, is determined as follows. $N_{kl} = 1$, if node l is the start node of kth feeder. $N_{kl} = -1$, if node l is the end node of kth feeder. $N_{kl} = 0$, if node l is neither start node nor end node of kth feeder. A fault vector $[F]$ is further defined, with its element F_l being determined as follows. $F_l = 1$ if the current TW direction, which can be decided by its polarity, is the same as the designated feeder section direction; $F_l = -1$ if the current TW direction is opposite to the designated feeder section direction; $F_l = 0$ if the current TW does not reach node l. The faulted section identification vector is defined as $[FS] = [N][F]$. The criterion for identifying the faulted section is stated as follows. The value of FS_k, the kth element in $[FS]$, indicates whether the section is faulted or not. If no section is faulted, then it means there is no fault on the system.

In a centralised protection scheme, the data captured by all the IEDs are transmitted to a central controller that processes the TW information, decides the faulted segment and trips related breakers. In a decentralised scheme, the IEDs at the breaker location exchanges TW information and permits or blocks tripping operation of neighbouring breakers. The IEC 61850 standard may be used for data exchange between distributed IEDs and the central IED.

In addition, since an IED may misoperate, Bayes' theorem is employed to increase the fault tolerance of the fault location method [19]. Accuracy of IED operation is characterised by:

$P(X = 1|, F = 1)$: probability of correct operation when there is a fault, c, say 95%;
$P(X = 0|, F = 1)$: probability of incorrect operation when there is a fault, $1 - c$;
$P(X = 0|, F = 0)$: probability of correct operation when there is no fault, d, say 92%;
$P(X = 1|, F = 0)$: probability of incorrect operation when there is no fault, $1 - d$.

Suppose there are N IEDs and K sections or areas. When a fault occurs, each IED will report fault information, i.e. TW direction, to a central controller. Let S_{IED} be the set of all IEDs, $R_{S_{IED}}$ be the set of all IED reports and A be the set of sections. The Bayes' theorem is given by

$$P(A_k|R_{S_{IED}}) = P(R_{S_{IED}}|, A_k)P(A_k)/P(R_{S_{IED}}) \tag{5.76}$$

$$P(R_{S_{IED}}|, A_k) = \prod_{\forall i \in S_{IED}} P(R_i|, A_k) \tag{5.77}$$

$P(A_k|R_{S_{IED}})$ is the probability of fault occurring in section A_k given IED reports, $P(R_{S_{IED}}|, A_k)$ is the probability of IED reports if fault occurs in section A_k, $P(R_{S_{IED}})$ is the probability of IED reports, $P(A_k)$ is the probability of fault occurring in section A_k and $k = 1, \ldots, K$. The faulted section A_{k_f} is determined as the section that has the maximum probability given the IED reports:

$$A_{k_f} = \underset{\forall k}{\operatorname{argmax}}(P(A_k|R_{S_{IED}}) \tag{5.78}$$

Moreover, TW information from multiple IEDs will help pinpoint the exact location of the fault and speed up maintenance and restoration efforts. Depending on the number and location of IEDs, single-end or two-end fault location method can be used, after extracting pertinent wave front arrival times from the series of wave front information. In single-end method, fault location is obtained as $L_{T1R} = \Delta T v/2$, where ΔT is the appropriate time difference and v the TW speed. In two-end method, we have $L_{T1R} = \left\{ L_{T1T2} - \left(\Delta T + \sum_{i \in I} \frac{L_i}{v_i} - \sum_{i \in J} \frac{L_i}{v_i} \right) v_{T1T2} \right\} /2$, where, L_i, v_i are length and wave speed of segment i. I, J are set of non-faulted feed segments between measuring terminals and faulted segment terminals.

The methods will be applicable to all types of faults including balanced and unbalanced faults and will be suitable to radial or meshed distribution networks with unbalanced, heterogenous feeders and loads. A modal analysis method can be applied to cope with unbalances in the system. Advanced signal processing techniques including wavelet analysis and correlation analysis can be applied to precisely identify the different types of wave fronts, decide the wave arrival time and pinpoint the fault location.

For economic considerations, it may be desirable to perform fault detection and location observability analysis and develop an optimal IED placement scheme to precisely detect and locate each fault across the network. Mathematically, the problem is formulated as a mixed integer linear programming problem that minimise the total cost of IED installation and operation, i.e. $\min \sum_i c_i x_i$, S.T. $f(x) \geq 0$, where c_i, x_i denotes cost associated with installing an IED at bus i, and installation decision variable, respectively. Constraint function $f(x)$ will be constructed based on network data and fault analysis.

Although the TW method is expected to work for almost all fault scenarios, its performance degrades when the fault occurs around zero voltage. Thus, it is necessary to have in place a fundamental frequency voltage-based method as a supplemental approach for enhanced reliability. This method does not require detailed DER modelling and the current measurements from the DER. So, this method has the advantage of being immune to DER output variation and varied short circuit current. This method can use rather low sampling frequency, say 12 samples per cycle. The relationship between the voltage measurements and the bus impedance matrix of the network can be derived. The bus impedance matrix is analytically deduced as a function of the unknown fault location, where the driving point and transfer impedances are particularly useful for fault location. Then, the voltage measurements at any buses can be derived as a function of the fault location. Therefore, we can compute the fault location based on voltage measurements. The method can also handle feeder unbalances. It does not require fault type information, is applicable to any type of fault and the fault type can be decided as a by-product.

It is important to ensure the security of the method. DER resources operation variation and other utility routine operations such as load switching may cause temporary voltage excursions outside normal ranges. The protection method should avoid tripping during these conditions. One way to achieve this is that the method will not only use voltage magnitude changes but also use angle changes, which typically have more drastic variation during a fault than during non-fault utility operations. The voltage magnitude variation pattern may be utilised for fault detection, since the pattern may differ for a fault than for a non-fault operation.

Besides, it is possible to apply machine learning-based methods for detecting and locating faults in the integrated power systems. IEDs installed across the system continuously collect voltage and current measurements from the system, and pertinent signal signatures will be extracted for training the network and predicting future faults.

5.5 Summary

The unique advantages of PMUs have brought numerous opportunities to enhance power grid monitoring and protection. This chapter presents potential applications of PMUs in areas of fault detection, location and load modelling. While applying the data collected by PMUs, various techniques including analytical method, signal processing and machine learning-based approaches may be harnessed. In addition to the topics discussed above, many other applications can be enhanced by PMUs such as frequency stability monitoring, power oscillations and voltage monitoring, etc. In summary, PMU has potential to greatly improve system operators' ability to monitor the grid, and the real-time operation and control can be conducted with more confidence. In the future, increasingly more PMU-based applications will be expected to be developed, tested and deployed on power grids.

References

[1] IEEE Standard C37.118 2005. *IEEE Standard for Synchrophasors for Power Systems*, New York, 2006.

[2] IEEE Guide C37.244-2013. *Guide for Phasor Data Concentrator Requirements for Power System Protection, Control and Monitoring*, New York, 2013.
[3] Abur A., and Gomez-Exposito A. *Power System State Estimation Theory and Implementation*. New York: Marcel Dekker, Inc.; 2004.
[4] Monticelli A. 'Electric power system state estimation'. *Proceedings of the IEEE*. 2000, vol. 88(2), pp. 262–282.
[5] Huang Y.-F., Werner S., Huang J., Kashyap N., and Gupta V. 'State estimation in electric power grids: meeting new challenges presented by the requirements of the future grid'. *IEEE Signal Processing Magazine*. 2012, vol. 29(5), pp. 33–43.
[6] North American Electric Reliability Corporation (USA). *Power System Model Validation*. 1 vol. Princeton, NJ: NERC; 2010.
[7] Siemens Energy, Inc. (USA). *PSS/E Program Operation Manual*. Schenectady, NY: Siemens; 2010.
[8] North American Synchrophasor Initiative (USA). *Model Validation Using Phasor Measurement Unit Data*. 1 vol. Richland, WA: NASPI; 2015.
[9] U.S.-Canada Power System Outage Task Force (USA & Canada). *Final Report on the August 14, 2003 Blackout in the United States and Canada: Causes and Recommendations*. 1 vol. Washington: DOE; 2004.
[10] He R., Ma J., and Hill D. 'Composite load modeling via measurement approach'. *IEEE Transaction on Power Systems*. 2006, vol. 21(2), pp. 663–672.
[11] Du Y., and Liao Y. 'Online estimation of power transmission line parameters, temperature and sag'. *North American Power Symposium*; Boston, USA, August 2011. New York: IEEE; 2011. pp. 1–6.
[12] Liao Y., and Kezunovic M. 'Optimal estimate of transmission line fault location considering measurement errors'. *IEEE Transaction on Power Delivery*. 2007, vol. 22(3), pp. 1335–1341.
[13] Zhang Y., and Liao Y. 'Kalman filter based method for tracking dynamic transmission line parameters'. *SN Applied Science*. 2021, vol. 3, p. 117.
[14] Electric Power Research Institute (USA). *Measurement-Based Load Modeling*. 1 vol. Palo Alto, CA: EPRI; 2006.
[15] Utts J.M., and Heckard R.F. *Mind on Statistics*. 4th edn. Boston, MA: Brooks/Cole, Cengage Learning; 2012.
[16] Zhang Y., Liao Y., and Yadav G. 'Experiences about calculating ZIP and exponential load model parameters'. *International Journal of Emerging Electric Power Systems*. 2024. https://doi.org/10.1515/ijeeps-2024-0011.
[17] Zhang Y., Liao Y., Jones E., Jewell N., and Ionel D. 'ZIP load modeling for single and aggregate loads and CVR factor estimation'. *International Journal of Emerging Electric Power Systems*. 2022, vol. 23, no. 6, pp. 839–858.
[18] Jiao X., and Liao Y. 'Accurate fault location for untransposed/transposed transmission lines using sparse wide area measurements'. *IEEE Transaction on Power Delivery*. 2016, vol. 31, no. 4, pp. 1797–1805.
[19] Liao Y. 'Transients based faulted feeder identification for ungrounded power distribution systems with distributed generations'. *Kansas Power and Energy Conference, Kansas State University*, April 19–20, 2021.

Chapter 6

Micro-phasor measurement unit-based energy management system

Jennie Angela Jose Shirley[1], *Prashant Khare*[1], *Harini Manivelan*[1] *and Maddikara Jaya Bharata Reddy*[1]

6.1 Introduction

The power system has become increasingly complex to satiate the ever-growing needs of the economies. This has caused a shift from the conventional power networks to smart power grids with advanced control and monitoring schemes. Renewable energy sources have become an integral part of smart grids, in order to meet the increasing power demand and to make the energy generation process more sustainable. However, the incorporation of such resources has introduced variability and uncertainty in the networks, which need to be properly addressed. To achieve complete observability of these power grids, wide area monitoring systems based on phasor measurement units (PMUs) and micro-phasor measurement units (μPMUs) have been adopted [1,2]. In such a scenario, an energy management system (EMS) plays an important role in the optimization of electricity distribution and efficient integration of new energy sources [3]. This chapter discusses an EMS for maintaining supply–demand balance in the distribution system using μPMUs.

6.2 Evolution of the power grid

The conventional power grid transmits the generated power over a long distance and provides supply to the end-user. It allows a uni-directional flow of power from the generator to the consumer. These systems mainly rely on conventional energy sources such as fossil fuels that are environmentally polluting. They seldom use renewable energy sources, which facilitates control and monitoring with fewer sensors and actuators [4].

With the incorporation of renewables in the system and the increased need for luxury, smart grids are being adopted. These modern grids encourage localized

[1]Department of Electrical and Electronics Engineering, National Institute of Technology Tiruchirappalli, India

Table 6.1 Differences between conventional and modern power grids

Conventional power grid	Modern power grid
Bulk energy generation	Distributed energy generation
Requires few sensors	Requires large number of sensors
Uni-directional information and power flow	Bi-directional information and power flow
Renewable energy resources are rarely integrated	Renewable energy resources are integrated often
Passive system	Active system
High environmental pollution	Low environmental pollution

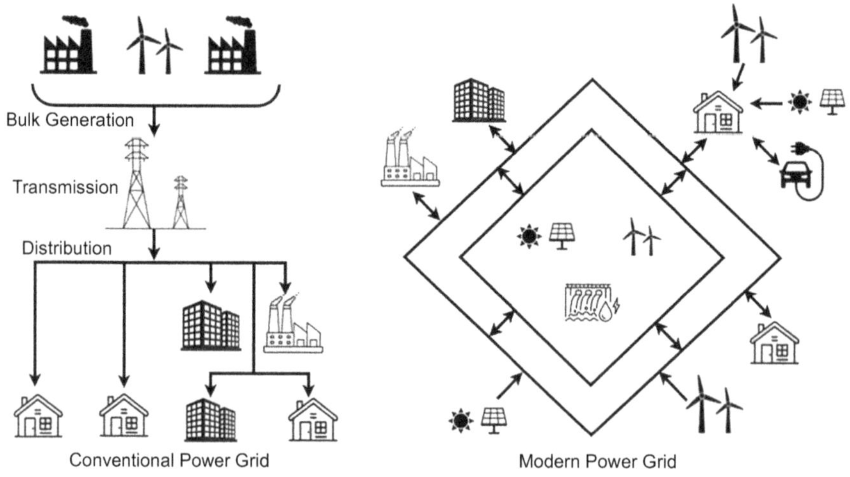

Figure 6.1 Conventional and modern power grids

power generation and advanced control techniques. These systems facilitate two-way flow of power. They enlist the use of smart distribution systems, smart sub-stations, smart meters and various smart measuring devices to provide enhanced reliability and efficiency. The modern systems rely on real-time data to take energy decisions [5]. The differences between a conventional power system and a modern power grid are listed in Table 6.1. Figure 6.1 shows a pictorial representation of the conventional and modern power grids.

6.2.1 Challenges in the modern power grid

As the functionalities and size of the power grid increases, the challenges faced in the grid operation also increase. These challenges may be caused by large-scale usage of renewable energy sources, integration of electric vehicles, grid modernization, transmission losses, and power outages. Renewable energies are intermittent in nature and cannot provide a stable power profile. Further, their integration into the system creates a bi-directional power flow, as against the uni-directional power flow in conventional systems [6]. This necessitates a well-developed control scheme without which there may be grid imbalance.

Electric vehicles have gained popularity owing to their sustainable nature; however, their long charging periods pose a major setback. To speed up the charging, superchargers are used that take up a huge amount of energy. Hence, when an electric vehicle is charged, the power demand shoots up, which may cause disturbances in the grid [7]. Quick and robust monitoring and control techniques need to be employed to retain the stability of the grid.

As the years progress, the power lines and other equipment age, preventing them from operating to their full capabilities. The capacities of these equipment need to be increased or they need to be replaced. This would be expensive. Local generation and usage of electricity would lower the amount of energy transmitted through the power lines, enabling them to last longer.

Distribution of power over long distances causes significant energy losses in the form of heat. This may be avoided by generating energy locally such that energy is consumed as close to the source as possible. Frequent power outages prevent the end-users from having a reliable supply of power. Blackouts refer to a complete outage of power in an area. They may be caused by poor system maintenance, voltage collapse, equipment failure, overloading, and maloperation of protective systems. Blackouts have far-reaching impacts on day-to-day-activities and may affect educational institutions, hospitals, industries, banks, and businesses hindering social, economic, and political activities, thereby causing financial losses. It is necessary to maximize the use of energy storage devices and conduct pre-disturbance studies to mitigate such outages [8,9]. Microgrids provide a means of backup in case of such power outages.

6.2.2 Microgrids

Microgrids are a combination of loads, distributed energy generators, and storage devices. They operate separately as well as in combination with the grid. When the grid fails, microgrids operate as independent entities, thereby providing power supply to the consumers. An important feature of a microgrid is islanding, which provides continuous supply of power during outages, reducing the impact of power system disturbances. When a microgrid is disconnected from the utility, as in Figure 6.2, it is said to operate in the islanded mode. Operation of a microgrid in islanded mode can improve system efficiency. When the system is islanded, the generated power must be capable of ensuring stability and reliability, maintaining the voltage and frequency within acceptable levels. System abnormalities may lead to a deficit between the generation and demand, which, if not addressed in time, would lead to system collapse [10–13].

In a microgrid, in the islanded mode, supply–demand imbalance would be a frequent issue due to the intermittent nature of renewables. A commonly used method to mitigate this issue is load shedding. Load shedding is a means of direct

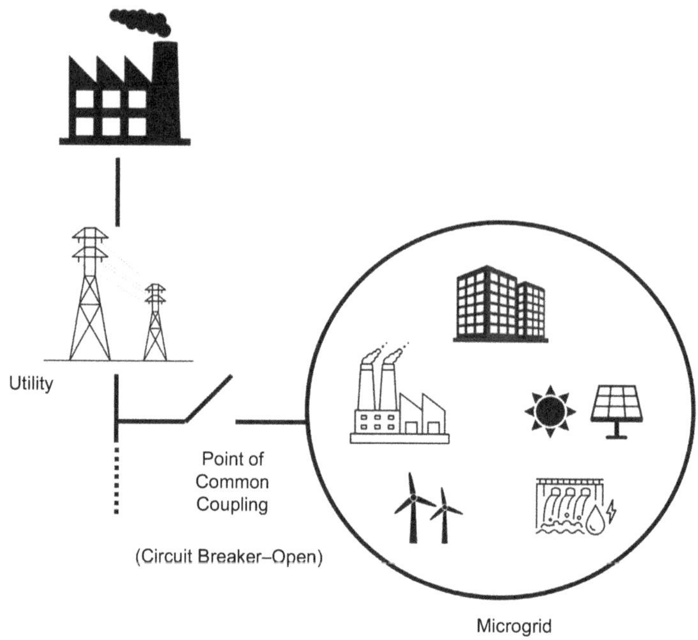

Figure 6.2 Microgrid in the islanded mode

load control, where unnecessary loads are shed based on the instantaneous power generation and demand. The system has to be brought back to its optimal operating condition, once a fault or disturbance is cleared. This is achieved through system restoration [14,15]. An EMS can be used to achieve load shedding and system restoration, to ensure the stable operation of the power system.

6.3 Energy management system

An EMS is a framework for energy consumers to manage their energy use. It enables implementation of cost-effective, energy-saving technologies in commercial, industrial, and public sector organizations [16]. It optimizes a set of objective functions based on specified constraints and provides the best solution possible. With the ever-changing energy scenario, EMS plays an important role in overcoming global energy challenges. It promotes reduction in waste and intelligent resource utilization. In a smart grid, EMS can be used for real-time monitoring, collection of weather data, control of renewable energy sources, and energy storage systems (ESSs). It can also be used to develop smart homes and smart neighborhoods [3].

6.3.1 Evolution of EMS

The first occurrence of an EMS dates back to the 1960s. Back then, it was called a control center and was tasked with the control of power generation and load demand. In the 1970s, this name was changed to energy control centers. Here, the control of energy was the main objective. It was further changed to EM systems in the 1990s. The aim here was to manage energy through various methods like demand side management and load management [17].

6.3.2 Objectives of EMS

An EMS may have one or multiple objectives. In a power system, it would play a crucial role in maintaining its reliable, stable, and efficient operation. The technical objectives of the EMS are to improve system performance, power quality, and to reduce downtime. Its economic objective would be to minimize the operating cost and the cost paid by the end-user. An EMS would also aim to minimize the carbon footprint of the system, in terms of greenhouse gases.

From the point of view of the end-user, an EMS would help to plan and implement energy-saving infrastructure. EMS enables the operators to have a complete overview of the system, by monitoring each equipment in real time. This would help in taking protective measures quickly in case of any system disturbances or faults. New employment opportunities may be created in places where energy efficiency and renewable energy sources are promoted [17,18]. The benefits of using an EMS are presented in Figure 6.3.

Sustainable and reduced usage of electricity are the major concerns of the current power systems. These concerns may be addressed by the EMS at the generation, transmission, and distribution levels. An EMS monitors, supervises, optimizes, and controls the various components of a power system. It aids in the optimal allocation of power from renewable energy sources, ESSs, and power generators to the end-users. The EMS can thus be used to maintain the supply–demand balance efficiently,

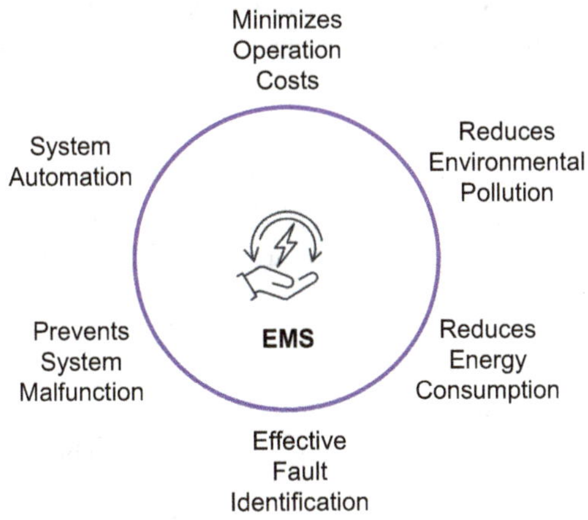

Figure 6.3 Benefits of energy management system

while following specific system constraints, in an economic manner [16]. The real-time data that the EMS needs to implement this can be obtained through synchrophasor technology.

6.4 Synchrophasor technology

Synchrophasor technology is an emerging technology that is being used all around the world for monitoring and controlling power systems. It involves phasor measurements of parameters such as current and voltage. All the devices in the synchrophasor system measure simultaneously. They obtain data that are time-synchronized with the global positioning system (GPS). This allows the users to obtain very accurate time-stamped data that can be utilized for time-critical applications. Synchrophasor measurements can thus be used to obtain the real sequence of events in a power system in case of a blackout or failure [19]. PMUs and μPMUs are the devices used to obtain synchrophasor data.

6.4.1 Phasor measurement units

PMUs are devices that are used to obtain synchrophasor data in the transmission system. They enable the measurement of both phase angle and difference in angle, which allows the calculation of impedance. Phase angle also provides information about the direction of power flow in the system. The ability of PMUs to continuously monitor electrical networks have enabled their application in a wide range of real-time applications. PMUs have an accuracy of $\pm 1^\circ$. This allows the system analysts to obtain accurate, time-stamped data, that can be used to effectively control, monitor, and protect the system. PMUs are placed at various locations of the power system and the data obtained is sent to the phasor data concentrators (PDC) for processing. The PDC contains algorithms that are implemented to achieve the system objectives [20,21]. PMUs are expensive and may not be deployed in distribution networks, since they have a large number of nodes that are placed at short distances. μPMUs are used for measurement in the distribution networks.

6.4.2 Micro-phasor measurement units

In case of distribution networks, μPMUs are used to obtain synchrophasor measurements. μPMUs have a higher sampling rate (120/s) and accuracy ($\pm 0.01^\circ$), when compared to PMUs. Unlike in case of transmission systems, a higher level of accuracy is required for measurements in the distribution systems. Since the power flow in a distribution network is smaller than in case of transmission networks, the reactance and the voltage difference between two points on the distribution network will also be significantly smaller than that of the transmission networks [21].

A block diagram of the μPMU is presented in Figure 6.4. Voltage and current measurements from the field are passed through an anti-aliasing filter. It removes

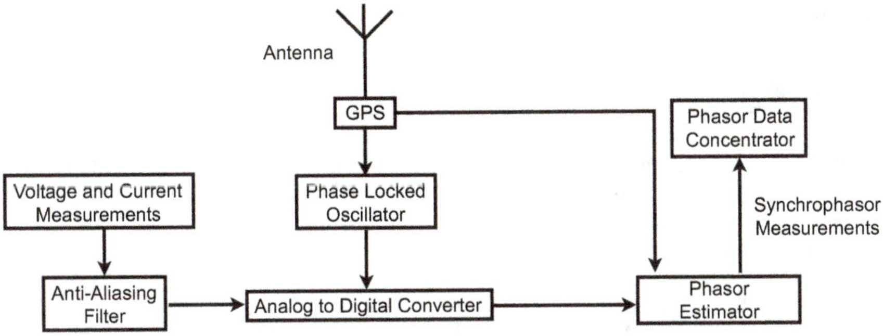

Figure 6.4 Block diagram—micro-phasor measurement unit

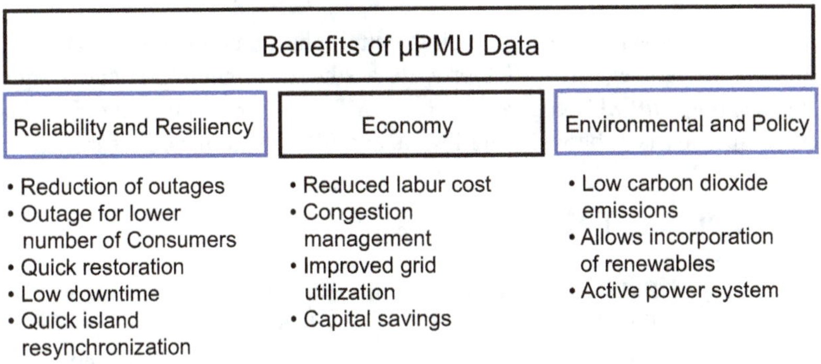

Figure 6.5 Benefits of micro-phasor measurement unit data

the components of the signal that has frequency greater than or equal to half of the Nyquist rate. The current and voltage phasor waveforms are analog in nature. They are converted into digital signals by passing them through an analog to digital converter. A phase-locked oscillator is used to convert the GPS signal into a sequence of timing pulses used to sample the waveform. The phasor estimator calculates the phasors, which are then time tagged. The final synchrophasor measurement is then sent to the PDC [22].

A μPMU obtains four fundamental measurements on three phases, namely, current magnitude, voltage magnitude, current phase angle, and voltage phase angle per phase. μPMUs use the GPS clock to provide accurate time synchronized data. μPMU data help in improving reliability, efficiency, resiliency, and economy of the power systems. They also provide environmental and policy benefits by allowing integration of renewable energy sources, as shown in Figure 6.5. μPMUs find application in power distribution monitoring, diagnostics, and control. They may be used for fault detection, state estimation, power quality monitoring, phase identification, adaptive protection, and microgrid coordination [22,23].

Table 6.2 Comparison between PMUs and μPMUs

Characteristics	PMU	μPMU
Application	Transmission network	Distribution network
Advantage	Better phasor computation and response speed	Smoother magnitude and frequency tracking response
Cost	Higher	Lower
Angle measurement accuracy	±1%	±0.01%
Angle resolution	±0.1%	±0.002%
Data rate	15 readings per second	100–120 readings per second

6.4.3 Comparison between PMUs and μPMUs

PMUs have the advantage of providing better phasor computation with a high response speed. However, μPMUs provide a smoother magnitude and frequency tracking response. μPMUs cost lesser than PMUs and are used in the distribution network. A comparison between the PMUs and μPMUs is made in Table 6.2 [24].

6.5 EMS using μPMUs

A well-developed EMS can contribute significantly to improving power system stability and reliability. The use of μPMUs in the distribution system would enable a complete observability of the system, thereby facilitating quick and effective mitigation of power system disturbances. This section outlines a method for EM in a distribution system using μPMUs at each node [25]. An EMS using μPMUs is depicted in Figure 6.6.

6.5.1 Energy management algorithm

An EMS to facilitate dynamic load shedding and system restoration using synchrophasor measurements is discussed here. The algorithm considers assigned priority indices and number of consumers for each load category to preferentially shed or restore loads, based on data obtained from the μPMUs. When islanding occurs, power imbalance is a frequent issue that needs to be addressed. Load shedding is a commonly adopted method of direct load control that is used to restore supply–demand balance. While it is essential for critical loads to run without a break, it is also necessary for loads that stand next in the line of importance to be given reliable power supply. Based on their importance of operation, loads may be classified as high, medium, and low priority loads. These loads participate in load shedding and restoration. It is also crucial to support as many consumers as possible. This is significant to reduce consumer complaints and improve customer satisfaction. Electricity has become an expensive commodity, and the reliability of power provided to consumers plays a major role in improving quality of life.

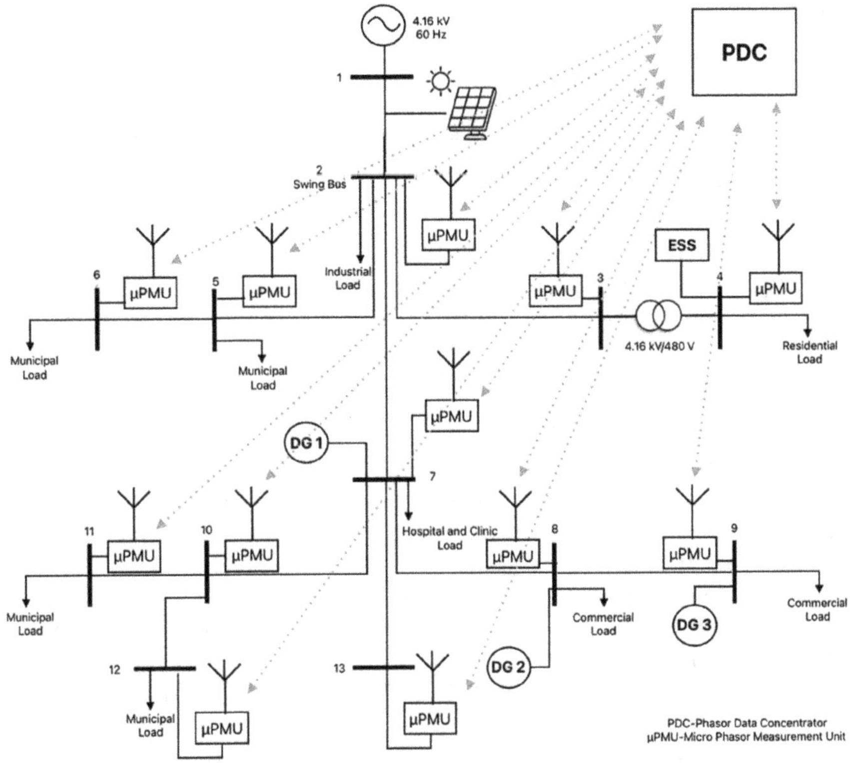

Figure 6.6 Energy management system using micro-phasor measurement units. Reprinted, with permission, from [25].

The objectives of the EMS presented here are:

- The load shed or restored is a linear function of the supply–demand imbalance.
- Load shedding and restoration considering both priority indices and number of consumers.
- Frequency and voltage profile lies within acceptable limits.

6.5.2 Constraints for EM algorithm

The μPMUs collect real-time data and sends them to the PDC, which has the EM algorithm for processing. The results from the PDC are then communicated to each μPMU to shed/restore each load. The EMS was tested on a modified IEEE 13 bus distribution system, in which the slack bus at 4.16 kV, 60 Hz rated was used as the reference.

Voltage, frequency, and current constraints were enforced at each node during the design of the system. At each node, the voltage profile is bounded by a defined lower limit (0.95 pu) and upper limit (1.05 pu).

$$V_l \leq V_n \leq V_u \tag{6.1}$$

The system operates at a frequency between a defined lower threshold (59.7 Hz) and upper threshold (60.3 Hz):

$$f_l \leq f \leq f_u \tag{6.2}$$

The current flowing through the transmission lines should lie below the rated value:

$$I \leq I_{\text{line limit}} \tag{6.3}$$

6.5.3 Implementation of EMS

The EM algorithm begins with island detection, followed by determination and comparison of power demand and supply, based on data obtained from μPMUs. The required loads are shed/restored till the terminating criteria are met. Two factors, namely, the number of consumers using each load and the priority index of each load, were considered for load shedding and restoration. A solar photovoltaic (PV) system, an energy storage system, and diesel generators (DGs) were considered to be a part of the system.

The EM algorithm, as depicted in Figure 6.7, is carried out in the following steps:

Step 1—Island Detection: The status of the microgrid is assessed through a circuit breaker (CB) placed at the point of common coupling (PCC) that connects the utility to the microgrid. The CB is open when the microgrid is islanded and is closed when the microgrid is operated in the grid-connected mode.

$$CB_{Micro\ grid,\ PCC} = \begin{cases} 0\ Islanded\ Mode \\ 1\ Grid\ Connected\ Mode \end{cases} \tag{6.4}$$

The process continues only when the microgrid is in the islanded mode.

Step 2—Assessment of state of charge (SOC): The SOC of the ESS is checked. For safe charge and discharge, lower (20%) and upper (80%) SOC limits are defined for the ESS.

$$SOC_l \leq SOC \leq SOC_u \tag{6.5}$$

The operation of the DGs and ESS depend on the SOC of the ESS.

Step 3—Estimation of generated power: The power generated in the system is a sum of the power derived from each of the connected power source.

$$\begin{aligned} P_G = {} & (\alpha \cdot P_{uti}) + (\beta \cdot P_{spv}) \\ & + (\gamma \cdot P_{es}) + (\delta \cdot P_{dg}) \end{aligned} \tag{6.6}$$

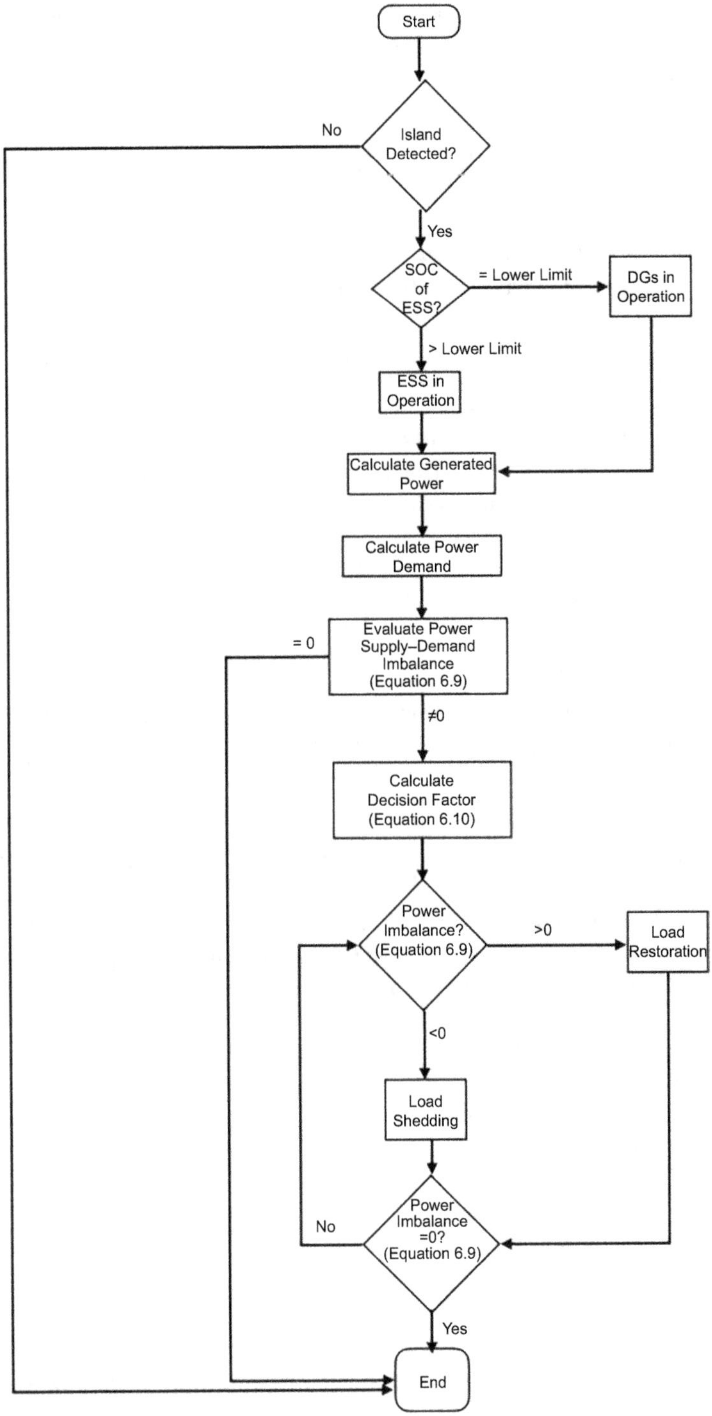

Figure 6.7 Energy management algorithm

where P_{uti} represents the power obtained from the grid, P_{spv} is the power from the solar PV system, P_{es} is the power obtained from the ESS, and P_{dg} is denotes the power from the DGs. The coefficients α, β, γ, and δ indicate if the corresponding source is connected to the system.

$$\alpha, \beta, \gamma, \delta = \begin{cases} 0 \; Source\ disconnected \\ 1 \; Source\ connected \end{cases} \tag{6.7}$$

Step 4—Evaluation of power demand: The power demand is equal to the sum of the demand of each load in the system.

$$P_D = \sum_{l=1}^{i} P_l \tag{6.8}$$

Step 5—Check power supply–demand imbalance: Compare the power generated and the power demanded in the system. Load shed/restoration occurs according to the difference in the islanded mode.

$$P_{imbalance} = P_G - P_D \begin{cases} > 0 \; Load\ restoration \\ = 0 \; No\ load\ shedding/restoration \\ < 0 \; Load\ Shedding \end{cases} \tag{6.9}$$

Move to step six if the power imbalance is not equal to zero.

Step 6—Evaluate decision factor: The decision factor is determined as a product of the assigned priority index and number of consumers of each load:

$$d.f. = (p_l \cdot n_l) \tag{6.10}$$

Move to step 7 if the power imbalance in step 5 was less than zero. Else, move to step 8.

Step 7—Load Shedding: Arrange the loads in the increasing order of their decision factor. Compare the $P_{imbalance}$ obtained from (6.9) with the power demand of each load in the set order. Shed the load that has a demand less than or equal to the power imbalance. Continue shedding loads till the power imbalance is equal to zero. Move to step 9.

Step 8—Load Restoration: Arrange loads in the decreasing order of their decision factor. Compare the $P_{imbalance}$ obtained from (6.9) with the power demand of each load in the set order. Restore the load that has a demand less than or equal to the power imbalance. Continue restoring loads till the power imbalance is equal to zero. Move to step 9.

Step 9—End: The terminating criteria are met and the process stops.

The EM algorithm was tested on a modified IEEE 13 bus system in MATLAB® Simulink®, which is presented in the following subsection.

6.5.4 Modified IEEE 13 bus distribution system

A modified IEEE 13 bus system was used to test the working of the EM algorithm. The reference is provided by the utility connected at node 1, which functions as the

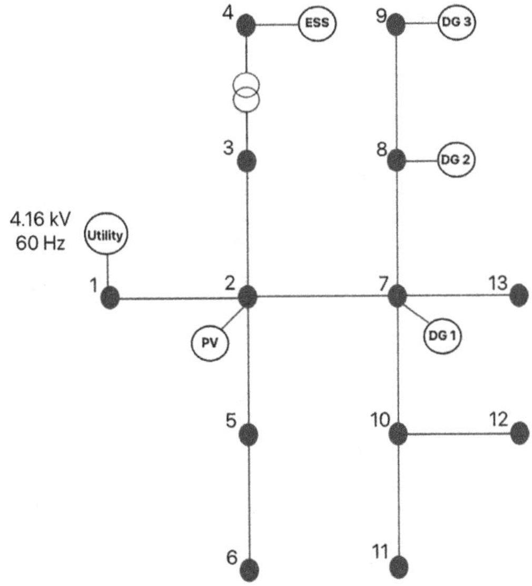

Figure 6.8 Modified IEEE 13 bus distribution system

slack bus. The reference voltage and frequency for the system are 4.16 kV and 60 Hz, respectively. A solar PV system, DGs, and ESS have been appended to the system. A graphical representation of the modified IEEE 13 bus system is provided in Figure 6.8.

The solar PV system at node 2 has a rated output of 1 MW when optimal conditions of irradiance and temperature are provided (1,000 W/m^2, 25°C). The ESS is rated for 500 kW. The minimum and maximum levels of ESS SOC considered for safe operation of the ESS are 20% and 80%, respectively. Three DGs rated for 700, 750, and 800 kW are considered.

The DGs and ESS operate depending on the SOC of the ESS. In the grid-connected mode, the ESS is charged when the SOC is less than 80%. In the islanded mode, the ESS discharges when the SOC is greater than 20%, and the DGs operate when the SOC of the ESS is at 20%.

To test the working of the EM algorithm, five categories of loads, each having four levels of priority, were considered. The priority indices range from 1 to 3.8, in increments of 0.2. A low priority load is assigned a lower priority index. The priority index and number of consumers is dependent on the kind of loads connected to each load category. Critical loads are not assigned priority indices since they need to run always. A total of 250 consumers are assumed to use the load. μPMUs are emulated at each bus to obtain real-time data [25]. The loads connected, along with their priority index and number of consumers, are listed in Table 6.3.

Table 6.3 Loads in the modified IEEE 13 bus system

Type of load	Bus	Priority	Priority index	Number of consumers
Residential	4	Critical	–	1
		High	3	5
		Medium	2.4	10
		Low	1.2	25
Hospitals	7	Critical	–	15
		High	3.8	15
		Medium	2.8	20
		Low	1.8	17
Commercial	8,9	Critical	–	3
		High	3.4	10
		Medium	2.2	17
		Low	1	13
Industrial	2	Critical	–	5
		High	3.6	10
		Medium	2.6	15
		Low	1.6	20
Municipal	5,6,11,12	Critical	–	4
		High	3.2	12
		Medium	2	15
		Low	1.4	18

The decision factor for load shedding and system restoration in the EM algorithm is obtained as a product of the entries on the last two columns of Table 6.3. The critical loads do not participate in load shedding and restoration.

6.6 Intelligent load shedding and system restoration

Based on the instantaneous data available from the μPMUs, the EM algorithm in the PDC determines whether a load has to be shed or restored to maintain supply–demand balance in the islanded mode. This section presents the operation of the EMS for effective load shedding and system restoration.

6.6.1 Operating conditions of the system

The microgrid comprising the modified IEEE 13 bus system may operate together with the utility or independently. In the former case, i.e., the grid-connected mode, the utility would provide the voltage and frequency reference, functioning as the master. During this time, the other sources operate as current-controlled sources. In the latter case, i.e., the islanded mode, one of the distributed energy sources is assigned the master, while the others operate as current-controlled sources. The operation of the DGs and ESS depends on the SOC of the ESS, as depicted in Figure 6.9.

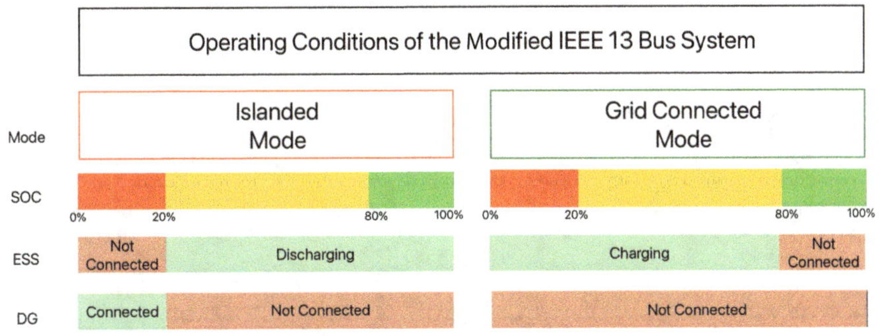

Figure 6.9 Operating conditions of the modified IEEE 13 bus system

In the grid-connected mode, the ESS is charged when the SOC of the ESS lies below 80%. The DGs are not connected to the system. When the SOC of the ESS is above 80%, it is not connected to the system. However, in the islanded mode, the ESS and DG are connected or disconnected from the system depending on the SOC of the ESS. If the SOC is below 20%, the ESS is not connected, while DGs are connected to the system. Above 20% SOC, ESS is discharged while DGs are not in operation. The operating conditions of the system are tested over a period of 6 s when all the three DGs are connected to the circuit, as depicted in Figure 6.10.

6.6.1.1 Time interval 1 (t = 0 s to t = 1 s)

The microgrid is connected to the grid. The grid is thus the master, while the other distributed generation sources are in the current-controlled mode. The initial SOC of the ESS is 60%. The ESS is charged till the SOC reached 80%, after which it remains stagnant. The DGs are not in operation in this interval. The frequency and voltage profiles stay within the specified limits.

6.6.1.2 Time interval 2 (t = 1 s to t = 3 s)

The microgrid now functions independent of the grid, i.e., it operates in the islanded mode. Since the SOC lies at 80%, the ESS discharges during this period, till it reaches 20%. The DGs continue being disconnected from the system. The ESS provides the voltage and frequency reference in this case. The frequency and voltage profiles stay within the specified limits after a short transient.

6.6.1.3 Time interval 3 (t = 3 s to t = 5 s)

The microgrid continues to operate in the islanded mode. Since the SOC lies at 20%, the ESS is disconnected from the system. Instead, the DGs are connected to the system. The DGs function as the master in this case. After a short period of transients, the frequency and voltage profiles stay within the specified limits.

6.6.1.4 Time interval 4 (t = 5 s to t = 6 s)

The utility grid is back in operation, i.e., the microgrid is back to operation in the grid-connected mode. ESS begins charging while the DGs are disconnected. After a

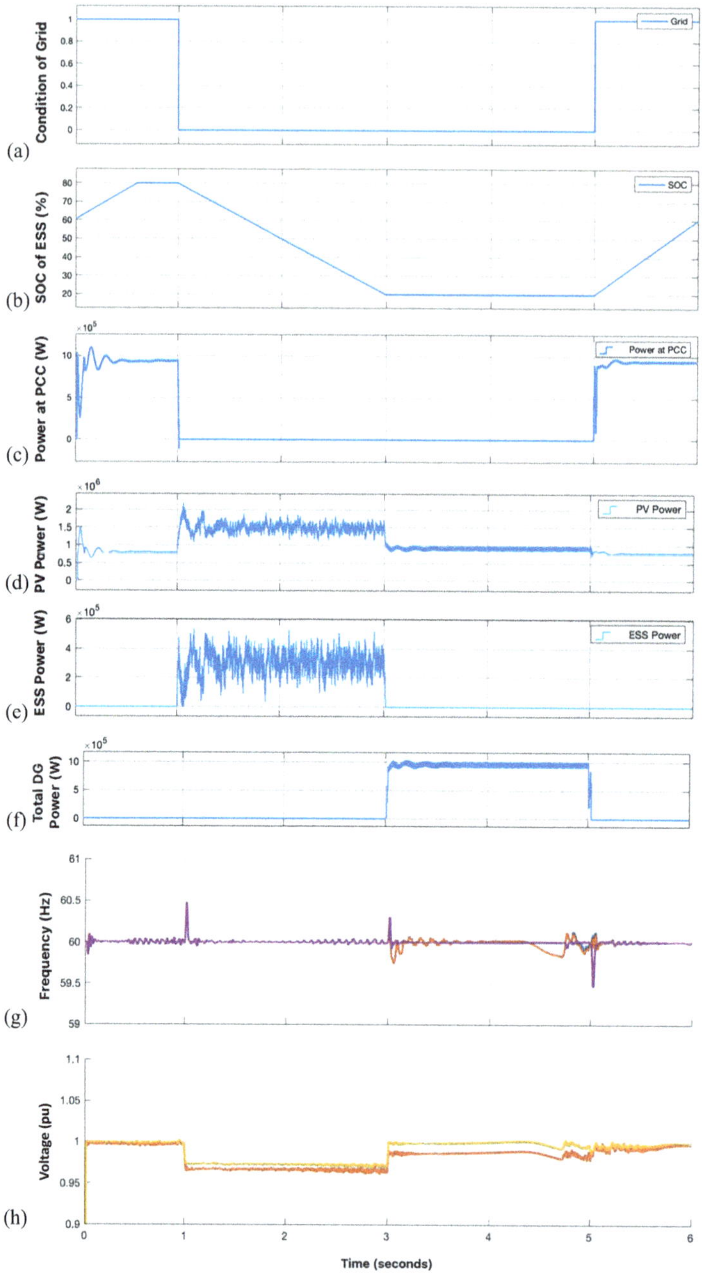

Figure 6.10 (a) Condition of the grid, (b) SOC of ESS, (c) power at PCC, (d) power from solar PV system, (e) ESS power, (f) total DG power, (g) frequency profile, and (h) voltage profile for a duration of 6 s (Note: The horizontal axis is same for all the graphs.)

short transient at $t = 5$ s, the voltage and frequency profiles settle down at 1 pu and 60 Hz, respectively.

The waveforms obtained from Figure 6.10 conform with the operating conditions laid down in Figure 6.9. Load shedding and dynamic system restoration occur only in the islanded mode, i.e., the time interval of $t = 1$ s to $t = 5$ s, in this case. Complete system restoration occurs during the interval of $t = 5$ s to $t = 6$ s. It is seen that the voltage and frequency profile lie in their acceptable ranges.

6.6.2 Intelligent load shedding

To demonstrate load shedding in the EMS, two situations were considered. The loads shed during day time and night time were studied. Based on the operating conditions of the DGs, nine cases were observed in each situation. The results obtained for each case are presented in Table 6.4.

Load shedding occurs when there is a deficit in power generation, when compared to the power demand in the system. The load with a lower priority and a lower number of connected consumers is shed. “S” represents that the load is shed. On the other hand, “C” represents that the load is connected to the system. The last column shows the number of consumers that receive power, out of the 250 consumers considered. The critical loads are always provided power. However, the high, medium, and low priority loads participate in load shedding.

6.6.2.1 Day time

During the day time, solar irradiation is present. The loads that are supplied power and shed from the system are presented in Table 6.4.

When the initial SOC of the ESS is 20%, and DG3 is in operation, the commercial high and low priority loads are shed, whereas the commercial medium priority load remains connected. In the hospital category, only the high priority load remains connected, while the rest of the loads are shed. In the industrial loads category, only the high priority load is shed. Only the low priority load is shed in the municipal category. In the residential load category, only the high priority load remains connected. In this case, 127 consumers receive supply.

When the initial SOC lies between 20% and 80%, and DG2 is connected, 145 consumers receive power. In the commercial category, only the medium priority load is connected. Only the high priority load is connected in the hospital category. Only the high priority load is shed in the industrial category. All the loads in the municipal category remain connected to the system. In the residential load category, only the high priority load is connected. The power profile for this case is presented in Figure 6.11.

Consider the case when the initial SOC of the ESS is greater than or equal to 80% and DG1 is connected. In the commercial category, only the medium priority load remains connected. In the hospital category, only the medium priority load is shed. Only the high priority load is shed in the industrial category. No load is shed in the municipal category. Only the high priority load remains connected in the residential category. A total of 162 consumers are provided power supply in this case.

Table 6.4 Load shedding in the energy management system

Situation	Initial ESS SOC (%)	Connected diesel generator	Status (C—connected, S—shed) of each connected load (H—high priority, M—medium priority, L—low priority)															Number of connected consumers
			Commercial			Hospital			Industrial			Municipal			Residential			
			H	M	L	H	M	L	H	M	L	H	M	L	H	M	L	
Day Time	SOC = 20	DG1	S	C	S	C	S	S	S	C	C	C	C	S	C	S	S	127
		DG2	S	C	S	C	S	C	S	C	C	C	C	C	C	S	S	162
		DG3	S	C	S	C	S	S	S	C	C	C	C	S	C	S	S	127
	20 < SOC < 80	DG1	S	C	S	C	S	S	S	C	C	C	C	C	C	S	S	145
		DG2	S	C	S	C	S	S	S	C	C	C	C	C	C	S	S	145
		DG3	S	C	S	C	S	S	S	C	C	C	C	C	C	S	S	145
	SOC ≥ 80	DG1	S	C	S	C	S	C	S	C	C	C	C	C	C	S	S	162
		DG2	S	C	S	C	S	C	S	C	C	C	C	C	C	S	S	162
		DG3	S	C	S	C	S	C	S	C	C	C	C	C	C	S	S	162
Night Time	SOC = 20	DG1	S	C	S	C	S	S	S	S	S	C	S	S	C	S	S	77
		DG2	S	C	S	C	S	S	S	S	S	C	S	S	C	S	C	102
		DG3	S	C	S	C	S	S	S	S	S	C	S	S	C	S	S	77
	20 < SOC < 80	DG1	S	C	S	C	S	S	S	S	S	C	S	C	C	S	S	95
		DG2	S	C	S	C	S	S	S	S	S	C	S	C	C	S	S	95
		DG3	S	C	S	C	S	S	S	S	S	C	S	C	C	S	S	95
	SOC ≥ 80	DG1	S	C	S	C	S	C	S	S	S	C	S	C	C	S	S	112
		DG2	S	C	S	C	S	C	S	S	S	C	S	C	C	S	S	112
		DG3	S	C	S	C	S	C	S	S	S	C	S	C	C	S	S	112

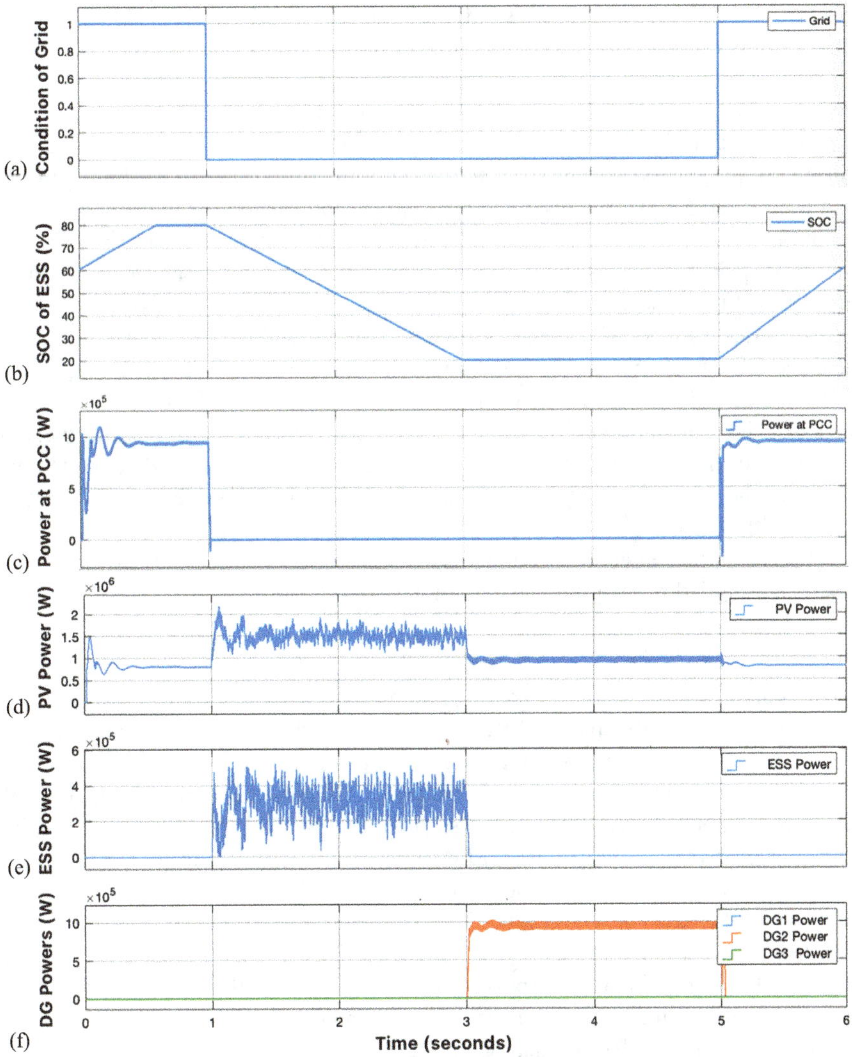

Figure 6.11 (a) Condition of the grid, (b) SOC of ESS, (c) power at PCC, (d) power from solar PV system, (e) ESS power, (f) power from each DG when initial SOC lies between 20% and 80% and DG2 is connected in day time (Note: The horizontal axis is same for all the graphs.)

6.6.2.2 Night time

During the night time, solar irradiation is absent. The loads that are supplied power and shed from the system are presented in Table 6.4.

When the initial SOC of the ESS is 20%, and DG1 is in operation, the commercial high and low priority loads are shed, whereas the commercial medium priority load remains connected. In the hospital category, only the high priority load remains connected, while the rest of the loads are shed. In the industrial loads category, all the loads are shed. Only the high priority load is connected in the municipal category. In the residential load category, only the high priority load remains connected. In this case, 77 consumers receive supply.

When the initial SOC lies between 20% and 80%, and DG3 is connected, 95 consumers receive power. In the commercial category, only the medium priority load is connected. Only the high priority load is connected in the hospital category. All the loads are shed in the industrial category. Only the medium priority load in the municipal category is shed. In the residential load category, only the high priority load is connected.

Consider the case when the initial SOC of the ESS is greater than or equal to 80% and DG2 is connected. In the commercial category, only the medium priority load remains connected. In the hospital category, only the medium priority load is shed. All the loads are shed in the industrial category. Only the medium priority load is shed in the municipal category. Only the high priority load remains connected in the residential category. A total of 112 consumers are provided power supply in this case.

The cases presented above demonstrate the ability of the EMS to shed loads based on assigned priority indices and number of consumers.

6.6.3 Dynamic load restoration

The EM algorithm facilitates dynamic restoration of load when additional power is available in the islanded mode. To test this, a case where each DG is connected at different timings was considered, as presented in Figure 6.12.

The test was performed for a duration of 6 s, where the microgrid operates in the islanded mode from $t = 1$ s to $t = 5$ s. The initial SOC of the ESS is 30%, which gradually increases to 60% in the grid-connected mode, i.e., from $t = 0$ s to $t = 1$ s.

At $t = 1$ s, in the islanded mode, the ESS begins discharging. Once the SOC of the ESS reaches 20% at $t = 2$ s, the ESS is disconnected and the DGs come into operation. DG1, DG2, and DG3 come into operation at $t = 2$ s, $t = 3$ s, and $t = 4$ s, respectively. At $t = 5$ s, the microgrid is connected back to the main grid. The load shed and restored in this case are presented in Table 6.5.

The loads that undergo dynamic restoration in the islanded mode when additional power is available in the system are highlighted in Table 6.5. When the ESS is in operation, commercial high and low priority loads, hospital medium and low priority loads, industrial high priority load, municipal low priority load, and residential medium and low priority loads are shed.

When DG1 comes into operation, commercial low priority load, hospital low priority load, industrial high priority load, municipal low priority load, and residential low priority load are restored. When DG2 is also connected, no load is restored. This indicates that the power demand of none of the shed loads is met by DG2. When DG3 also comes into operation, the hospital medium priority load is restored. Complete restoration of the loads occurs when the grid is connected back at $t = 5$ s.

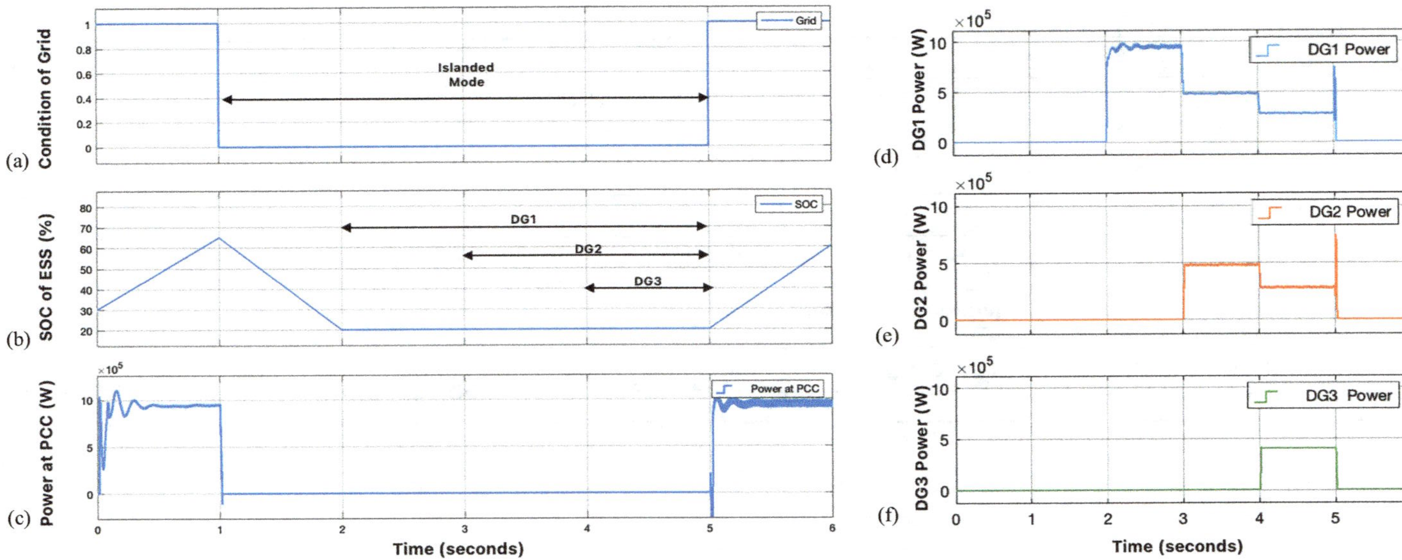

Figure 6.12 *Dynamic load restoration: (a) condition of the grid, (b) SOC of ESS, (c) power at PCC, (d) DG1 power, (e) DG2 power, and (f) DG3 power (Note: The horizontal axis is same for all the graphs.)*

Table 6.5 Dynamic load restoration

Mode of operation	**Time interval (s)**	**Distributed generator connected**	**Status (C—connected, S—shed) of each connected load (H—high priority, M—medium priority, L—low priority)**														
			Commercial			**Hospital**			**Industrial**			**Municipal**			**Residential**		
			H	**M**	**L**	**H**	**M**	**L**	**H**	**M**	**L**	**H**	**M**	**L**	**H**	**M**	**L**
Grid connected	$t = 0–1$	–	C	C	C	C	C	C	C	C	C	C	C	C	C	C	C
Islanded	$t = 1–2$	ESS	S	C	S	C	S	S	S	C	C	C	C	S	C	S	S
	$t = 2–3$	DG1	S	C	C	C	S	C	C	C	C	C	C	C	C	S	C
	$t = 3–4$	DG1 and DG2	S	C	C	C	S	C	C	C	C	C	C	C	C	S	C
	$t = 4–5$	DG1, DG2, and DG3	S	C	C	C	C	C	C	C	C	C	C	C	C	S	C
Grid connected	$t = 5–6$	–	C	C	C	C	C	C	C	C	C	C	C	C	C	C	C

The EMS is therefore capable of dynamic load shedding and restoration in the islanded mode, depending on the power demand and requirements of the system.

6.7 Conclusion

Supply–demand balance is integral in maintaining the stability of a power system. The use of μPMUs in the distribution system has enabled efficient monitoring and control of the system, thereby ensuring reliable power supply to its end-users. An EMS using μPMUs for load shedding and system restoration in the islanded mode was presented in this chapter. The power generation and demand were constantly measured through μPMUs at each node. Based on the instantaneous synchrophasor measurements, the EMS implemented load shedding or load restoration in a modified IEEE 13 bus system. Load shedding and restoration were dependent on the decision factor, which is obtained from assigned priority indices and number of consumers. This system provides power to a higher number of consumers, thereby reducing consumer complaints. There is a quick transition between the sources in the islanded and the grid-connected modes. From the case studies, it was observed that the number of loads shed was higher during night time, in the absence of solar irradiation. When additional power sources are connected to the system, load restoration occurs based on the available power. Further, the voltage and frequency profiles settle to the safe limits after short transients, when there is a change in the source connected. This EMS can thus be adopted in real power systems to maintain supply–demand balance and thereby prevent collapse of grid in the islanded mode.

References

[1] Gopakumar P, Reddy MJ, and Mohanta DK. Adaptive fault identification and classification methodology for smart power grids using synchronous phasor angle measurements. *IET Generation, Transmission & Distribution*. 2015;9(2):133–45.

[2] Von Meier A, Stewart E, McEachern A, Andersen M, and Mehrmanesh L. Precision micro-synchrophasors for distribution systems: A summary of applications. *IEEE Transactions on Smart Grid*. 2017;8(6):2926–36.

[3] Zhang J. Energy management system: The engine for sustainable development and resource optimization. *Highlights in Science, Engineering and Technology*. 2023;76:618–24.

[4] Kalkal P, and Garg VK. Transition from conventional to modern grids: Modern grid include microgrid and smartgrid. In *2017 4th International Conference on Signal Processing, Computing and Control (ISPCC)*, September 21, 2017 (pp. 223–8). IEEE.

[5] Aroua S. Spectrum Resource Assignment in Cognitive Radio Sensor Networks for Smart Grids (Doctoral dissertation, Université de La Rochelle).

[6] Shahsavari A, Farajollahi M, Stewart E, Roberts C, and Mohsenian-Rad H. A data-driven analysis of lightning-initiated contingencies at a distribution

grid with a PV farm using micro-PMU data. In *2017 North American Power Symposium (NAPS)*, September 17, 2017 (pp. 1–6). IEEE.

[7] Ghodusinejad MH, Noorollahi Y, and Zahedi R. Optimal site selection and sizing of solar EV charge stations. *Journal of Energy Storage*. 2022; 56:105904.

[8] Raza MA, Khatri KL, Hussain A, Khan MH, Shah A, and Taj H. Analysis and proposed remedies for power system blackouts around the globe. *Engineering Proceedings*. 2022;20(1):5.

[9] Haes Alhelou H, Hamedani-Golshan ME, Njenda TC, and Siano P. A survey on power system blackout and cascading events: Research motivations and challenges. *Energies*. 2019;12(4):682.

[10] Hosseinzadeh M, and Rajaei Salmasi F. Islanding fault detection in microgrids—A survey. *Energies*. 2020;13(13):3479.

[11] Karimi M, Wall P, Mokhlis H, and Terzija V. A new centralized adaptive underfrequency load shedding controller for microgrids based on a distribution state estimator. *IEEE Transactions on Power Delivery*. 2016;32(1):370–80.

[12] Abo-Khalil AG, Abdalla M, Bansal RC, and Mbungu NT. A critical assessment of islanding detection methods of solar photovoltaic systems. *Case Studies in Thermal Engineering*. 2023;52:103681.

[13] Sapari NM, Mokhlis H, Laghari JA, Bakar AH, and Dahalan MR. Application of load shedding schemes for distribution network connected with distributed generation: A review. *Renewable and Sustainable Energy Reviews*. 2018;82:858–67.

[14] Keles C, Alagoz BB, and Kaygusuz A. A note on demand side load management by maximum power limited load shedding algorithm for smart grids. In *2015 3rd International Istanbul Smart Grid Congress and Fair (ICSG)*, April 29, 2015 (pp. 1–5). IEEE.

[15] Sharma D, Lin C, Luo X, Wu D, Thulasiraman K, and Jiang JN. Advanced techniques of power system restoration and practical applications in transmission grids. *Electric Power Systems Research*. 2020;182:106238.

[16] Energy Management System (EnMS) Energy Efficiency Solutions Series Industrial Energy Accelerator [Internet]. [cited August 9, 2024]. Available from: https://www.industrialenergyaccelerator.org/wp-content/uploads/EnMS_brochure-1.pdf [accessed 9 Aug 2024].

[17] Rathor SK, and Saxena D. Energy management system for smart grid: An overview and key issues. *International Journal of Energy Research*. 2020;44(6):4067–109.

[18] Amaral J, Reis C, and Brandão RM. Energy management systems. In *2013 48th International Universities' Power Engineering Conference (UPEC)*, September 2, 2013 (pp. 1–6). IEEE.

[19] De La Ree J, Centeno V, Thorp JS, and Phadke AG. Synchronized phasor measurement applications in power systems. *IEEE Transactions on Smart Grid*. 2010;1(1):20–7.

[20] Krishna RV, Ashok S, and Krishnan MG. Synchronised phasor measurement unit. In *2014 International Conference on Power Signals Control and Computations (EPSCICON)*, January 6, 2014 (pp. 1–6). IEEE.

[21] Jain A, and Bhullar S. Micro-phasor measurement units (μPMUs) and its applications in smart distribution systems. In *ISGW 2017: Compendium of Technical Papers: 3rd International Conference and Exhibition on Smart Grids and Smart Cities*, 2018 (pp. 81–92). Springer Singapore.

[22] Dusabimana E, and Yoon SG. A survey on the micro-phasor measurement unit in distribution networks. *Electronics*. 2020;9(2):305.

[23] Das HP, and Pradhan AK. Development of a micro-phasor measurement unit for distribution system applications. In *2016 National Power Systems Conference (NPSC)*, December 19, 2016 (pp. 1–5). IEEE.

[24] Ravi A, Saranathan M, Achuthan PH, Lavanya MC, and Rajini V. A comprehensive review on the current trends in micro-phasor measurement units. In *IOP Conference Series: Materials Science and Engineering*, October 1, 2022 (vol. 1258, no. 1, p. 012045). IOP Publishing.

[25] Jose Shirley JA, Manivelan H, Khare P, and Reddy MJ. Dynamic load shedding and system restoration using wide area distribution management system. *IET Generation, Transmission & Distribution*. 2024;18(17):2860–74.

Chapter 7

A resilient framework to facilitate seamless ancillary power support in a microgrid cluster

T.P. Vishnu[1], *P. Gopakumar*[1], *R. Sunitha*[1] *and Maddikara Jaya Bharata Reddy*[2]

7.1 Introduction

The penetration of renewable sources has transformed the operation of power grids. The power grid now accommodates multiple small-scale distributed generators (DGs) that penetrate the low-voltage distribution systems. Advancements in renewable energy and reduced costs of renewable generators have enabled many countries to invest more in renewable power generation [1]. In 2022, global renewable energy generation increased by 295 GW, reaching a capacity of 3,372 GW. Solar energy added 192 GW, the largest share of increased renewable energy generation, followed by wind, renewable hydropower, bio, and geothermal energy sources [2]. The penetration of renewable sources into the distribution system has led to the formation of microgrids. Microgrids are self-sufficient controllable entities with well-defined electrical boundaries. Microgrids can be operated in either islanded or grid-connected mode. The islanded operation of a microgrid offloads some of the generation capacity and load from the main grid. Therefore, it reduces the dependence on the transmission infrastructure of the main grid. The microgrid is immune to disturbances in the main grid, thereby improving its power quality. In addition, during grid emergencies, the microgrid can provide backup power to the main grid, thereby improving reliability.

However, the intermittent and variable nature of power generation in renewable-based DGs creates uncertainty in microgrid operation. This necessitates maintaining sufficient ancillary and reserve power for microgrid operation. Renewable sources are underutilised when used for ancillary and reserve power support. Hence, dispatchable sources which can be controlled promptly to meet the real-time electricity demand in the microgrid are necessary to ensure self-sufficiency in microgrids. Dispatchable sources in microgrids can have energy

[1]Department of Electrical Engineering, National Institute of Technology (NIT) Calicut, India
[2]Department of Electrical and Electronics Engineering, National Institute of Technology (NIT) Tiruchirappalli, India

storage elements and are capable of plug-and-play (PnP) operation depending on the demand in the microgrid. Ancillary power support in this study is facilitated through such PnP sources. Electric vehicles, battery energy storage systems, ultracapacitor banks, and other smart electrical appliances with energy storage elements constitute PnP sources [3,4]. PnP sources provide microgrids with flexibility to meet the load demand, reserve, and ancillary power requirements. However, the participation of PnP sources in microgrid operations must ensure both economic viability and operational stability. This chapter elaborates on the need for a resilient framework to provide ancillary power support to microgrids using PnP sources. The role of synchrophasor-based measurements in enabling the operation of microgrids and their participation in the ancillary power market is also elaborated.

7.2 Grid interactive sources: opportunities and challenges in microgrid clusters

The energy requirements of microgrids vary from one another. Microgrids can be energy-rich (energy-surplus) or energy-deficient in nature depending on the availability of sources. Obtaining the desired power support from the main AC grid is a common method for ancillary power support in microgrids. However, the erratic nature of the power flow between the microgrid and AC grid burdens the AC grid, making its control difficult. Alternatively, interconnecting microgrids to form a microgrid cluster can facilitate ancillary power support for energy-deficient microgrids from energy-rich ones. It also provides the PnP sources in the microgrid cluster with an opportunity to transact with multiple microgrids, thereby improving their utilisation. This also reduces the stochastic nature of the availability of PnP sources for market participation. Moreover, a conscious selection can be made from a larger pool of PnP sources to satisfy the operating constraints of the energy storage system. Therefore, a microgrid cluster enhances the reliability, resiliency, and economics of microgrid operation [5]. A schematic representation of a test system depicting the clustered operation of several microgrids is shown in Figure 7.1. Each microgrid is interconnected to other microgrids in the cluster

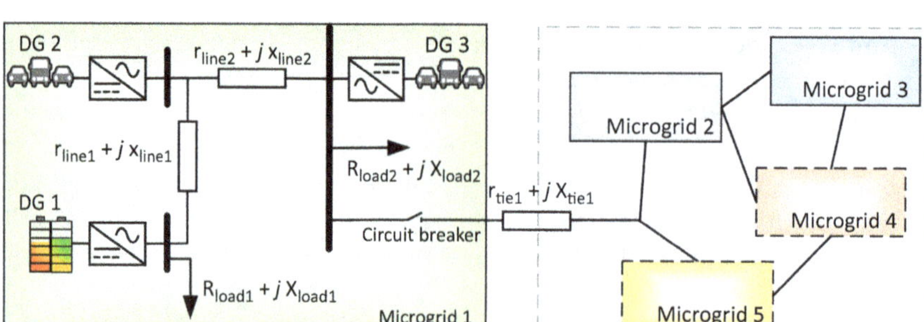

Figure 7.1 Schematic representation of a microgrid cluster

through tie lines. The power flow through the tie line is controlled by varying the voltage and frequency of individual microgrid through proper control of its available resources.

7.2.1 Transactive market in a microgrid cluster

Many markets offer incentives for PnP sources to participate in market operations. They can either participate in the ancillary/reserve power markets to provide stability to the grid or compete with conventional sources in the energy market. The various roles of PnP sources in ancillary and reserve power markets include congestion management, reactive power compensation and voltage regulation, maintaining spinning and standby reserves, voltage asymmetry reduction, and voltage harmonics mitigation. However, integrating and controlling a PnP source to provide various services to the conventional central electricity market pose several challenges. The sheer number of such sources adds a computational burden to scheduling and market-clearing operations. Therefore, all electricity markets have a minimum bid size and a high market fee to discourage the direct participation of small-scale generators. Hence, restructuring the centralised conventional energy market is necessary to utilise the capabilities of PnP sources. This has led to the restructuring of market operations from the grid-level central electricity market to distribution-system-level electricity markets. The operation of electricity markets at the distribution level in a microgrid cluster provides the following advantages [6]:

1) It can accommodate the increasing penetration of PnP sources into the distribution system without burdening the central market.
2) It enables PnP sources to participate in peer-to-peer and PnP-source-to-grid market operations, apart from their services to the local microgrid.
3) The deployment of PnP sources for various services can be monetarily compensated.
4) It encourages the formation of independent microgrids and microgrid clusters, decongesting the main grid.

The distribution-system-level market structure has two hierarchical control levels. Each microgrid in the cluster has a local energy market, whereas the microgrid cluster has a distribution energy market to supervise various local markets. The local energy market, operated by a local market operator (LMO), schedules resources available within a microgrid and ensures sufficiency to meet the varied demands in the microgrid. The LMO facilitates the participation of diverse PnP sources and their varied services in the microgrid. PnP sources can participate in market operations independently or through an aggregator. The distribution energy market facilitates individual microgrids to transact energy with other microgrids within the cluster. A schematic representation of the market operation of a microgrid cluster is shown in Figure 7.2. Here, the LMO obtains the available reserve power capacities from individual PnP sources and performs preliminary scheduling of resources [7]. In the preliminary scheduling, an economic dispatch solution is obtained for the local market. It also identifies the bidirectional power

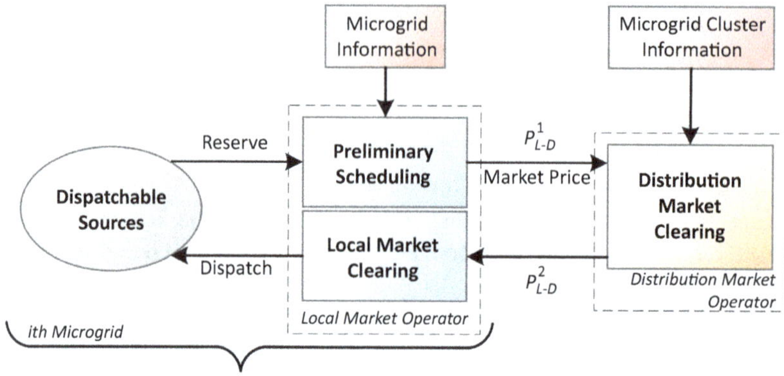

Figure 7.2 Schematic representation of distribution-system-level electricity market for the participation of plug-and-play sources

support between the microgrid and cluster (P^1_{L-D}) and updates the electricity price to the distribution market operator (DMO). The DMO on obtaining the bids from all LMOs clears the optimised schedule for ancillary power support in the microgrid cluster and passes on the cleared scheduled power (P^2_{L-D}) to the individual LMOs. The necessary interconnections between the microgrids in the cluster should be ensured to facilitate the power flow between the microgrids. Individual LMOs perform market clearing by dispatching the necessary power from PnP sources. The PnP sources can also schedule their charging to assist the operation of the microgrid, which is not considered in this study.

7.2.2 Challenges for sustainable use of plug-and-play sources

PnP sources are storage-enabled distributed resources capable of dispatching power support when desired. PnP sources should ramp up and curtail their generation, according to the requirements of the microgrid. Several studies have elaborated on the role of PnP sources in ensuring power adequacy and improving microgrid stability. In Ref. [8], the authors used a battery energy storage system to meet the resource uncertainty and deficiency in microgrids. This improved the stability and reliability of the microgrid. Several energy management models using energy storage elements are elaborated in Ref. [9]. PnP sources with different types of chemical and electrical energy storage elements are used in grid operations. These include lead-acid batteries, lithium-ion batteries, nickel-metal hydride batteries, and ultracapacitors [10].

Battery energy storage systems can provide a rapid response to meet the diverse demands of microgrids, and their capacity can be easily scaled. Moreover, the participation of PnP sources with battery energy storage, such as electric vehicles, is becoming more prevalent in microgrids. However, all battery energy storage systems have a useful life cycle; hence, every participation in a grid operation will reduce the life and degrade them. Subjecting battery sources to extreme operations, such as a lower depth of discharge (DoD), higher state of charge (SoC), higher charging and discharging limits, and higher cell temperatures (T_C), will increase the rate of degradation. The operation of battery energy storage systems is also limited by the thermal constraints imposed by the battery. The remaining battery life (L) of a lithium-ion battery subjected to N cyclic charge–discharge operation can be empirically estimated using (7.1) [11].

$$L = 1 - \alpha_{sei} e^{-N\beta_{sei} f_d} - (1 - \alpha_{sei}) e^{-N f_d} \tag{7.1}$$

$$f_d = [S_\delta(DoD) + S_t(t_c)] S_\sigma(SoC) S_T(T_C) \tag{7.2}$$

where f_d is the cyclic degradation of the battery over one cycle given by (7.2) and α_{sei} and β_{sei} are the fitting functions of the nonlinear degradation model. S_δ, S_σ, S_T, and S_t represent the stress factor models for cyclic DoD, cycle average SoC, cell temperature (T_C), and time (t_c), respectively. Subjecting the battery to extreme operating conditions aggravates the degradation process and reduces the number of operating cycles.

On the other hand, ultracapacitors are suitable for delivering short-term, high-intensity power. Ultracapacitors can help reduce the uncertainties of renewable generators, fluctuations in the grid, perform peak shaving, ramp-up fossil fuel generators, provide frequency and voltage stabilisation, and reduce the load on batteries by forming a hybrid system. The performance of an ultracapacitor depends on its SoC, state of health, and state of power [12].

To ensure a sustainable and safe operation, every storage-based PnP source requires a management system. The management system should alter the dispatching capabilities of the PnP sources to meet the operating constraints of the storage system. It should also monitor the operating conditions and respond to electricity market signals, prioritising the health and life of the storage system. A battery management system can impose limitations on the ramp rates for charging and discharging operations, vary the dispatchable power capacity, and select the nature of the service provided to the grid. Along with these operating constraints, the PnP sources also alter the network topology through their participation, making power scheduling a computationally complex process. Nevertheless, modern power grid will be dominated by PnP sources, and their participation in grid operations should be utilised to enhance the stability of the microgrid.

7.2.3 Challenges with diverse characteristics of distributed generators

A power flow controller based on the droop control of active and reactive power is commonly used to control PnP sources. It requires only local measurements for execution. The droop-based power-flow controller mimics the droop characteristics of a synchronous generator and enables power sharing between various sources in a microgrid cluster. The power-flow controller ensures that the power demand in the microgrid is shared between its sources, adhering to their operating limits. The droop-based power flow controllers have two droop controls: the

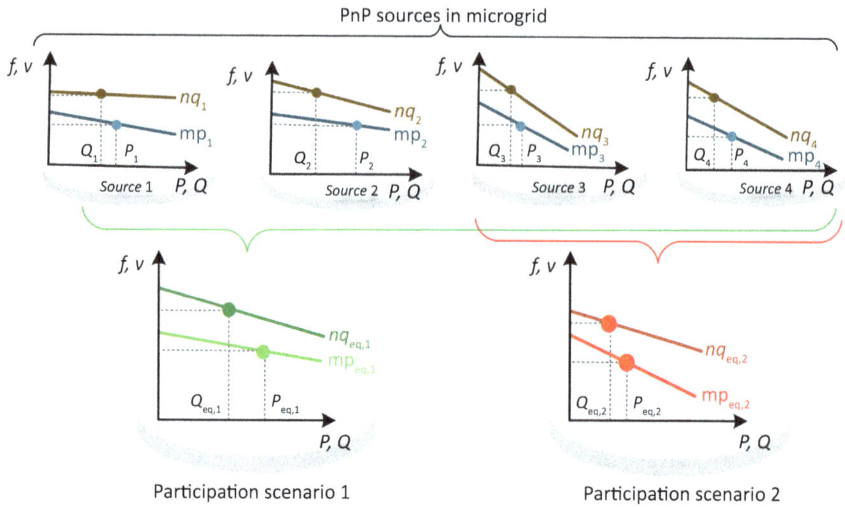

Figure 7.3 The variation of equivalent droop with varying number of participating sources

active power-frequency droop ($P-f$) and reactive power–voltage magnitude droop ($Q-v$). In this chapter, the $P-f$ and $Q-v$ droop coefficients are represented using *mp* and *nq*, respectively. The droop coefficients of the PnP sources may require modifications during market operations [13]. The dispatchable capacity of an aggregator can vary owing to the inherent uncertainty associated with the availability of the PnP sources. The equivalent droop of an aggregator with *n* participating sources can be computed by using $mp_{eq}= 1/(\sum_{i=1}^{n}(1/mp_i)$ and $nq_{eq}= 1/(\sum_{i=1}^{n}(1/nq_i)$ [14]. Two scenarios encountered by an aggregator that enable PnP sources for market participation are shown in Figure 7.3. In Scenario 1, the aggregator has four sources to participate in the market operation, whereas in Scenario 2, the dispatchable capacity is lower because of the unavailability of the two PnP sources. The equivalent droop for both scenarios is presented. As the number of participating sources decreases, the equivalent droop is modified and becomes steeper. In addition, individual PnP sources have several operating constraints, as mentioned in Section 7.2.2, which can vary as the source participates in the market operations. Therefore, the droop coefficients of the participating sources must be modified to satisfy the operating constraints.

In addition to droop-based power flow controllers, the PnP source can have other controllers such as a voltage controller and a current controller. In general, the primary droop control layer generates reference set points for the voltage and current control loops by using the terminal measurements of each source. Active and reactive power thus computed are often passed through a low-pass filter to avoid unnecessary oscillations in voltage and frequency references. A secondary control layer restores the voltage and frequency deviations created by the primary control layer, which is subject to the availability of generating sources [15].

A droop-based power controller with a low-pass filter, voltage controller, and current controller ensures the desired power delivery from the PnP source. However, this operation can introduce low-frequency oscillatory modes into microgrid dynamics [16]. The choice of the droop coefficients in power controllers primarily affects the dynamic stability margin of the microgrid. Therefore, it is imperative for a PnP-rich microgrid to ensure that the participation of PnP sources in market operations does not lead to unstable operating conditions. A viable operating scenario for each participating source should be obtained to ensure seamless ancillary power support in a microgrid cluster.

In summary, the use of storage-based PnP sources is necessary in microgrids to provide ancillary power support and meet the reserve power capacity. However, they also add the following three major complexities to microgrid operation:

1) Monitoring and control of battery resources for their safe, reliable, and sustainable operation.
2) Facilitating energy market participation for small-scale PnP sources to obtain financial benefits.
3) Ensuring a stable operation of the microgrid accommodating various constraints of PnP sources.

Therefore, a resilient ancillary power support framework should generate a viable and stable operating scenario for each participating source, simultaneously satisfying the power requirements of the microgrid.

7.3 Dynamic analysis of microgrid cluster

The operation of a microgrid cluster relies on efficient control of its generating sources and controllable load. As elaborated in Section 7.2.3, when PnP sources participate in market operations, low-frequency oscillatory modes are introduced into the microgrid cluster dynamics. These oscillatory modes drift to new positions depending on the droop coefficients, connected loads, and operating conditions. It is essential to identify such low-frequency modes for each load condition and study their movements with variations in the ancillary power demand. The poor damping performance of the power-sharing mechanism can lead to the collapse of microgrid clusters [16]. This necessitates an analysis based on the mathematical model of the microgrid cluster to ensure a stable operating scenario.

7.3.1 Modelling of microgrid cluster

In this chapter, a systematic approach for modelling inverter-based microgrid clusters is presented. The PnP sources are considered inverter-based generators connected to the AC microgrid and are controlled using droop-based power controllers. A DC–DC converter for batteries and ultracapacitor-based PnP sources maintains a constant DC-link voltage for the inverter [17]. This study assumes a constant DC-link voltage at the input of the inverter, and the dynamics of the inverter is modelled for analysis. Each inverter-based DG has an outer power controller based on droop control to share the real and reactive powers with other sources. They also have voltage and current controllers designed to reject high-frequency disturbances and damp the output to avoid resonance with the external network. A small-signal state-space model of a single inverter was obtained by placing the controllers, output filter, and coupling inductor on a synchronous reference frame. The inverter power controller sets the frequency of the rotating reference frame. An arbitrary choice is made to select one inverter frame as the common reference frame, and all other inverters are translated to this common reference frame. Hence, a dynamic state-space model of the network is formed on the common reference frame. The microgrid modelling given in Ref. [16] is adapted for a microgrid cluster in this chapter, and similar variables are used here to maintain congruity.

The inverter model of the PnP source has 13 state variables, as given in (7.3). The subscript dq represents the d and q state variable components. Here δ_i is the angle between the individual reference frame (dq) of the PnP source and the common reference frame (DQ) of the microgrid cluster. Each PnP source is modelled in their individual dq reference frame. $\Delta\delta_i$, ΔP_i, and ΔQ_i are the state variables of the droop-based power controller. $\Delta\phi_{dq,i}$ and $\Delta\gamma_{dq,i}$ are the state variables for the voltage and current controller of the inverter, respectively. $\Delta i_{ldq,i}$ and $\Delta v_{odq,i}$ are the state variables for the LC filter in dq reference frame. Similarly, $\Delta i_{odq,i}$ is the state variable of the coupling inductance. To connect the individual PnP sources to the microgrid, the measured voltage and current are transformed into the common DQ reference frame using the frequency of the common reference frame ω_{com}. This is performed on the steady-state node voltage V_b using the transformation given in (7.4), where matrices T_S and T_C are computed using (7.5) and (7.6), respectively. The inputs to the small-signal state-space representation of the individual inverters in a microgrid are the node voltages $\Delta V_{bDQ,i}$ and frequency of common reference frame $\Delta\omega_{com}$, and the outputs are the current injected into node, $\Delta i_{oDQ,i}$, and the frequency of inverter in the individual reference frame $\Delta\omega_i$.

$$\Delta x_{inv,i} = [\Delta\delta_i\ \Delta P_i\ \Delta Q_i\ \Delta\phi_{dq,i}\ \Delta\gamma_{dq,i}\ \Delta i_{ldq,i}\ \Delta v_{odq,i}\ \Delta i_{odq,i}]^T \tag{7.3}$$

$$[\Delta V_{bDQ,i}] = [T_S][\Delta V_{bdq,i}] + [T_C][\Delta\delta_i] \tag{7.4}$$

$$T_S = \begin{bmatrix} \cos(\delta_0) & -\sin(\delta_0) \\ \sin(\delta_0) & \cos(\delta_0) \end{bmatrix} \tag{7.5}$$

$$T_C = \begin{bmatrix} -V_{bd,i}\sin(\delta_0) - V_{bq,i}\cos(\delta_0) \\ V_{bd,i}\cos(\delta_0) - V_{bq,i}\sin(\delta_0) \end{bmatrix} \tag{7.6}$$

The state equations for each inverter can be formed, and the overall state-space representation combining all the inverters in a microgrid can be written as shown in (7.7). The distribution network is modelled as an RL network in the common DQ reference frame, and the state equation is represented by (7.8). Similarly, the loads in the microgrid are also modelled as RL loads, and their state equation in the common reference frame is given by (7.9).

$$[\Delta\dot{x}_{INV}] = A_{INV}[\Delta x_{INV}] + [B_{INV}][\Delta V_{bDQ}] + [B_{\omega com}][\Delta\omega_{com}] \tag{7.7}$$

$$[\Delta\dot{i}_{lineDQ}] = A_{NET}[\Delta i_{lineDQ}] + [B_{1NET}][\Delta V_{bDQ}] + [B_{2NET}][\Delta\omega_{com}] \tag{7.8}$$

$$[\Delta\dot{i}_{loadDQ}] - A_{LOAD}[\Delta i_{loadDQ}] + [B_{1LOAD}][\Delta V_{bDQ}] + [B_{2LOAD}][\Delta\omega_{com}] \tag{7.9}$$

The state-space models of inverter, network, and load in a microgrid can be combined using the relationship in (7.10) to obtain the state equation of the microgrid. M_{INV} is a mapping matrix that maps the position of the PnP sources in the microgrid. Similarly, M_{load} and M_{NET} map the location of the loads and the interconnection between the nodes in the network, respectively. R_N is a matrix consisting of virtual resistances connected at each node [16].

$$[\Delta v_{bDQ}] = R_N(M_{INV}[\Delta i_{oDQ}] + M_{load}[\Delta i_{loadDQ}] + M_{NET}[\Delta i_{lineDQ}]) \tag{7.10}$$

The state equations thus formed can be written as given by (7.11), where the state matrix and input matrix are given by (7.12) and (7.13).

$$\begin{bmatrix} \Delta\dot{x}_{INV} \\ \Delta\dot{i}_{lineDQ} \\ \Delta\dot{i}_{loadDQ} \end{bmatrix} = [A_{mg}]\begin{bmatrix} \Delta x_{INV} \\ \Delta i_{lineDQ} \\ \Delta i_{loadDQ} \end{bmatrix} + [B_{mg}]\begin{bmatrix} \Delta V_{bDQ} \\ \Delta\omega_{com} \end{bmatrix} \tag{7.11}$$

$$[A_{mg}] = \begin{bmatrix} A_{INV} & 0 & 0 \\ B_{1NET}\,R_N\,M_{INV}\,C_{INVc} & A_{NET} + B_{1NET}\,R_N\,M_{NET} & B_{1NET}\,R_N\,M_{LOAD} \\ 0 & 0 & A_{LOAD} \end{bmatrix} \tag{7.12}$$

$$[B_{mg}] = \begin{bmatrix} B_{INV} & B_{\omega com} \\ B_{1NET} & B_{2NET} \\ B_{1LOAD} & B_{2LOAD} \end{bmatrix} \tag{7.13}$$

The output equations are formulated to represent the ancillary power support from an energy-rich microgrid to an energy-deficient microgrid. In this chapter, the ancillary power support for an energy-deficient microgrid is studied by considering the energy-deficient microgrid as a load connected to the energy-rich microgrid. The ancillary power support to the energy-deficient microgrid is the output of the overall state-space model. The active power balance equation within a microgrid is formed by equating the total power generation for all s generators with the total load demand of p loads and power lost during transmission in the n connected lines, given by (7.14). The reactive power balance equation is obtained by equating the reactive power injected by the sources with the power flowing to the loads as represented by (7.15).

$$\sum_{i=1}^{s} \Delta P_{inj,i} = \sum_{i=1}^{p} \Delta P_{load,i} + \sum_{i=1}^{n} \Delta P_{line,i} \tag{7.14}$$

$$\sum_{i=1}^{s} \Delta Q_{inj,i} = \sum_{i=1}^{p} \Delta Q_{load,i} \tag{7.15}$$

The power injected by each inverter in the microgrid is represented in its small-signal model through (7.16) and (7.17), and the transmission line copper losses are given by (7.18).

$$\begin{aligned}\Delta P_{inj,i} = \frac{3}{2}[& \underbrace{(-V_{bD,i}\, i_{od,i}\ \sin(\delta_i) - V_{bD,i}\, i_{oq,i}\ \cos(\delta_i) + V_{bQ,i}\, i_{od,i}\ \cos(\delta_i) - V_{bQ,i}\, i_{oq,i}\ \sin(\delta_i))}_{\mathrm{P}^{(1)}_{\mathrm{inj,i}}}\Delta\delta_i \\ & \underbrace{+(V_{bD,i}\ \cos(\delta_i) + V_{bQ,i}\ \sin(\delta_i))}_{\mathrm{P}^{(2)}_{\mathrm{inj,i}}}\Delta i_{od,i}\underbrace{+(V_{bQ,i}\ \cos(\delta_i) - V_{bDi}\ \sin(\delta_i))}_{\mathrm{P}^{(3)}_{\mathrm{inj,i}}}\Delta i_{oq,i} \\ & \underbrace{+(i_{od,i}\ \cos(\delta_i) - i_{oq,i}\ \sin(\delta_i))}_{\mathrm{P}^{(4)}_{\mathrm{inj,i}}}\Delta V_{bD,i}\underbrace{+(i_{od,i}\ \sin(\delta_i) + i_{oq,i}\ \cos(\delta_i))}_{\mathrm{P}^{(5)}_{\mathrm{inj,i}}}\Delta V_{bQ,i}]\end{aligned} \tag{7.16}$$

$$\begin{aligned}\Delta Q_{inj,i} = \frac{3}{2}[& \underbrace{(-V_{bQ,i}\, i_{od,i}\ \sin(\delta_i) - V_{bQ,i}\, i_{oq,i}\ \cos(\delta_i) - V_{bD,i}\, i_{od,i}\ \cos(\delta_i) + V_{bD,i}\, i_{oq,i}\ \sin(\delta_i))}_{\mathrm{Q}^{(1)}_{\mathrm{inj,i}}}\Delta\delta_i \\ & \underbrace{+(V_{bQ,i}\ \cos(\delta_i) - V_{bD,i}\ \sin(\delta_i))}_{\mathrm{Q}^{(2)}_{\mathrm{inj,i}}}\Delta i_{od,i}\underbrace{+(-V_{bQ,i}\ \sin(\delta_i) - V_{bD,i}\ \cos(\delta_i))}_{\mathrm{Q}^{(3)}_{\mathrm{inj,i}}}\Delta i_{oq,i} \\ & \underbrace{+(-i_{od,i}\ \sin(\delta_i) - i_{oq,i}\ \cos(\delta_i))}_{\mathrm{Q}^{(4)}_{\mathrm{inj,i}}}\Delta V_{bD,i}\underbrace{+(i_{od,i}\ \cos(\delta_i) - i_{oq,i}\ \sin(\delta_i))}_{\mathrm{Q}^{(5)}_{\mathrm{inj,i}}}\Delta V_{bQ,i}]\end{aligned} \tag{7.17}$$

$$\Delta P_{line,i} = 3[\underbrace{i_{lineD,i}\, r_{line,i}}_{\mathrm{P}^{(1)}_{\mathrm{line,i}}}\ \Delta i_{lineD,i} + \underbrace{i_{lineQ,i}\, r_{line,i}}_{\mathrm{P}^{(2)}_{\mathrm{line,i}}}\ \Delta i_{lineQ,i}] \tag{7.18}$$

Similarly, the active power delivered to a load connected to the *i*th bus in the microgrid can be represented by its small-signal representation using (7.19) and (7.20).

$$\Delta P_{load,i} = \frac{3}{2}[\underbrace{V_{bD,i}}_{\mathrm{P}^{(1)}_{\mathrm{load,i}}}\ \Delta i_{loadD,i} + \underbrace{V_{bQ,i}}_{\mathrm{P}^{(2)}_{\mathrm{load,i}}}\ \Delta i_{loadQ,i} + \underbrace{i_{loadD,i}}_{\mathrm{P}^{(3)}_{\mathrm{load,i}}}\ \Delta V_{bD,i} + \underbrace{i_{loadQ,i}}_{\mathrm{P}^{(4)}_{\mathrm{load,i}}}\ \Delta V_{bQ,i}] \tag{7.19}$$

$$\Delta Q_{load,i} = \frac{3}{2}[\underbrace{V_{bQ,i}}_{\mathrm{Q}^{(1)}_{\mathrm{load,i}}}\ \Delta i_{loadD,i} \underbrace{-V_{bD,i}}_{\mathrm{Q}^{(2)}_{\mathrm{load,i}}}\ \Delta i_{loadQ,i} \underbrace{-i_{loadQ,i}}_{\mathrm{Q}^{(3)}_{\mathrm{load,i}}}\ \Delta V_{bD,i} + \underbrace{i_{loadD,i}}_{\mathrm{Q}^{(4)}_{\mathrm{load,i}}}\ \Delta V_{bQ,i}] \tag{7.20}$$

The output equation given by (7.21) is formed by substituting (7.16)–(7.20) in (7.14) and (7.15) to represent the desired ancillary power support flowing into an energy-deficient microgrid. The output matrix (C_{mg}) and feedforward matrix (D_{mg}) thus formed are given by (7.22), and (7.23). The variables in these matrices are marked using the underbraces in (7.16)–(7.20).

$$\begin{bmatrix}\Delta P_{load,i} \\ \Delta Q_{load,i}\end{bmatrix} = [C_{mg}]\begin{bmatrix}\Delta x_{INV} \\ \Delta i_{lineDQ} \\ \Delta i_{loadDQ}\end{bmatrix} + [D_{mg}]\begin{bmatrix}\Delta V_{bDQ} \\ \Delta\omega_{com}\end{bmatrix} \tag{7.21}$$

$$[C_{mg}] = \frac{3}{2}\begin{bmatrix} P_{inj,1}^{(1)} & [0]_{1\times10} & P_{inj,1}^{(2)} & P_{inj,1}^{(3)} \cdots & P_{line,1}^{(1)} & P_{line,1}^{(2)} \cdots & P_{load,1}^{(1)} & P_{load,1}^{(2)} \cdots \\ Q_{inj,1}^{(1)} & [0]_{1\times10} & Q_{inj,1}^{(2)} & Q_{inj,1}^{(3)} \cdots & [0]_{1\times1} & [0]_{1\times1} \cdots & Q_{load,1}^{(1)} & Q_{load,1}^{(2)} \cdots \end{bmatrix} \tag{7.22}$$

$$[D_{mg}] = \frac{3}{2}\begin{bmatrix} P_{inj,1}^{(4)} + P_{load,1}^{(3)} & P_{inj,1}^{(5)} + P_{load,1}^{(4)} \cdots & [0]_{1\times1} \\ Q_{inj,1}^{(4)} + Q_{load,1}^{(3)} & Q_{inj,1}^{(5)} + Q_{load,1}^{(4)} \cdots & [0]_{1\times1} \end{bmatrix} \tag{7.23}$$

The overall model of the microgrid participating in an ancillary service by delivering the output power of $P_{load,i}$ and $Q_{load,i}$ to an energy-deficient microgrid in the cluster can be represented using (7.11) and (7.21).

7.4 Framework for real-time ancillary power support

7.4.1 Framework overview

In this section, a general framework necessary to facilitate ancillary power flow between interconnected microgrids considering the diverse dynamics of the emerging distributed sources is presented. In this framework, the reserve power capacity of each source is identified and a viable economic dispatch schedule is generated for all participating sources in the cluster. The viability of the new operating condition involving the participating PnP sources is studied using eigenvalue analysis. Upon obtaining an unstable operating condition, the participation of the PnP sources is rescheduled to obtain an economic and stable operating scenario. The ancillary power support cleared by the framework is realised in the microgrid cluster through necessary network reconfigurations. This framework encourages the participation of electric vehicles and other PnP sources in the transactive market in a PnP fashion. The ancillary power support framework consists of the following four operating modules: (1) Preparatory Data Processing Module, (2) Economic Dispatch Synthesizer Module, (3) Viable Performance Facilitator Module, and (4) Schedule Execution and Administration Module. A schematic representation of the various modules operating in a microgrid cluster is shown in Figure 7.4. The necessary variables used in the framework are shown in Figure 7.5. The operation of these modules is elaborated further for the microgrid cluster model in Section 7.3.1.

7.4.2 Preparatory data processing module

The *Preparatory Data Processing Module* receives time-synchronised phasor data from the cluster-level phasor data concentrator (PDC), various parameters of the PnP sources, and their operating constraints for market participation. It identifies the reserve power limit available for ancillary support of each participating source. The various steps involved in the operation of this module are enumerated in detail below:

Step 1: Read the voltages V_{bdq} and v_{odq}, voltage angle δ, currents i_{odq} and i_{ldq}, and droop coefficients m_p and n_q from all the inverter sources. The subscripts o and b

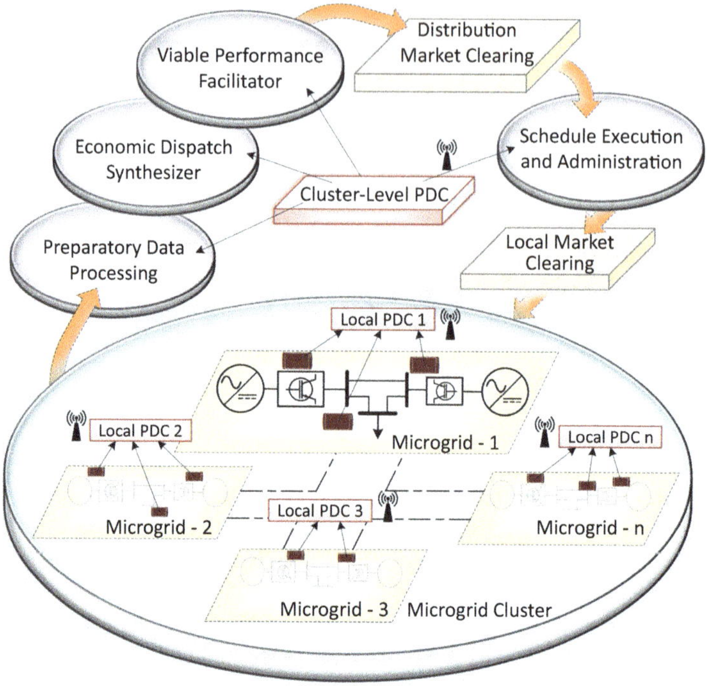

Figure 7.4 Schematic representation of the framework for real-time ancillary power support

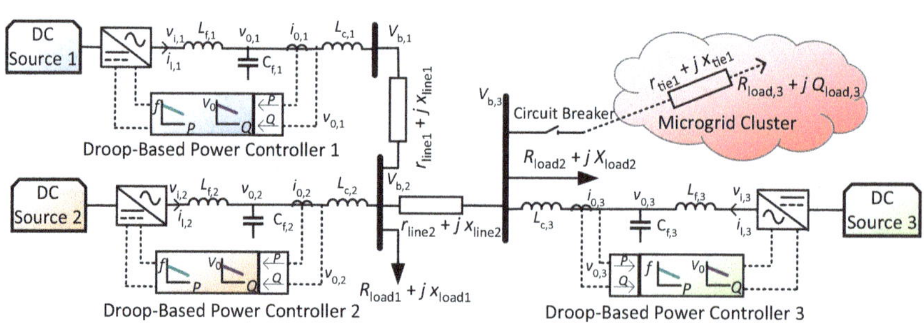

Figure 7.5 Test system depicting the plug-and-play sources in a microgrid cluster

refer to the readings from the inverter terminal and the bus terminal, respectively as shown in Figure 7.5.

Step 2: Obtain the values of R_{load} and X_{load} using the load flow information of the load flow buses.

Step 3: Identify the network topology of each microgrid and the microgrid interconnections within the microgrid cluster. Hence, r_{line}, x_{line}, r_{tie}, and x_{tie} are computed.

Step 4: Obtain the list of PnP sources from each microgrid that would provide ancillary power support to the microgrid cluster.
Step 5: Obtain the control parameters of the PnP sources and other necessary model information.
Step 6: The reserve power limit of each PnP source is obtained by estimating its operating power.

7.4.3 Economic dispatch synthesizer module

The *Economic Dispatch Synthesizer Module* receives the bids from individual microgrids and generates an economic schedule to meet the demand in the microgrid cluster. This module also identifies probable changes in the active and reactive droop parameters necessary to participate in the market operations. The step-by-step operation of this module is described in detail below:

Step 1: Obtain the probable ancillary power support necessary for the microgrid cluster from DMO. Read the bids P^1_{L-D} and the price of energy from each microgrid.
Step 2: Perform economic load dispatch and obtain a schedule for each microgrid as elaborated in Section 7.2.1. Identify the generation schedule for all the participating PnP sources. Check if the active and reactive power droops of the PnP sources require modification to satisfy the new dispatch schedule. Ensure that the droops are maintained in such a way that the reserve power is dispatched with minimum changes to the grid voltage and frequency.
Step 3: If necessary, compute the new active and reactive power droops m_p and n_q.
Step 4: Populate the mapping matrices M_{INV}, M_{load}, and M_{NET} shown in (7.10) based on the network information and the new dispatch schedule.

7.4.4 Viable performance facilitator module

The *Viable Performance Facilitator Module* first forms the model for the microgrid cluster based on the updated information from the economic dispatch synthesizer module. By performing eigenvalue analysis, this module ensures that the microgrid cluster is stable when it migrates to this new operating point. The detailed step-by-step operation of this module is as follows:

Step 1: Form the state, input, output, and feedforward matrices for the microgrid cluster model given by (7.11) and (7.21). Perform eigenvalue analysis.
Step 2: Ensure that the new operating condition for the microgrid cluster is stable. In addition, ensure that the damping ratio (ζ) and the damped natural frequency of oscillation (ω_n) are within the permissible limits.
Step 3: If the new operating schedule is not viable, identify the PnP sources whose eigenvalues are out of range.
Step 4: If the non-viable operating condition is due to any newly participated PnP sources, the participation of such sources is limited by redistributing power to other sources. If the non-viable operating condition is due to existing participating

sources, then the market participation P^1_{L-D} is reduced, and check for the viability of the solution. If the non-viable operating condition persists, this module withdraws the sources from current market participation. Inform Economic Dispatch Synthesizer Module.
Step 5: Generate the list of sources whose droop needs to be modified, and update the Economic Dispatch Synthesizer Module.
Step 6: Upon obtaining a viable operating scenario, generate P^2_{L-D} and price of energy for each microgrid, and communicate with the LMO.

7.4.5 Schedule execution and administration module

The *Schedule Execution and Administration Module* ensures that the viable operating scenario developed by the viable performance facilitator module is executed, and that the market obligations are met by each microgrid in the cluster. The operational steps are described below:

Step 1: The LMO ensures that the PnP sources are dispatched as per the schedule P^2_{L-D} and ensures the tie-line power flow in the microgrid cluster by using measurements from the distribution-level phasor measurement units (D-PMU).
Step 2: Ensure that market obligations are always satisfied. Perform load shedding if necessary.

7.4.6 Synchrophasor measurement-based realisation of ancillary power support framework

The operation of a microgrid cluster involves monitoring available resources and grid conditions, detection of contingencies, control of resources, and management of energy through market operations. In addition, improving the observability of microgrid clusters is important in the context of increasing PnP sources and flexible loads. This can only be realised through a synchrophasor-measurement-based measurement and monitoring system. This system includes D-PMU units, intelligent devices, local and cluster-level PDCs, communication systems, and various tools for extracting microgrid conditions [18]. D-PMUs are strategically placed in a microgrid to ensure complete observability. The time-stamped measurements from the D-PMU include the voltage and current phasors, frequency, and rate of change of frequency. The measurements from the individual D-PMUs in a microgrid are sent to a local PDC at the microgrid level through a wired or wireless communication system. The data acquisition and communication requirements for phasor measurements should follow the standard IEC/IEEE 60255-118-1:2018 [19]. In the PDC, the database management system aggregates and time-synchronises the streams of data. These data are stored in a database, and the applications can then be used to raise awareness of the microgrid. The minimum analytical functionalities that the PDC should offer are defined in IEEE C37.247-2019 [20]. At the local PDC, analytical applications include situational awareness, control, and protection of individual microgrids. The cluster-level PDC can receive time-aligned data from local PDCs or stream directly from individual D-PMUs in the microgrid. Compared to local PDCs, cluster-level PDCs have several additional responsibilities. It should

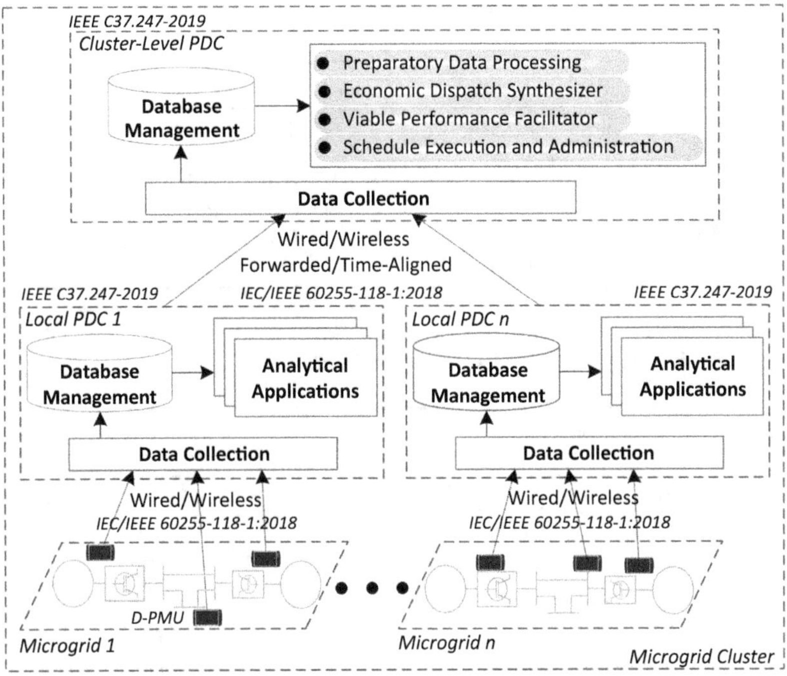

Figure 7.6 Data flow in a synchrophasor measurement-based operation of microgrid cluster

ensure energy sufficiency, ancillary power support, and reserve power support to all the microgrids in the cluster. This is achieved through in-build applications in the PDC or through third-party applications that use time-synchronised data streams from the PDC. The various modules elaborated in Section 7.4.1 for ancillary power support will obtain data from the cluster-level PDC and ensure seamless ancillary power support to various microgrids in the cluster. The flow of data in a synchrophasor-measurement-based monitoring system in a microgrid cluster is shown in Figure 7.6.

7.5 Case studies and discussion

A test system based on the schematic shown in Figure 7.1 is used to elaborate the necessity of the framework for seamless ancillary power support. In an operating scenario shown in Figure 7.5, three PnP sources and two constant-impedance loads are considered in the microgrid. This microgrid is connected to the cluster through a tie line. The various parameters used to simulate the operating conditions are listed in Table 7.1. By performing a load flow study on the test system, necessary parameters were obtained to define the initial operating conditions for a base voltage of 380 V and a base power of 10 kVA. The initial values are listed in Table 7.2.

Table 7.1 Parameters for the plug-and-play source-based microgrid cluster

Par.	Value	Par.	Value	Par.	Value	Par.	Value
mp	3.6733e−5	nq	1.5516e−4	L_f	1.35e-3 H	r_{Lf}	0.1 Ω
C_f	50e−6 F	F	0.75	L_c	0.35e−3 H	r_{Lc}	0.03 Ω
K_{pc}	10.5	K_{ic}	16e3	K_{pv}	0.05	K_{iv}	390
ω_c	314 rad/s	ω_n	314 rad/s	ω_0	31.41 rad/s	r_N	1000 Ω
r_{line1}	0.23 Ω	l_{line1}	3.185e−4 H	r_{line2}	0.35 Ω	l_{line2}	1.847e−3 H
r_{load1}	24.85 Ω	l_{load1}	6.822e−3 H	r_{load2}	19.79 Ω	l_{load2}	4.33e−3 H

Table 7.2 Initial values for the state-space model

Variable	Value (pu)	Variable	Value (pu)
v_{od}	[0.9998 0.9998 0.9996]	v_{oq}	[0 0 0]
i_{od}	[0.3275 0.2214 0.3246]	i_{oq}	[−0.0428 0.0716 −0.10]
V_{bD}	[0.9992 1.0 0.9983]	V_{bQ}	[−0.003 −0.002 −0.015]
i_{ld}	[0.3275 0.2214 0.3246]	i_{lq}	[−0.0428 0.0716 −0.1]
$i_{lineD,1}$	−0.0587	$i_{lineQ,1}$	−0.0081
$i_{lineD,2}$	−0.1626	$i_{lineQ,2}$	−0.0063
$i_{loadD,1}$	0.3862	$i_{loadQ,1}$	−0.0346
$i_{loadD,2}$	0.48612	$i_{loadQ,2}$	−0.0406

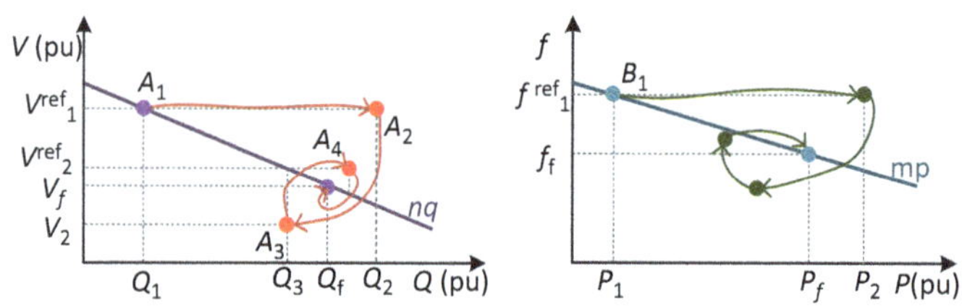

(a) (b)

Figure 7.7 A probable locus for the operating points following an increase in load demand: (a) reactive power on Q–V droop curve and (b) active power on P–f droop curve

Droop-based control of the participating PnP resources is one of the popular approaches to ensure the sharing of power demand among various sources in a microgrid. Droop-based controllers ensure that the steady-state operating point of the source lies on the droop characteristics while providing the desired participation. To understand the operation of droop-based power control, the probable locus of the operating point for active and reactive power following a load change is shown in Figure 7.7. Here, the PnP source is initially delivering P_1 and Q_1 active and reactive power, at a voltage magnitude of V_1^{ref} and a frequency of f_1^{ref}. When the load demand on the source increases to P_2 and Q_2, the operating points deviate from their respective droop characteristics. The inner voltage and current control loops try to maintain the voltage magnitude at V_1^{ref}, while trying to deliver the desired demand. However, owing to the increased power flow, the terminal voltage decreases, which reduces the reactive and active power flow. When the reactive power flow is Q_3, the droop-based power controller operates and modifies the voltage reference to V_2^{ref}. A probable locus for the operating point of the reactive power is $A_1 - A_2 - A_3$. The locus of the operating point of the active power owing to the variation in voltage is similar to that of the reactive power. This process continues until the voltage, current, and power controller converge to a stable operating point on the droop characteristics. This is achieved only when, at a voltage V_f and frequency f_f, the demand on the inverter equals the delivered power P_f and Q_f. When the action of the controllers cannot simultaneously converge both the active and reactive power operating points on the droop characteristics, the operating point continuously encircles the droop line, creating an oscillatory response in the microgrid cluster. Furthermore, for certain values of the control parameters, particularly the droop coefficients, the operating point tends to diverge from the droop characteristics. This creates unstable operating conditions in the microgrid cluster. For a given set of parameters and operating conditions, viable load sharing of PnP sources can be ensured by performing eigenvalue analysis on the microgrid cluster.

The dominant eigenvalues of the test system are shown in Figure 7.8(a). The eigenvalues of the microgrid can be grouped into three different frequency-mode clusters. The high-frequency modes ($\geq$1,500 Hz), represented as Cluster 3, are caused by the LCL filters and current loops in the inverter controller. The medium-frequency modes (200–500 Hz), represented as Cluster 2, are contributed by the voltage-control loop in the inverter controller. The oscillatory modes in Cluster 1, the low-frequency mode (2–10 Hz), are due to the power-sharing controllers and the power filter. These low-frequency oscillatory modes from Cluster 1 primarily affect the ancillary power sharing. The eigenvalues in Cluster 1 and their corresponding damped frequencies are shown in Figure 7.8(b). The eigenvalues of the test system are listed in Table 7.3 along with their damping ratio (ζ) and natural frequency of

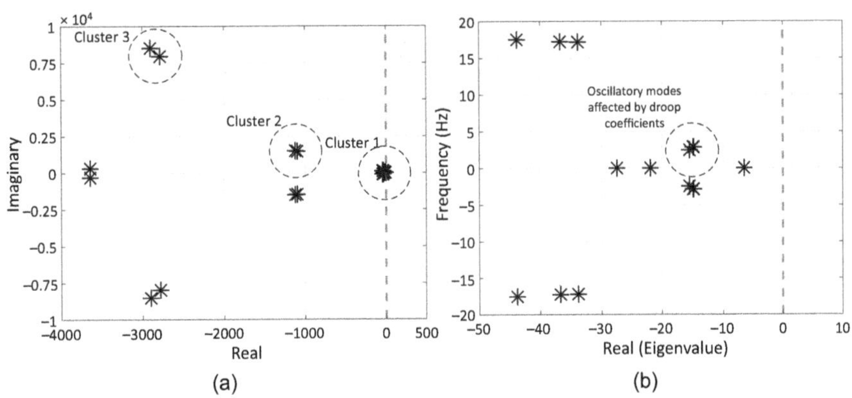

Figure 7.8 (a) The eigenvalues of the microgrid under a stable operating scenario and (b) the frequency bandwidth for eigenvalues in cluster 1

Table 7.3 Eigenvalues and eigenvectors of the microgrid under a stable operating scenario

Eigenvalue	ζ	ω_n (rad/s)	Eigenvalue	ζ	ω_n (rad/s)	Eigenvalue	ζ	ω_n (rad/s)
-6.48	1.00	6.48	$-15.48 \pm 15.48i$	0.71	21.90	-21.94	1.00	21.95
$-14.81 \pm 18.08i$	0.63	23.38	$-14.89 \pm 18.14i$	0.63	23.48	-27.49	1.00	27.50
$-33.82 \pm 107.97i$	0.30	113.15	$-36.69 \pm 108.227i$	0.32	114.27	$-43.78 \pm 110.16i$	0.37	118.54
$-1097.23 \pm 1474.01i$	0.60	1837.56	$-1097.52 \pm 1473.97i$	0.60	1837.71	$-1097.63 \pm 1473.96i$	0.60	1837.76
$-1122.63 \pm 1508.49i$	0.60	1880.39	$-1122.72 \pm 1508.47i$	0.60	1880.43	$-1122.99 \pm 1508.42i$	0.60	1880.54
$-3642.62 \pm 314.00i$	1.00	3656.14	$-4570.43 \pm 314.00i$	1.00	4581.21	$-2784.16 \pm 7962.28i$	0.33	8435.02
$-2784.16 \pm 7962.28i$	0.33	8435.02	$-2784.16 \pm 7962.30i$	0.33	8435.04	$-2902.08 \pm 8514.33i$	0.32	8995.33
$-2902.08 \pm 8514.35i$	0.32	8995.35	$-2902.08 \pm 8514.35i$	0.32	8995.35	$-774283.31 \pm 314.00i$	1.00	774283.38
$-6588900.20 \pm 314.00i$	1.00	6588900.21						

Eigenvectors corresponding to the eigenvalue $-15.48 + 15.48i$

0.972379, 0.207426 − 0.106891i, 0.002242 + 0.002177i, 0.000716 + 0.000219i, 0.002242 + 0.002177i, 0.000717 + 0.000218i, 0.001495 + 0.001452i, 0.000483 + 0.000151i, 0.000747 + 0.000726i, 0.000241 + 7.53237e-05i

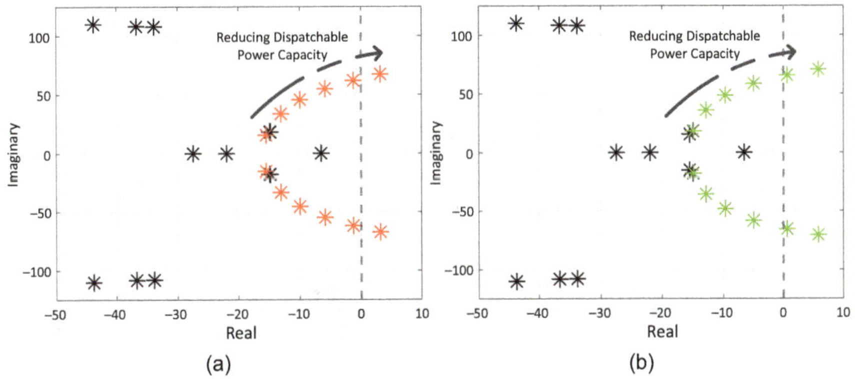

Figure 7.9 The locus of eigenvalue when the dispatchable power capacity of participating sources reduces: (a) in generator 1 and (b) in generator 2

oscillation (ω_n). The eigenvector corresponding to an eigenvalue $-15.48 + 15.48i$ participating in the low-frequency oscillation is also shown.

The droop of PnP sources will change based on their availability, market participation, and health of the PnP sources. This in turn modifies the equivalent droop of the aggregator, as explained in Section 7.2.3. When the droop coefficients mp and nq are varied, the oscillatory modes in the cluster migrate to a new operating point. For example, when a few PnP sources withdraw from market operations, the effective power capacity of the aggregator reduces. For an aggregator with 10 kVA capacity for market participation, loss of PnP sources reduces its capacity and makes its equivalent droops steeper. This is simulated in the test system by varying the droop coefficients of droop-based power controllers 1 and 2. As shown in Figure 7.9(a) and (b), the variation in the active power droop triggered the movement of low-frequency oscillatory modes. This shows that the system can move into oscillatory instability based on the nature of the PnP sources participating in the market operation.

Similarly, the ancillary power support through the tie line from microgrid 1 to the microgrid cluster was simulated by varying mp for droop-based power controller 1. This is achieved by providing a frequency change $\Delta\omega_{com}$ as the input to the state-space model simulating a load variation. As shown in Figure 7.10, limitations on the dispatchable capacity of the PnP sources can cause the cluster to experience oscillatory instability. Hence, the droop coefficient of the power controllers must be chosen appropriately to meet the ancillary power demand. This can be ensured by following the ancillary power support framework described in Section 7.4.1.

For any ancillary power demand on the microgrid cluster, the viable performance facilitator ensures that the operating conditions to be deployed in the system are stable. In case the participation of PnP sources creates a non-viable power dispatch scenario, the viable performance facilitator and economic dispatch

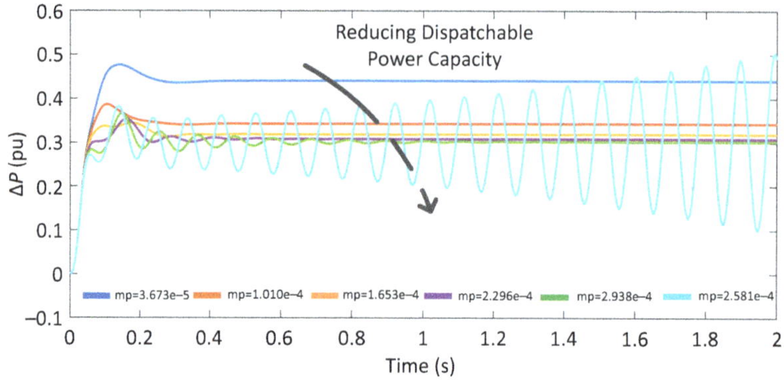

Figure 7.10 Oscillatory instability in the microgrid when the droop of the droop-based power controller 1 is varied

synthesizer coordinate to settle at a viable set of droop coefficients. This prevents the system from transitioning to unstable operating conditions.

7.6 Conclusion

The integration of battery-based PnP sources in a microgrid cluster is an emerging solution to satisfy the ancillary power demand. However, this poses a major drawback by jeopardising the stable operation of the microgrid cluster. This chapter elaborates on the various operating constraints of PnP sources that can lead to unfavourable operating conditions in the microgrid cluster. This chapter also highlights the difficulties faced by market operators when integrating such sources into a microgrid cluster. The need for a synchrophasor-assisted resilient framework for providing ancillary support in a microgrid cluster is explained in detail. This framework ensures that the participation of PnP sources in market operations is economically viable and operationally stable. The operating conditions leading to oscillatory instability in a microgrid cluster can be identified and avoided by following this framework. A dedicated reactive power market that encourages reactive power support in a microgrid cluster can aid in the stable operation of the microgrid cluster.

References

[1] IRENA (2021), Renewable power generation costs in 2020, *Abu Dhabi, Tech. Rep.*, 2021.

[2] IRENA (2023), Renewable capacity statistics 2023, *Abu Dhabi, Tech. Rep.*, 2023.

[3] S. Tirunagari, M. Gu, and L. Meegahapola, "Reaping the benefits of smart electric vehicle charging and vehicle-to-grid technologies: Regulatory, policy and technical aspects," *IEEE Access*, vol. 10, pp. 114657–114672, 2022.

[4] J. J. C. Mancera, J. L. Saenz, E. López, "Experimental analysis of the effects of supercapacitor banks in a renewable DC microgrid," *Applied Energy*, vol. 308, p. 118355, 2022.

[5] L. Che, X. Zhang, M. Shahidehpour, A. Alabdulwahab, and A. Abusorrah, "Optimal interconnection planning of community microgrids with renewable energy sources," *IEEE Transactions on Smart Grid*, vol. 8, no. 3, pp. 1054–1063, 2017.

[6] F. Alfaverh, M. Denai, and Y. Sun, "A dynamic peer-to-peer electricity market model for a community microgrid with price-based demand response," *IEEE Transactions on Smart Grid*, vol. 14, no. 5, pp. 3976–3991, 2023.

[7] M. Farrokhseresht, N. G. Paterakis, H. Slootweg, and M. Gibescu, "Enabling market participation of distributed energy resources through a coupled market design," *IET Renewable Power Generation*, vol. 14, no. 4, pp. 539–550, 2020.

[8] P. B. L. Neto, O. R. Saavedra, and L. A. de Souza Ribeiro, "A dual-battery storage bank configuration for isolated microgrids based on renewable sources," *IEEE Transactions on Sustainable Energy*, vol. 9, no. 4, pp. 1618–1626, 2018.

[9] X. Liu, T. Zhao, H. Deng, P. Wang, J. Liu, and F. Blaabjerg, "Microgrid energy management with energy storage systems: A review," *CSEE Journal of Power and Energy Systems*, vol. 9, no. 2, pp. 483–504, 2023.

[10] S. Sahoo, and P. Timmann, "Energy storage technologies for modern power systems: A detailed analysis of functionalities, potentials, and impacts," *IEEE Access*, vol. 11, pp. 49689–49729, 2023.

[11] B. Xu, A. Oudalov, A. Ulbig, G. Andersson, and D. S. Kirschen, "Modeling of lithium-ion battery degradation for cell life assessment," *IEEE Transactions on Smart Grid*, vol. 9, no. 2, pp. 1131–1140, 2018.

[12] F. Naseri, S. Karimi, E. Farjah, and E. Schaltz, "Supercapacitor management system: A comprehensive review of modeling, estimation, balancing, and protection techniques," *Renewable and Sustainable Energy Reviews*, vol. 155, p. 111913, 2022.

[13] S. S. Guggilam, C. Zhao, E. Dall'Anese, Y. C. Chen, and S. V. Dhople, "Optimizing power–frequency droop characteristics of distributed energy resources," *IEEE Transactions on Power Systems*, vol. 33, no. 3, pp. 3076–3086, 2018.

[14] Q. Zhou, M. Shahidehpour, Z. Li, and X. Xu, "Two-layer control scheme for maintaining the frequency and the optimal economic operation of hybrid AC/DC microgrids," *IEEE Transactions on Power Systems*, vol. 34, no. 1, pp. 64–75, 2019.

[15] W. Du, Z. Chen, K. P. Schneider, *et al.*, "A comparative study of two widely used grid-forming droop controls on microgrid small-signal stability," *IEEE Journal of Emerging and Selected Topics in Power Electronics*, vol. 8, no. 2, pp. 963–975, 2020.

[16] N. Pogaku, M. Prodanovic, and T. C. Green, "Modeling, analysis and testing of autonomous operation of an inverter-based microgrid," *IEEE Transactions on Power Electronics*, vol. 22, no. 2, pp. 613–625, 2007.

[17] J. Shen, and A. Khaligh, "A supervisory energy management control strategy in a battery/ultracapacitor hybrid energy storage system," *IEEE Transactions on Transportation Electrification*, vol. 1, no. 3, pp. 223–231, 2015.

[18] A. Meier, M. Brown, R. Arghandeh, and E. Stewart, "Synchrophasor monitoring for distribution systems – Technical foundations and applications," *NASPI-2018-TR-001*, 2018.

[19] "IEEE/IEC international standard – Measuring relays and protection equipment – Part 118-1: Synchrophasor for power systems – Measurements," *IEC/IEEE 60255-118-1:2018*, pp. 1–78, 2018.

[20] "IEEE standard for phasor data concentrators for power systems," *IEEE Std C37.247-2019*, pp. 1–44, 2019.

Chapter 8

PMU-based vulnerability assessment of power systems

Guanqun Wang[1], Chen-Ching Liu[1], Evangelos Farantatos[2] and Navin Bhatt[2]

8.1 Introduction

8.1.1 Cascaded events and vulnerability assessment

Since 1960s, the scale and complexity of interconnected power grids have increased to such an extent that the vulnerability of power systems with respect to large-scale blackouts has increased significantly. For example, on November 9, 1965, the blackout in Northeast USA caused a loss of 21 GW load and left 30 million people without power for up to 13 hours [1]. The 2003 blackout in North America affected more than 50 million people, costing about $3 billion economic losses and contributed to 11 reported fatalities [2]. On July 30 and 31, 2012, two consecutive blackouts took place in India. The second one is the largest power outage that has ever happened in the world. A total of 600 million people were affected, nearly 10% of the world's population. An estimated 48 GW of generating capacity was lost during the outage [3,4]. Table 8.1 lists some major blackouts in the world in recent years.

Analysis of these large-scale blackouts indicates that these outages were triggered by a cascading sequence of events [8]. Generally, these sequences of events follow a common process [8,9]. First, the cascaded events are initiated by a single or multiple events such as line outages or loss of generators. Due to relay actions, one initiating contingency may trigger another, leading to successive tripping of transmission lines and generators if the contingencies are not cleared in time. The subsequent events can cause power flow rerouting, line overloading, voltage problems, and reactive power deficiency, which further weaken the system. Frequency instability, voltage instability, and rotor angle instability can take place during this process, leading to an uncontrolled system partitioning, and eventually a widespread blackout.

As a result, it is important to identify the vulnerable patterns of power systems. That is, events that may trigger other events. In Ref. [8], authors report the patterns

[1]Department of Electrical Engineering and Computer Science, Washington State University, USA
[2]Grid Operations & Planning R&D Group, Electric Power Research Institute, USA

Table 8.1 Some major blackouts in recent years [3–7]

Date	Location	Effects
2012-7-30 and 7-31	India	Affected more than 600 million people. 48 GW of generating capacity was taken off.
2011-9-8	USA and Mexico	2.7 million people were affected, including 1.4 million customers in San Diego County and 1.1 million in Mexico.
2009-11-10	Brazil Paraguay	60 million people were affected and 28.8 GW of load was lost.
2006-11-04	West Europe	Eight countries and approximately 50 million people were affected. 17 GW of load was lost. The blackout was restored after 1.5 hours.
2003-09-28	Italy	180 GW of load was lost, 57 million people were affected. The blackout was restored after 20 hours.
2003-09-23	North Europe	1800 MW of load was lost, 5 million people were affected. The blackout was restored after 6.5 hours.
2003-08-14	USA and Canada	One province and eight states were out of power and more than 50 million people were affected. The blackout last for up to 29 hours, cost the loss of 61.8 GW of power and economic loss of $6 billion.

of cascaded events as line tripping due to overloading, generator tripping due to over-excitation, line tripping due to loss of synchronism, generator tripping due to abnormal voltage, and frequency system condition and under-frequency/voltage load shedding. In Ref. [10], it is stated that zone 3 relay operation serves as a basic pattern for cascaded events. A fuzzy inference system is developed to identify contingencies that are likely to trigger cascaded zone 3 relay operations. Furthermore, the concept of strategic power infrastructure defense (SPID) systems is proposed in Refs. [11,12], which involves real-time sensing and communication, failure analysis, vulnerability assessment, and self-healing control features. This system is a real-time, wide-area, adaptive protection, and control system that depends on the power, communication, and computer infrastructures. It is intended to prevent catastrophic failures that may lead to large-scale power outages.

Rotor angle instability that can further cause frequency instability and voltage instability is one of the main vulnerable patterns of power systems. It is important to predict rotor angle instability before the system loses synchronization. For transient stability, the power system dynamics are divided into three stages, represented by three sets of nonlinear dynamic equations

$$\text{Pre}-\text{fault}\begin{cases}\dot{x}=f_0(x,y,u)\\ 0=g_0(x,y,u)\end{cases}\quad 0<t\le t_0^- \tag{8.1}$$

$$\text{Fault}-\text{on}\begin{cases}\dot{x}=f_F(x,y,u)\\ 0=g_F(x,y,u)\end{cases}\quad t_0^+\le t\le t_{cl}^- \tag{8.2}$$

$$\text{Post}-\text{fault}\begin{cases}\dot{x}=f_{PF}(x,y,u)\\ 0=g_{PF}(x,y,u)\end{cases}\quad t_{cl}^+\le t \tag{8.3}$$

where x represents state variables that describe the dynamics of generators, speed governors and excitation systems, etc. The symbol y represents algebraic variables such as node voltage amplitudes and angles, and u represents control variables such as tripping and restoration of generators and load. The time instants t_0 and t_{cl} represent the time at which a fault occurs and is cleared, respectively. After severe disturbances, the rotor angle speed of generators will oscillate due to imbalance between mechanical power inputs and electrical power outputs of generators. If the system does not have the ability to eliminate such oscillations, it will lose synchronism and become unstable. As a result, it is important to monitor vulnerable patterns and assess transient stability of power system on line so that controls can be taken to prevent the system from entering an emergency operating state.

There are two major approaches to transient stability analysis for power systems. One is time-domain simulation methods [13,14], which is based on numerical integration of system dynamic equations. With an accurate component model, this method can provide details of the dynamic process for a large-scale system. However, the high computation burden and lack of a stability index limit the online applicability of this method. The other one is the energy functions-based direct methods such as relevant unstable equilibrium point (RUEP) method [15], controlling UEP (CUEP) method [16], potential energy boundary surface (PEBS) [17], BCU method [18], and extended equal area criteria (EEAC) method [19]. These methods have a relatively fast computational speed and can provide a quantitative index for stability assessment, and thus there are advantages for online application. A shortcoming of direct methods is that the assessment result may not incorporate sufficient details due to the simplification of system models.

8.1.2 PMU-based monitoring of power systems

The concept of phasor measurement units (PMUs) was proposed in mid-1980s [20]. Algorithms for the application of PMUs in static state estimation and adaptive protection were proposed [21–23]. The early prototype of PMU was built in 1992 at Virginia Tech [24]. In recent years, due to the increasing need for wide area monitoring and control of the power grid, the number of PMUs installed on the grid has increased significantly. The new and accurate measurement device is an enabling technology for power grids to perform wide area monitoring and control of system dynamics.

The PMU-based out-of-step relay is proposed and applied to the Florida-Georgia system [25–28]. The system is modeled as two interconnected equivalent generators and they are measured by two PMU sets to monitor the angular difference between them. Based on the monitored angular difference, the equal-area criterion is utilized to assess transient stability of the power system after a power swing. In Ref. [29], synchronized phasor measurements are used for solving the states of a post-fault power system based on differential-algebraic equation model. The numerical results are used to evaluate system stability. In Ref. [30], the authors investigate the application of extended Kalman filter techniques for tracking dynamic states of power systems using phasor measurement data. In Refs. [31,32], PMU measurements at the substation level are applied to a distributed frequency-domain optimization method for monitoring of oscillations. PMUs are also applied to the energy function-based direct methods and decision tree methods for online dynamic security assessment [33,34]. In Ref. [35], PMU is used to monitor long-term voltage stability by calculating voltage stability assessment index in a real-time operational environment. When combined with machine learning technique, PMUs can also be applied for advanced power system situational awareness, including event identification [36].

In this research, a new method is proposed for monitoring of wide-area rotor angle stability based on PMU data in an on line operational environment [37,38]. The maximum Lyapunov exponent (MLE), which is applied as a tool for prediction of out-of-step conditions [39], is connected with rotor angle stability analysis. An efficient computational algorithm is proposed for the calculation of MLE based on dynamic waveforms acquired from PMU data within a finite time window.

Furthermore, the observability properties of the system with the limited availability of PMUs are discussed. Consider a large-scale power system with only a small number of PMUs. With the available PMUs, is it possible to reconstruct all states through the available PMU measurements? This research provides the results to clarify different observability concepts for static and dynamic models. Nonlinear dynamic observability which is based on a rigorous theoretical foundation is applied to evaluate the level of observability for dynamic state calculation with partial PMU measurements. The proposed method is illustrated by a case study with a three-bus system.

8.2 PMU-based monitoring and assessment of rotor angle stability

8.2.1 Maximum Lyapunov exponent method

In Ergodic theory [40], Lyapunov exponent is used to characterize whether a given system is "chaotic" and how chaotic it is. For two trajectories that start out very close to each other

$$\begin{aligned} x(t) &= f^t(x_0) \\ x(t) + \Delta x(t) &= f^t(x_0 + \Delta x_0) \end{aligned} \tag{8.4}$$

all points in a neighborhood of $x(t)$ may converge toward it, or separate exponentially as time goes on, which is called chaotic. Sensitivity of the system trajectory toward initial conditions can be quantified by

$$||\Delta x(t)|| \approx e^{\lambda t}||\Delta x_0|| \tag{8.5}$$

which is shown in Figure 8.1. The index λ is called the Lyapunov exponent which distinguished between exponential divergence or convergence of nearby trajectories.

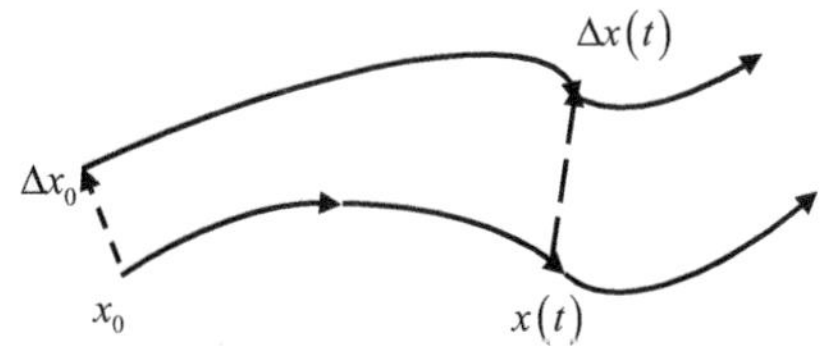

Figure 8.1 Perturbation of trajectories of a dynamic system

For a N-dimensional dynamical system represented by nonlinear differential equations

$$\dot{x} = f(x) \tag{8.6}$$

Based on (8.5), MLE can be calculated as

$$\lambda = \lim_{t \to \infty} \frac{1}{t} \ln \frac{\|J^t(x_0)\Delta x_0\|}{\|\Delta x_0\|} \tag{8.7}$$

where $J^t(x_0)$ is the Jacobian matrix of the system, and Δx_0 is a randomly chosen vector in small scale, as shown in Figure 8.1. The relationship between MLE and the asymptotic behavior of the dynamical system is established in Ref. [41]. If MLE is found to be negative after the fault, the waveform following the disturbance is asymptotically stable. Otherwise, it is considered unstable.

8.2.2 MLE calculation algorithm

The flowchart of the proposed MLE calculation algorithm of a power system is shown in Figure 8.2.

In the flowchart, the procedure for calculation of MLE can be described in the following steps:

1. Spectral analysis

   MLE is estimated over a specified, finite time window T with the length of seconds. It should be pre-determined properly so that waveforms of the PMU measurements can reliably represent the features of system dynamics under a disturbance. In the algorithm, the spectral analysis is used to compute an appropriate time interval length. Credible disturbances can be simulated in the offline environment to obtain the power swing curves following each disturbance. The spectral analysis is performed for the power swing curves to determine a corresponding frequency range. Based on the result of the spectrum analysis, the window size T is selected as the inverse of the lower limit of the frequency range so that it covers at least one cycle of the power swing. The MLE estimated based on T, $\lambda_{t=T}$, is an approximate value of λ, the theoretical MLE over an infinite time interval.
2. State calculator

   The complete trajectories of all state variables over the time window T are required to calculate MLE. That is, rotor angles and angular speeds of all

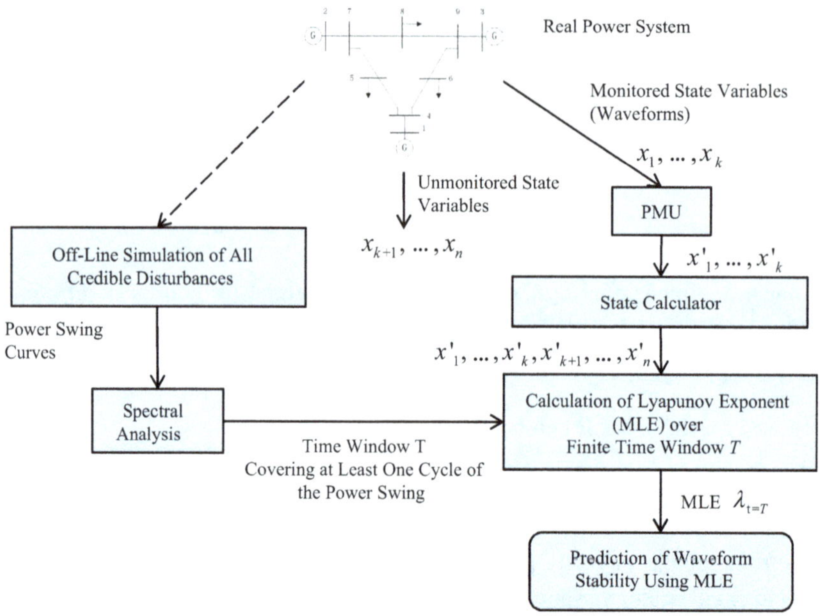

Figure 8.2 Flowchart of MLE calculation algorithm

generators are to be observed by PMU data. However, at present, there are only limited numbers of PMUs in the power system. To overcome the difficulty, the state calculator (SC) algorithm is developed to estimate the unmonitored state variables from the initial condition to the post-fault period.

The rotor angle dynamics of power system is formulated as

$$\begin{cases} \dot{x}(t) = f(x(t)) \\ x(t_0) = x_0 \end{cases} \tag{8.8}$$

where $x(t) = [x_1(\mathrm{t}), x_2(\mathrm{t}), \ldots, x_n(\mathrm{t})]^T$ are state variables, x_0 is the initial state. Assuming the waveforms of $x_1(t), \ldots, x_k(t)$ are monitored by PMUs, it is necessary to estimate the unmonitored state variables $x_{k+1}(t), \ldots, x_n(t)$ at a sequence of discrete points, $t = t_0, t_0 + \Delta t, t_0 + 2\Delta t, \ldots$. Numerical methods are needed for the solution of the above problem.

In Ref. [37], the implicit integration method with a trapezoidal rule is used to solve this problem. The state variables at instant $t = t_0 + \Delta t$ are estimated by integration of the dynamical system, as shown below:

$$x(\Delta t) = x(0) + \frac{\Delta t}{2}\{f(x(0)) + f(x(\Delta \mathrm{t}))\} \tag{8.9}$$

The above equation can be formulated as $G[x(\Delta t)] = 0$. Since this is a set of nonlinear simultaneous equations, an estimated value of $x(\Delta t)$ is obtained by the Newton–Raphson method. Note that for the next time increments, $x(2\Delta t)$, $x(3\Delta t)$, ..., can be approximated by the same procedure.

Since $x_1, x_2, \ldots, x_k$ in $x(t)$ are monitored by PMUs at buses 1, 2, ..., k, $x_1(t), x_2(t), \ldots, x_k(t)$ are known at each discrete points $t = t_0, t_0 + \Delta t$, $t_0 + 2\Delta t, \ldots$. The differential equations in (8.8) of $x_1, x_2, \ldots, x_k$ can, therefore, be eliminated and $x_1(t), x_2(t), \ldots, x_k(t)$ in other equations can be replaced by known PMU measurements $x_1{}'(t), x_2{}'(t), \ldots, x_k{}'(t)$

$$\begin{cases} \dot{x}_i(t) = f(x_i(t), x_1{}'(t), x_2{}'(t), \ldots, x_k{}'(t)) & i = k+1, \ldots, n \\ x_i(t_0) = x_{i0} \end{cases} \tag{8.10}$$

Note that the order of the dynamic system is reduced from n to $(n-k)$ due to the direct measurements from the available PMUs at buses 1, 2, ..., k.

3. Computation of MLE

The standard method with Gram–Schmidt reorthonormalization (GSR) [42] is used to calculate MLE for the long time average, after the estimation of the dynamic states. By this method, MLE can be represented as a mean growth rate of the perturbation between neighboring trajectories, as shown in Figure 8.3. By calculating the waveform changes for each time increment Δt, an average of the changes over a selected time window will be the approximate value of MLE

$$\lambda = \lim_{t \to \infty} \frac{1}{t} \sum_i \Delta t_i \lambda_i \tag{8.11}$$

Based on a nonlinear dynamic model of the power system, the MLE can be calculated based on the waveforms resulting from the dynamics of system states. The corresponding control actions can be determined based on the prediction. As PMUs can measure the phasors accurately and with synchronized time stamps, their installation on a power grid can considerably improve the performance of monitoring for system dynamics.

4. Prediction of system stability

The relationship between MLE and rotor angle stability of power systems has been established and, therefore, MLE as an index of the waveform stability is founded on a solid theoretical basis. Note that waveform stability is different from the usual stability around an equilibrium point of a nonlinear dynamic system. The former is to evaluate whether the waveform can stay close to a

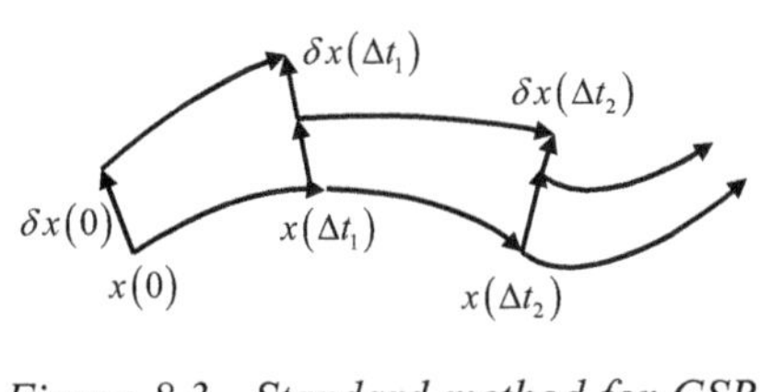

Figure 8.3 Standard method for GSR

given waveform from the collected PMU data if a perturbation occurs to the initial operating condition. The latter, however, is to evaluate whether the system will approach a stable equilibrium point after a disturbance for a given post fault system condition. The waveform stability does not assume the existence of a stable or unstable equilibrium point.

8.3 Observability for PMU-based monitoring of system dynamics

Observability analysis is a crucial task in model-based state monitoring and estimation. At this time, the total number of monitored buses is still modest compared with the large number of buses (tens of thousands) in a large interconnected grid such as the Western or Eastern Interconnection of North America. Since MLE is calculated from available PMU measurements, it is questionable whether these PMU measurements contain sufficient information for estimation of the entire set of states and how reliable the results are. This is the issue of observability one needs to deal with.

The concepts of observability in power systems are defined for different contexts and in different ways. We need to clarify these concepts and apply appropriate concept for PMU-based monitoring of system dynamics.

8.3.1 Static observability

For static state estimation, a network is said to be observable if any flow in the network can be observed by some indication in the set of measurements [43]. For simplicity, consider the DC power flow model of the power system in Figure 8.4.

By applying DC power flow, the measurement in the system could be written as

$$z = H\theta \tag{8.12}$$

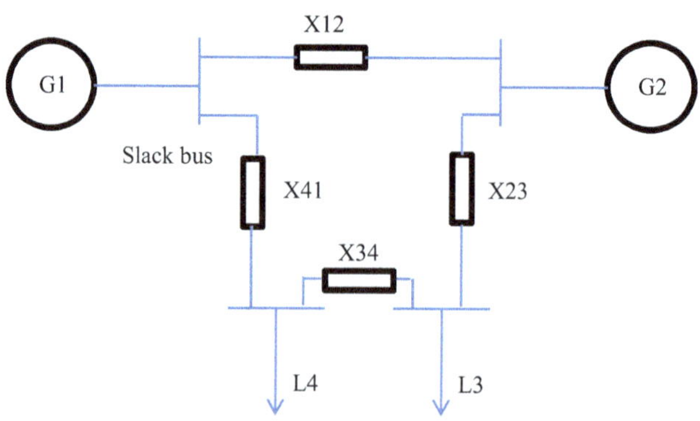

Figure 8.4 A simple power system

where z represents measurements, H represents the measurement matrix, and $\theta = \begin{bmatrix} \theta_2 & \theta_3 & \theta_4 \end{bmatrix}$ represents the state variables (bus angles) of the system. By the least-squared algorithm for static state estimation, θ can be estimated by

$$\theta^* = \left(H^T H\right)^{-1} H^T z \tag{8.13}$$

Therefore, if $H^T H$is nonsingular, that is, the rank of H equals to H's column number, the system is observable [44]. The observability for static state estimation is different depending on whether the number of measurements is greater than (or equal to) or less than the number of system states. Equation (8.13) applies to the case when the number of measurements is greater than or equal to the number of states. This solution also corresponds to the maximum likelihood estimate for jointly Gaussian noises of the measurement errors. Therefore, it is physically meaningful. If the number of measurements is lower than that of the state variables, it is possible to obtain a minimum norm estimate based on the available measurements. However, this minimum norm estimate is not physically meaningful.

8.3.2 Topology observability

For static state estimation with measurements provided by PMUs, however, it is possible to observe beyond the PMU buses (see Figure 8.5). There are 14 state variables, i.e., 7 voltage magnitudes and 7 angles, to be measured. Two PMUs are installed. However, a PMU installed at a specific bus is capable of measuring not only the bus voltage phasor but also the current phasors along all lines incident with the bus. As a result, in addition to the voltage phasor at this bus, the voltage phasors of all neighboring buses can be computed, hence they also become observable.

If the complete set of bus voltage phasors can be determined by the number and location of PMUs, the system with this PMU placement scenario is called the system topological observability [45]. For a topologically observable network with only PMU measurements, the static state estimation becomes linear, as shown in the following equation:

$$\begin{pmatrix} z_V \\ z_C \end{pmatrix} = \begin{pmatrix} I & 0 \\ Y_{CL1} & Y_{CL2} \end{pmatrix} \begin{pmatrix} \dot{V}_{L1} \\ \dot{V}_{L2} \end{pmatrix} + \begin{pmatrix} \varepsilon_V \\ \varepsilon_C \end{pmatrix} \tag{8.14}$$

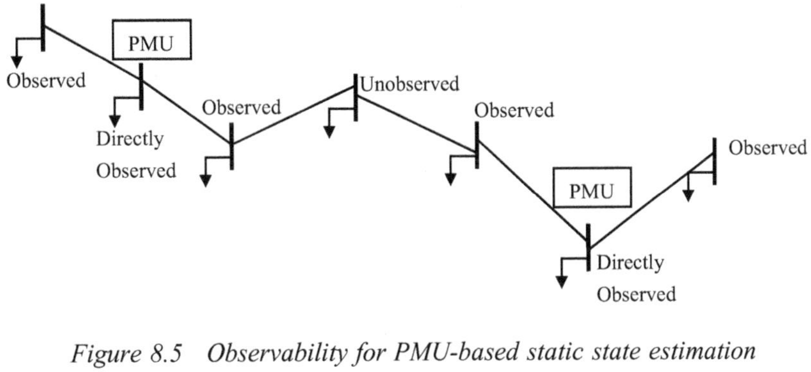

Figure 8.5 Observability for PMU-based static state estimation

where $\dot{V}_{L1}$, $\dot{V}_{L2}$ are state vectors representing the bus voltage phasor at buses with and without PMU measurements; I is identical matrix. Y_{CL1} and Y_{CL2} are the measurement matrices with the network admittances, respectively; z_V and z_C are the measurement vectors with the bus voltages and line currents, respectively; ε_V and ε_C are the vector of measurement errors. According to [46], a minimum number of 1/5 to 1/4 of the system buses to be installed with PMUs should ensure the complete topological observability of the system.

8.3.3 Dynamic observability

A dynamic system can be represented by differential equations

$$\begin{aligned} \dot{x}(t) &= f(x(t)) + u(x(t)) \\ z &= h(x(t)) \end{aligned} \tag{8.15}$$

where x and z represent state vector and measurement vector, respectively, f, u, and h are state function vector, control function vector, and measurement function vector, respectively. The system is said to be observable if, for any possible sequence of state and control vectors, the current state can be determined in finite time using the outputs [47]. In other words, for an observable system, it is possible to use the system measurements to determine the states of the dynamic system. If a system is not observable, the current values of some of its states may not be determined through the output values.

To illustrate the concept of observability, the mathematical tool of Lie derivative is needed [48], which is summarized as follows.

Let $f: \mathbb{R}^n \rightarrow \mathbb{R}^n$ be a vector field in $\mathbb{R}^n$, that is

$$f = \begin{bmatrix} f_1(x) \\ \cdots \\ f_n(x) \end{bmatrix} \tag{8.16}$$

Let $h: \mathbb{R}^n \rightarrow \mathbb{R}$ be a scalar function. Then the Lie derivative of h with respect to f is

$$L_f h = \nabla h \cdot f(x) = \frac{\partial h}{\partial x} f = \sum_{i=1}^{n} \frac{\partial h}{\partial x_i} f_i \tag{8.17}$$

The higher-order Lie derivatives can also be defined. By definition, the zero-order Lie derivative of h is

$$L_f^0 h = h \tag{8.18}$$

and the k-order derivative can be calculated by repeated derivative, i.e.,

$$L_f^k h = \frac{\partial \left(L_f^{k-1} h \right)}{\partial x} \cdot f \tag{8.19}$$

System (8.15) is locally observable at x_t, if the matrix

$$O(x_t) = \frac{\partial l(x)}{\partial x}|_{x=x_t} \tag{8.20}$$

has a full rank [46]. $O(x)$ is called the observability matrix, in which

$$l(x_t) = \begin{bmatrix} L_f^0 h_1(x) \\ \cdots \\ L_f^0 h_p(x) \\ L_f^1 h_1(x) \\ \cdots \\ L_f^{n-1} h_p(x) \end{bmatrix} \tag{8.21}$$

The rank condition of (8.21) can be checked by the singular value decomposition method [49]. Since the observability property is local, the smallest singular value of $O(x)$ will change along the trajectory of $x(t)$. Furthermore, the smallest singular value of $O(x)$ can be used to determine the level of observability of the system with different outputs.

If system (8.15) becomes linear, the observability matrix $O(x)$ will also reduce to linear case, which is well described in standard linear system textbooks, e.g., [50].

8.3.4 PMU-based monitoring of power system dynamics

The static state estimation accounts only for the static system variables, i.e., bus voltages and angles, without the dynamic system states. For control techniques that require information about these dynamic system states, the static state estimation methods and corresponding observability concepts are not sufficient.

It is important to monitor power system dynamic states such as the rotor speed to ensure a reliable operation and an efficient control of the power grid. Since PMU data has a high sampling rate with high accuracy and synchronized time stamps, it can be acquired to approximate the continuous time waveforms and thus considerably improve monitoring of the system dynamics. The dynamics of the power system can be modeled as (8.15) and the PMU-measured waveform can be treated as the measurement function $h(x)$. With state calculation methods such as implicit integration methods, the full states of the system are solvable no matter how many measurements are available on the system. However, the accuracy of monitoring is questionable when the number of measurements is small compared to the number of states. As a result, the theory of dynamic observability can be applied to the PMU-based dynamic monitoring problem. The observability index serves as a metric for quality of the reconstructed dynamic states based on snapshots of PMU measurements.

Considering observability issue, the flowchart of the MLE calculation algorithm is revised as shown in Figure 8.6.

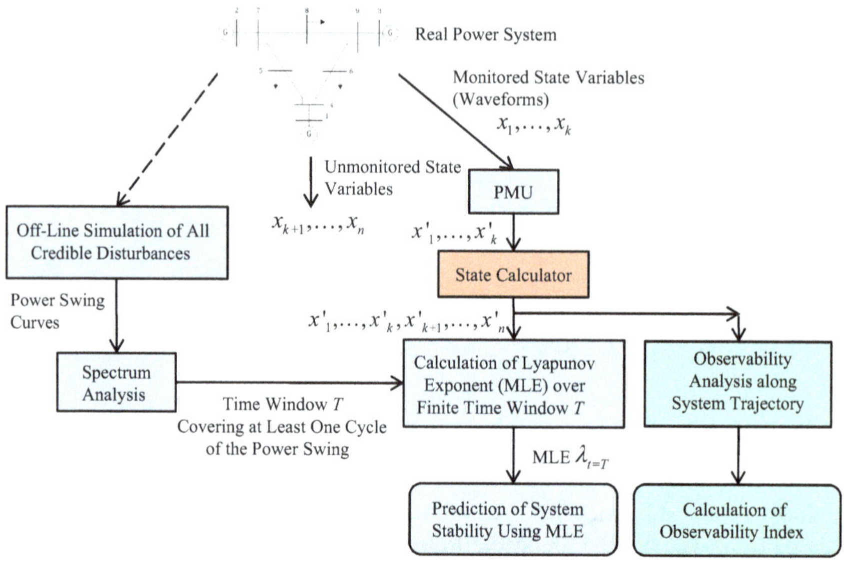

Figure 8.6 Flowchart of MLE calculation algorithm together with observability issue

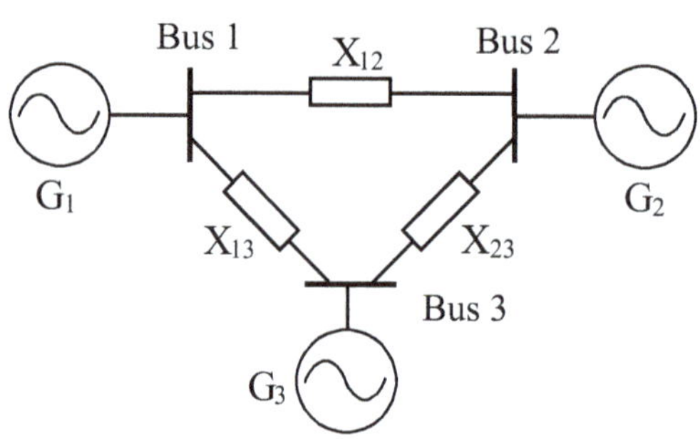

Figure 8.7 A three-bus power system

8.4 Example

The MLE calculation algorithm is illustrated by a three-bus system, as shown in Figure 8.7.

The classical rotor dynamic equations are as follows:

$$\begin{aligned}
P_{M1} &= M_1\ddot{\delta}_1 + D_1\dot{\delta}_1 + \frac{E_1E_2}{X_{12}}\sin(\delta_1 - \delta_2) + \frac{E_1E_3}{X_{13}}\sin(\delta_1 - \delta_3) \\
P_{M2} &= M_2\ddot{\delta}_2 + D_2\dot{\delta}_2 + \frac{E_1E_2}{X_{12}}\sin(\delta_2 - \delta_1) + \frac{E_2E_3}{X_{23}}\sin(\delta_2 - \delta_3) \\
P_{M3} &= M_3\ddot{\delta}_2 + D_3\dot{\delta}_3 + \frac{E_1E_3}{X_{13}}\sin(\delta_3 - \delta_1) + \frac{E_2E_3}{X_{23}}\sin(\delta_3 - \delta_2)
\end{aligned} \tag{8.22}$$

The parameters of the three-generator system are given by

$$D_1/M_1 = 0.5, D_2/M_2 = 0.3, D_3/M_3 = 0.2$$
$$P_{M1}/M_1 = P_{M2}/M_2 = 0.85, P_{M3}/M_3 = -1.7$$
$$E_1E_2/M_1 = E_1E_2/M_2 = 0.1, E_1E_3/M_1 = E_1E_3/M_3 = E_2E_3/M_2 = E_2E_3/M_3 = 1.0$$

By substitution of these parameters, a six-dimensional equation is obtained:

$$\begin{aligned}
\dot{\delta}_1 &= \omega_1 \\
\dot{\omega}_1 &= 0.85 - 0.1 \cdot \frac{\sin(\delta_1 - \delta_2)}{X_{12}} - \frac{\sin(\delta_1 - \delta_3)}{X_{13}} - 0.5\omega_1 \\
\dot{\delta}_2 &= \omega_2 \\
\dot{\omega}_2 &= 0.85 - 0.1 \cdot \frac{\sin(\delta_2 - \delta_1)}{X_{12}} - \frac{\sin(\delta_2 - \delta_3)}{X_{23}} - 0.3\omega_2 \\
\dot{\delta}_3 &= \omega_3 \\
\dot{\omega}_3 &= -1.7 - \frac{\sin(\delta_3 - \delta_1)}{X_{13}} - \frac{\sin(\delta_3 - \delta_2)}{X_{23}} - 0.2\omega_3
\end{aligned} \tag{8.23}$$

Assume that $X_{12} = X_{13} = X_{23} = 1$. A three-phase fault followed by clearing of the fault is applied to transmission line 2–3. The fault is cleared by relay and breaker operations, which will last for a fraction of 1 s. Then it is assumed that the breaker recloses successfully at 1 s. The relative rotor angles calculated by simulation are shown in Figure 8.8.

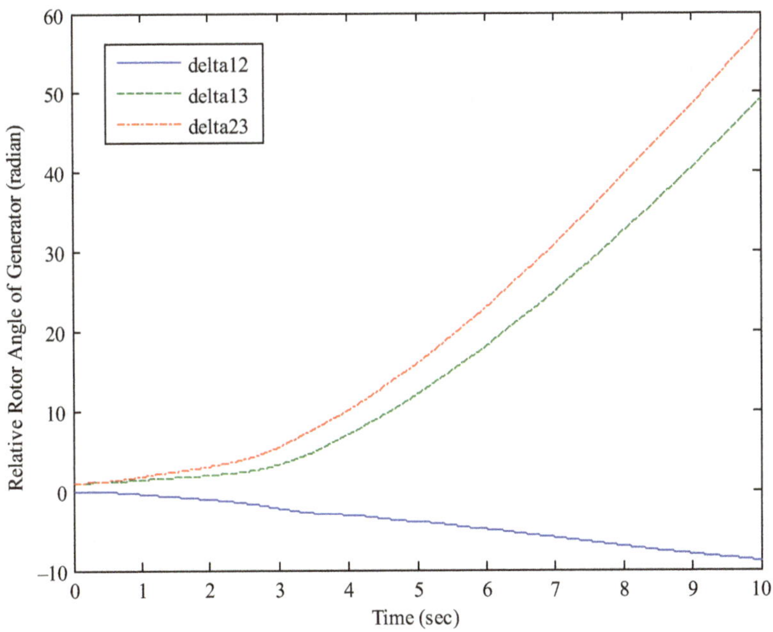

Figure 8.8 Simulation results of relative rotor angles

Assuming that only one PMU is installed and available at generator 1, so that δ_1, ω_1 are monitored as δ_1', ω_1', then the waveform of δ_1' is substituted into the last four equations and the first two equations in (8.23) are eliminated. The three-phase fault is applied to each transmission line, respectively. MLE is calculated by the above algorithm together with the dynamic trajectories. The observability indices along the trajectory are also computed. The results are shown in Table 8.2 and Figure 8.9.

Similarly, assuming that two PMUs are installed at generator 1 and 2. The MLE and observability calculation results are shown in Table 8.3 and Figure 8.10.

The results indicate that

1. The MLE calculation results can predict rotor-angle transient stability correctly.

Table 8.2 MLE of different time window T with one PMU placement

Three-phase fault	Stability	MLE of T = 1.25	MLE of T = 1.667	MLE of T = 2
Line 1–3	Unstable	0.5848	0.7448	0.7798
Line 2–3	Unstable	0.0236	0.1647	0.3086
Line 1–2	Stable	−0.0531	−0.0788	−0.0870

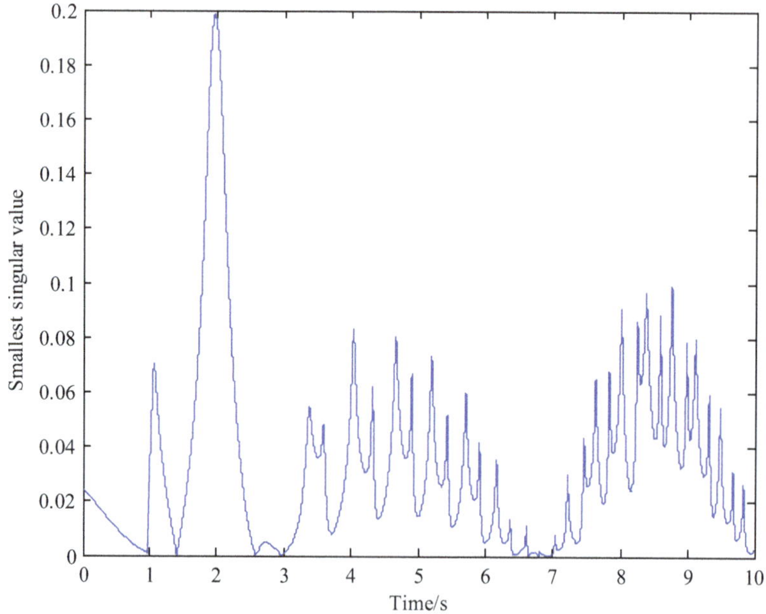

Figure 8.9 Smallest singular value of observability matrix with PMU placed at bus 1

Table 8.3 MLE of different time window T with two PMUs

Three-phase fault	Stability	MLE of $T = 1.25$	MLE of $T = 1.667$	MLE of $T = 2$
Line 1–3	Unstable	0.5848	0.7448	0.7798
Line 2–3	Unstable	0.0236	0.1647	0.3086
Line 1–2	Stable	−0.0531	−0.0788	−0.0870

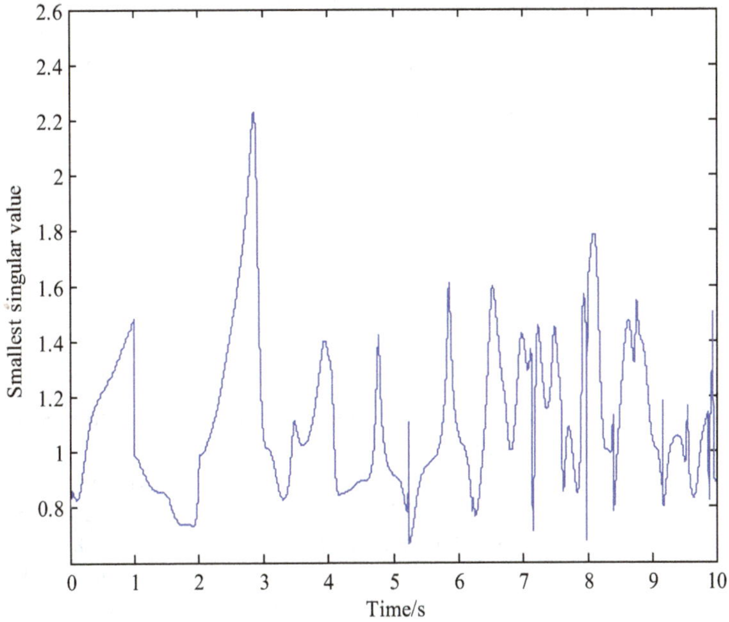

Figure 8.10 Smallest singular value of observability matrix with PMU placed at bus 1 and 2

2. When there is only one available PMU in the system, the smallest singular values of the observability matrices are very close to zero, indicating that the system is poorly observable.
3. As the numbers of PMUs increases, the level of observability is also improved.
4. When only one PMU is available in the system, the system is hardly observable. However, the MLE calculation results are the same with the scenario with two PMUs. It indicates that for this simple three-bus system, one PMU measurement is able to provide the necessary information for rotor angle stability assessment. However, this observation is not expected to be true for large-scale power systems.

8.5 Conclusion

The proposed MLE method is validated for the assessment of one of the vulnerable patterns of power systems: transient rotor angle instability. The main idea is to calculate MLE along the system dynamic trajectory in order to predict a loss of synchronism of power systems. Observability of PMU-based power system analysis is studied with a purpose of developing a method to quantify how well the available PMU measurements can be used to monitor dynamics of a power system. The research results in an innovative application of PMU data. Once an unstable power swing is captured and rotor angle instability is predicted by the above method, self-healing actions, such as power system partitioning control [51], can be initiated to reduce system vulnerability and avoid cascading failures that may result in a catastrophic blackout.

Acknowledgment

This research is sponsored by Electric Power Research Institute (EPRI), Palo Alto, CA, USA.

References

[1] Nye D. E. *When the Lights Went Out: A History of Blackouts in America.* Cambridge, MA: MIT Press, 2010.

[2] Parks B. Transforming the grid to revolutionize electric power in North America. *U.S. Department of Energy, Edison Electric Institute: Fall 2003 Transmission, Distribution and Metering Conference*, 2003.

[3] MoP. Report of the Enquiry Committee on Grid Disturbance in Northern Region on 30th July 2012 and in Northern, Eastern & North-Eastern Region on 31st July 2012. 2012. [Online]. Available: http://www.powermin.nic.in/pdf/GRID_ENQ_REP_16_8_12.pdf [Accessed 16 Aug 2012].

[4] Romero J. J. Blackouts illuminate India's power problems. *IEEE Spectrum*, 2012, 49(10): 11–12.

[5] FERC. Arizona-Southern California outages on September 8, 2011: Causes and recommendations. 2012. [Online]. Available: http://www.ferc.gov/legal/staff-reports/04-27-2012-ferc-nerc-report.pdf [Accessed 27 Apr 2012].

[6] Hines P., Balasubramaniam K., and Sanchez E. C. Cascading failures in power grids. *IEEE Potentials*, 2009, 28(5): 24–30.

[7] Pahwa S., Hodges A., Scoglio C., and Wood S. Topological analysis of the power grid and mitigation strategies against cascading failures. *IEEE 4th Annual Systems Conference*, San Diego, USA, 2010.

[8] Yamashita K., Joo S. K., Li J., Zhang P., and Liu C. C. Analysis, control, and economic impact assessment of major blackout events. *European Transactions on Electrical Power*, 2008, 18: 854–871.

[9] Yamashita K., Li J., Zhang P., and Liu C. C. Analysis and control of major blackout events. *IEEE PES. Power Systems Conference and Exposition*, Seattle, USA, 2009.

[10] Yamashita K., Li J., Liu C. C., Zhang P., and Hofmann M. Learning to recognize vulnerable patterns due to undesirable zone-3 relay operations. *IEEJ Transactions on Electrical and Electronic Engineering*, 2009, 4: 322–333.

[11] Liu C. C., Jung J., Heydt G., Vittal V., and Phadke A. G.The strategic power infrastructure defense (SPID) system: A conceptual design. *IEEE Control Systems Magazine*, 2000, 20(4): 40–52.

[12] Li H., Rosenwald G., Jung J., and Liu C. C. Strategic power infrastructure defense. *Proceedings of the IEEE*, 2005, 93(5): 918–933.

[13] Pai M. A., and Kulkarni A. Y. A simulation tool for transient stability analysis suitable for parallel computation. *IEEE 4th Conference on Control Applications*, Albany, USA, 1995.

[14] Suyono H., Nor K. M., and Yusof S. Component based development for transient stability power system simulation software. *Power Engineering Conference*, Thessaloniki, Greece, 2003.

[15] Fouad A. A., Vittal V., and Tae K. O. Critical energy for direct transient stability assessment of a multimachine power system. *IEEE Transactions on Power Apparatus and Systems*, 1984, PAS-103(8): 2199–2206.

[16] Chiang H. D. A theory-based controlling UEP method for direct analysis of power system transient stability. *IEEE International Symposium on Circuits and Systems*, Portland, USA, 1989.

[17] Kakimoto N., Ohnogi Y., Matsuda H., and Shibuya H. Transient stability analysis of large-scale power system by Lyapunov's direct method. *IEEE Transactions on Power Apparatus and Systems*, 1984, PAS-103(1): 160–167.

[18] Chiang H. D., Wu F. F., and Varaiya P. P. A BCU method for direct analysis of power system transient stability. *IEEE Transactions on Power Systems*, 1994, 9(3): 1194–1208.

[19] Xue Y., Van C. T., and Ribbens-Pavella M. A simple direct method for fast transient stability assessment of large power systems. *IEEE Transactions on Power Systems*, 1988, 3(2): 400–412.

[20] Phadke A. G., Thorp J. S., and Adamiak M. G. A new measurement technique for tracking voltage phasors, local system frequency, and rate of change of frequency. *IEEE Transactions on Power Apparatus and Systems*, 1983, PAS-102(5): 1025–1038.

[21] Thorp J. S., Phadke A. G., and Karimi K. Real-time voltage phasor measurements for static state estimation. *IEEE Transactions on Power Apparatus and Systems*, 1985, PAS-104(11): 3098–3106.

[22] Phadke A. G., Thorp J. S., and Karimi K. State estimation with phasor measurements. *IEEE Transactions on Power Systems*, 1986, PER-6(2): 233–241.

[23] Thorp J. S., Phadke A. G., Horowitz S. H., and Begovic M. M. Some applications of phasor measurements to adaptive protection. *IEEE Transactions on Power Systems*, 1988, 3(2): 791–795.

[24] Phadke, A. G. Synchronized phasor measurements: A historical overview. *IEEE/PES Transmission and Distribution Conference and Exhibition, Asia Pacific*, 2022, 1: 476–479.

[25] Centeno V., Phadke A. G., Edris A., Benton J., Gaudi M., and Michel G. An adaptive out-of-step relay for power system protection. *IEEE Transactions on Power Delivery*, 1997, 12(1): 61–71.

[26] Centeno V., Phadke A. G., and Edris A. Adaptive out-of-step relay with phasor measurement. *Sixth International Conference on Developments in Power System Protection*, 1997, pub 1, no. 434: 210–213.

[27] Centeno V., Phadke A. G., Edris A., Benton J., and Michel G. An adaptive out-of-step relay. *IEEE Power Engineering Review*, 1997, 17(1): 39–40.

[28] Centeno V., Ree J., de la, Phadke A. G., Michel G., Murphy R. J., and Burnett R. Adaptive out-of-step relaying using phasor measurement techniques. *IEEE Computer Application in Power*, 1993, 6(4): 12–17.

[29] Liu C. W., and Thorp J. New methods for computing power system dynamic response for real-time transient stability prediction. *IEEE Transactions on Circuits and Systems I: Fundamental Theory and Applications*, 2000, 47(3): 324–337.

[30] Huang Z., Schneider K., Nieplocha J., and Zhou N. Estimating power system dynamic states using extended Kalman filter. *IEEE PES General Meeting*, National Harbor, USA, 2014.

[31] Liu G., Ning J., Tashman Z., Venkatasubramanian V., and Trachian P. Oscillation monitoring system using synchrophasors. *IEEE PES General Meeting*, San Diego, USA, 2012.

[32] Ning J., Pan X., and Venkatasubramanian V. Oscillation modal analysis from ambient synchrophasor data using distributed frequency domain optimization. *IEEE Transactions on Power Systems*, 2013, 28(2): 1960–1968.

[33] Chow J. H., Chakrabortty A., Arcak M., Bhargava B., and Salazar A. Synchronized phasor data based energy function analysis of dominant power transfer paths in large power systems. *IEEE Transactions on Power Systems*, 2007, 22(2): 727–734.

[34] Sun K., Likhate S., Vittal V., Kolluri V. S., and Mandal S. An online dynamic security assessment scheme using phasor measurements and decision trees. *IEEE Transactions on Power Systems*, 2007, 22(4): 1935–1943.

[35] Zhu R., and Liu C. C. Online voltage stability monitoring and control using limited synchrophasor measurements. *IEEE Power & Energy Society Innovative Smart Grid Technologies Conference (ISGT)*, Washington DC, USA, 2020.

[36] Li H., Weng Y., Farantatos E., and Patel M. An unsupervised learning framework for event detection, type identification and localization using PMUs without any historical labels. *IEEE PES General Meeting*, Atlanta, USA, 2019.

[37] Yan J., Liu C. C., and Vaidya U. PMU based monitoring of rotor angle dynamics. *IEEE Transactions on Power Systems*, 2011, 26(4): 2125–2133.

[38] Yan J., Liu C. C., and Vaidya U. A PMU-based monitoring scheme for rotor angle stability. *IEEE PES General Meeting*, San Diego, USA, 2012.

[39] Liu C., Thorp J., Lu J., Thomas R. J., and Chiang H. D. Detection of transiently chaotic swings in power systems using real-time phasor measurements. *IEEE Transactions on Power Systems*, 1994, 9(3): 1285–1292.

[40] Osedelec V. I. Multiplicative ergodic theorem: Lyapunov characteristic exponent for dynamical systems. *Moscow Mathematical Journal*, 1968, 19: 539–575.

[41] Eckman J. P., and Ruelle D. Ergodic theory of chaos and strange attractors. *Reviews of Modern Physics*, 1985, 57: 617–656.

[42] Ramasubramanian K., and Sriram M. S. A comparative study of computation of Lyapunov spectra with different algorithms. *Physica D*, 2000, 139(1–2): 72–86.

[43] Monticelli A., and Wu F. F. Network observability: Theory. *IEEE Transactions on Power Apparatus and Systems*, 1985, PAS-104(5): 1042–1048.

[44] Monticelli A. Electric power system state estimation. *Proceedings of the IEEE*, 2000, 88(2): 262–282.

[45] Chakrabartiand S., and Kyriakides E. Optimal placement of phasor measurement units for power system observability. *IEEE Transactions on Power Systems*, 2008, 23(3): 1433–1440.

[46] Nuqui R. F., and Phadke A. G. Phasor measurement unit placement techniques for complete and incomplete observability. *IEEE Transactions on Power Delivery*, 2005, 20(4): 2381–2388.

[47] Kalman R. E. On the computation of the reachable/observable canonical form. *SIAM Journal of Control and Optimization*, 1982, 20: 258–260.

[48] Hermann R., and Krener A. J. Nonlinear controllability and observability. *IEEE Transactions on Automatic Control*, 1977, AC-2(5): 728–740.

[49] Röbenack K., and Reinschke K. J. An efficient method to compute Lie derivatives and the observability matrix for nonlinear systems. *Proceedings of 2000 International Symposium on Nonlinear Theory and Its Applications*, Dresden, 2000, 2: 625–628.

[50] Chen C. T. *Linear System Theory and Design* (3rd edn). Oxford: Oxford University Press, 1998.

[51] Li J., Liu C. C., and Schneider K. Controlled partitioning of a power network considering real and reactive power balance. *IEEE Transactions on Smart Grid*, 2010, 1(3): 261–269.

Chapter 9

Synchrophasor applications for load and generator parameter estimation

Sagnik Basumallik[1], *Tushar*[2], *Armina Foroutan*[3] *and Anurag K. Srivastava*[4]

The use of advanced real-time monitoring and control technologies has made it possible to transform the existing power grid into smart grid. With the development of phasor measurement units (PMUs) and the availability of synchronized real-time measurements, it has become possible to develop data-driven power system applications to enhance asset management capabilities and overall reliability of the power grid. This chapter presents two such applications of synchrophasor measurements: (1) real-time load parameter estimation and (2) generator parameter estimation. The load modeling tool used in the industry and comprehensive research effort in load modeling are discussed in this chapter. The adaptive search-based recursive least square algorithm is presented for improved online estimation of load parameters. Finally, the use of synchrophasor measurements for generator model validation is discussed, including anomaly detection and mitigation for data quality. The approach features Ljung modification for joint estimation and dual estimation with recurrent derivative modification for extended Kalman filter (EKF). The proposed algorithm enhances model calibration through non-intrusive validation, allowing proactive issue identification and timely mitigation, and is validated on a real-world Alaskan grid system.

9.1 Introduction

The electric power system is a highly complex man-made dynamical system. Monitoring the power system operating states is the first step toward stability and reliability of the power system. The power system is generally monitored by measuring key states, defined by the magnitude and phase angle of bus voltages. These states change continuously even during normal operation of the system and can be processed to track operating status of the grid given by (i) normal, (ii) alert, and (iii) emergency

[1]New York Power Authority, Albany, NY, USA
[2]GridAstra, Dallas, TX, USA
[3]GE Vernova, Bothell, WA, USA
[4]Lane Department of Computer Science and Electrical Engineering, West Virginia University, USA

states. Set of preventive and emergency control actions are initiated by the system operator, under alert and emergency operating states of the system, to keep the system in stable and secure state. Supervisory control and data acquisition system-based energy management systems (SCADA/EMS) have been deployed in the power system networks, which utilizes remote terminal units (RTUs) in the field to acquire analog and digital signals from the network. The RTUs are located at substations and provide the data of the amplitude of bus voltages, branch currents, and line power flows, typically at a scan rate of 2–10 s [1,2]. These data, received at control center, are processed by a state estimator (SE). The results obtained from the state estimator are used for security analysis, optimal power dispatch, automatic generation control, and auxiliary control actions. The data obtained from the SCADA measurements have very low sampling rate and are not time-stamped, and do not provide a direct measure of the angle of the bus voltages. These limitations of the SCADA system have been addressed by the introduction of PMUs, which can provide the voltage and the current phasors with respect to a global reference sinusoid, synchronized to the global positioning system (GPS) clock, and at approximately 200 times faster than SCADA. The phasor data estimated by the PMUs are time-stamped and archived by the phasor data concentrators (PDC), which are located at substations or at the control center. The archived phasor data in the control center are retrieved by different power system applications such as load parameter estimation and generator parameter estimation. The load parameter estimation is a quasi-steady state phenomenon, which needs to be updated in a few minutes and uses the high resolution fast rate of data. The high-resolution data can further be used to precisely estimate generator parameters such as inertia, damping, and reactance once events are detected.

9.1.1 Expected quality of synchrophasor data

The need for PMU data to be error-free for load and generator parameter estimation is important. Several sources of errors can compromise the quality of this data, ranging from instrument transformer inaccuracies to communication network issues, and even potential data compromise. The following points outline key areas where data quality must be maintained to prevent inaccuracies and maintain data integrity.

1. Free from errors introduced by instrument transformers: Errors in PMU measurements can arise from bias in current and potential transformers, often due to incorrect calibration, ratio, or phase errors.
2. Free from errors within PMUs: Errors in PMUs may include filtering anomalies, missing time-stamps, channel errors, or harmonic interference.
3. Free from errors in the communication network: Errors can result from packet loss, node failures, congestion, or latency within the communication network.
4. Free from errors in data concentrators: Errors during PMU data aggregation and storage can stem from time-stamping misalignment, packet loss, compression issues, or archival inconsistencies.
5. Free from errors due to data compromise: Errors from targeted cyber intrusions, like false data injection or GPS spoofing, can compromise the integrity of PMU data.

9.1.2 Bad data processing

The well-established synchrophasor anomaly detection (SyncAD) tool [3] is used to address synchrophasor data quality issues, specifically, those caused by (a) outliers and (b) missing data resulting from communication issues, hardware faults, and instrumentation errors. Pre-processing synchrophasor data involve several steps: (a) all data are processed in 0.5 s windows through an ensemble of detectors such as linear regression, Chebyshev, DBSCAN, and isolation forest to generate anomaly scores. These scores are normalized with expectation maximization and then analyzed by a maximum likelihood estimator (MLE) ensemble model. To reduce errors, prony analysis is further used to distinguish dynamic transients from true anomalies.

9.2 PMU-based load modeling

The representation of load model has a significant impact on the system stability analysis [4,5]. One of the major reasons for inaccuracies in simulation studies for planning and operations are load models [6]. Accurate load model is necessary to correctly estimate the operational state of the system. High-resolution measurement from PMUs make it possible to estimate load behavior models.

9.2.1 Definition of load modeling

Various definitions of load as recommended by IEEE Task force [5] are:

- A device, connected to a power system, that consumes power. The term used for this definition is "load device".
- The total power (active and/or reactive) consumed by all devices connected to a power system. The term used for this definition is "system load".
- A portion of the system that is not explicitly represented in a system model, but rather is treated as if it were a single power consuming device connected to a bus in the system model. The term used for this definition is "bus load".
- The power output of a generator or generating plant. The term used for this definition is "generator or plant load".

Following are some terms used to describe the composition of a load [5] are:

Load component: A load component is the aggregate equivalent of all devices of a specific or similar type, e.g., water heater, air conditioner, fluorescent lighting.
Load class: A load class is a category of load, such as residential, commercial, or industrial.
Load composition: A load composition is the fractional composition of the load by load components.
Load class mix: A load class mix is the fractional composition of the bus load by load classes.
Load characteristic: Load characteristic is determined by a set of parameters that characterize the behavior of a specified load such as the variation of the active power with voltage. This term may be applied to a specified load device, a load component, a load class or the total bus load.

9.2.2 Classification of load models

The load models are generally classified into two broad categories: static load models and dynamic load models [7] as shown in Figure 9.1.

9.2.2.1 Static load

A static load model describes the relation between active and/or reactive power at any time with voltage and/or frequency at the same instant of time. The static load model is time invariant. The most common forms of static load models for active and reactive power are polynomial load model and exponential load model. These models, if necessary, can also include a frequency-dependent term [8].

Polynomial model or ZIP model:

The static characteristics of a load can be classified as constant power (*P*), constant current (*I*), and constant impedance (*Z*) load, depending upon the relation of power to voltage [9]. Hence, a polynomial static load model is also known as ZIP model.

- Constant impedance load (Z): It is the part of load in which the active and reactive power of the load varies proportionally to the square of supply voltage magnitude, e.g., heaters, incandescent lamps.
- Constant current load (I): It is the part of load in which the active and reactive power of the load varies in direct proportion to supply voltage magnitude, e.g., motors used in paper mill industry.
- Constant power load (P): It is the part of load in which the active and reactive power of the load remain constant irrespective of changes in supply voltage magnitude, e.g., induction motors operating close to nominal voltage.

The polynomial form of static load model can be mathematically represented by the following equations:

$$P = P_0\left(Z_p\left(\frac{V}{V_0}\right)^2 + I_p\left(\frac{V}{V_0}\right) + P_p\right) \tag{9.1}$$

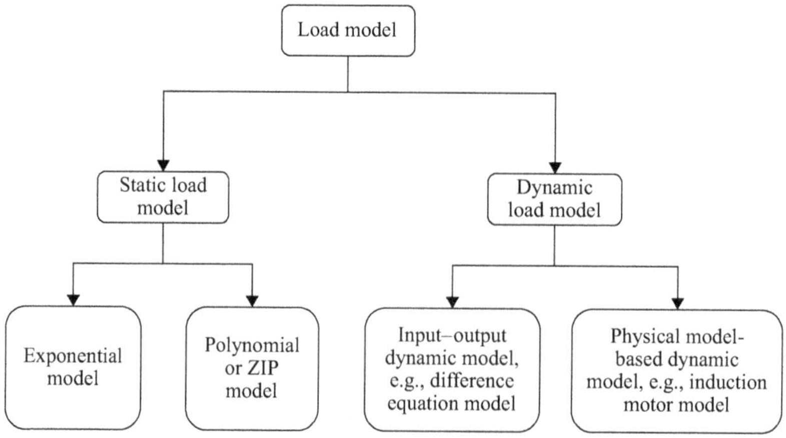

Figure 9.1 Classification of load model

$$Q = Q_0\left(Z_q\left(\frac{V}{V_0}\right)^2 + I_q\left(\frac{V}{V_0}\right) + P_q\right) \tag{9.2}$$

where

V_o is the rated base voltage
V is the operating voltage
P_o is the rated active base power at rated voltage
Q_o is the rated reactive base power at rated voltage
P is the active power at operating voltage
Q is the reactive power at operating voltage
Z_p, I_p, P_p are the ZIP coefficients for the active power with $Z_p + I_p + P_p = 1$
Z_q, I_q, P_q are the ZIP coefficients for the reactive power with $Z_q + I_q + P_q = 1$

Exponential model:

The static load models can also be represented as exponential model. The exponential model is represented by the following equations:

$$P = P_0\left(\frac{V}{V_0}\right)^a \tag{9.3}$$

$$Q = Q_0\left(\frac{V}{V_0}\right)^b \tag{9.4}$$

where

V_o, V, P_o, Q_o, P, and Q are same as ZIP model. "a" and "b" are non-integer exponents.

For special cases, where "a" or "b" is equal to 0, 1, and 2, the load model will represent a constant power, constant current, or constant impedance load, respectively.

The frequency dependency of load characteristics can be represented in both polynomial and exponential model by multiplying with a frequency factor as shown in (9.5)–(9.8).

For active power

$$P = P_0\left(Z_p\left(\frac{V}{V_0}\right)^2 + I_p\left(\frac{V}{V_0}\right) + P_p\right)(1 + k_{pf}\Delta f) \tag{9.5}$$

$$P = P_0\left(\frac{V}{V_0}\right)^a (1 + k_{pf}\Delta f) \tag{9.6}$$

For reactive power

$$Q = Q_0\left(Z_q\left(\frac{V}{V_0}\right)^2 + I_q\left(\frac{V}{V_0}\right) + P_q\right)(1 + k_{qf}\Delta f) \tag{9.7}$$

$$Q = Q_0\left(\frac{V}{V_0}\right)^b (1 + k_{qf}\Delta f) \tag{9.8}$$

where, Δf is the frequency deviation. Typically, k_{pf} ranges from 0 to 3 and k_{qf} ranges from −2 to 0 [8].

9.2.2.2 Dynamic load

The dynamic loads, unlike static loads, are dependent of time. The response of power system following a disturbance spans for several time intervals. The initial response that occurs immediately following the disturbance may be represented using a static load model. The dynamic loads, like induction motors, were approximated by static loads. This approximation is sufficient in some cases, but since load representation has critical effects in voltage stability studies, static loads need to be replaced by dynamic load models [7]. To predict a system response for longer time spans, like slow voltage recovery phenomena, etc., dynamic behavior of the loads has to be modeled using a dynamic load model.

A dynamic load is often expressed in the form of ordinary differential equations, or difference equations, or partial differential equations, or transfer functions. Dynamic load models are broadly classified as:

- **Input-output dynamic models**: This class of models include the difference equation-based dynamic models.

  The difference equation representation of the dynamic load is a numerical model that represents the time-varying nature of the load. It can be represented either as a first-order model or second-order model as follows:

  First-order model

$$\Delta P(k) = a_{p1}\Delta P(k-1) + c_{p0}\Delta V(k) + c_{p1}\Delta V(k-1) \tag{9.9}$$

$$\Delta Q(k) = a_{q1}\Delta Q(k-1) + c_{q0}\Delta V(k) + c_{q1}\Delta V(k-1) \tag{9.10}$$

  Second-order model

$$\begin{aligned}\Delta P(k) = {} & a_{p1}\Delta P(k-1) + a_{p2}\Delta P(k-2) + c_{p0}\Delta V(k) \\ & + c_{p1}\Delta V(k-1) + c_{p2}\Delta V(k-2)\end{aligned} \tag{9.11}$$

$$\begin{aligned}\Delta Q(k) = {} & a_{q1}\Delta Q(k-1) + a_{q2}\Delta Q(k-2) + c_{q0}\Delta V(k) \\ & + c_{q1}\Delta V(k-1) + c_{q2}\Delta V(k-2)\end{aligned} \tag{9.12}$$

- **Physical-model-based dynamic models**: The induction motor load model is considered as a physical-model-based load model.

The mathematical formulation of the induction motor model can be described by the following equations [10]:

$$\frac{dE_d'}{dt} = -\frac{1}{T'}[E_{d'} + (X - X')I_q] - (\omega - 1)E_q' \tag{9.13}$$

$$\frac{dE_q'}{dt} = -\frac{1}{T'}[E_{q'} + (X - X')I_d] - (\omega - 1)E_d' \tag{9.14}$$

$$\frac{d\omega}{dt} = -\frac{1}{2H}[T_0(A\omega^2 + B\omega + C + D\omega^E) - (E_d'I_d + E_q'Iq)] \tag{9.15}$$

$$I_d = \frac{1}{R_s^2 + X'^2}[R_s(U_d - E_d^{'}) + X'(U_q - E_q^{'})] \tag{9.16}$$

$$I_q = \frac{1}{R_s^2 + X'^2}[R_s(U_q - E_q^{'}) + X'(U_d - E_d^{'})] \tag{9.17}$$

$$T_L = T_0(A\omega^2 + B\omega + C + D\omega^E) \tag{9.18}$$

where

H is the rotor inertia constant
T_L is the load torque equation
T_0 is the steady state mechanical torque
ω is the rotor rotation speed
$E_{d'}$ is the d-axis transient EMF of motor
$E_{q'}$ is the q-axis transient EMF of motor
U_d is the d-axis bus voltage
U_q is the q-axis bus voltage
I_d is the d-axis stator current
I_q is the q-axis stator current

The A, B, C, D, E coefficients in load torque equation need to satisfy $C = 1 - A\omega^2 - B\omega - D\omega^E$ equation. In these equations X' is equivalent to $X_s + \left(\frac{X_m X_r}{X_m + X_r}\right)$ and is the transient reactance, whereas X is equivalent to $X_s + X_m$ which is the open circuit reactance.

9.2.2.3 Composite load (static load + dynamic load)

Generally, a composite load model approach is used by utilities. A composite load model is the combination of static load model and dynamic load model. Some of the composite models currently used [10] for research purposes are:

- ZIP model augmented with induction motor (three state).
- ZIP model augmented with difference equation model (second order).
- Exponential model augmented with induction motor (three state).
- Exponential model augmented with difference equation model (second order).

Since motors constitute about 60%–70% of the total load, most of the dynamic load is represented by an aggregate induction motor. Other load components that affect the stability studies are as follows [8]:

- Discharging the lamps and restart the lamp. The lamps that have mercury vapor, sodium vapor, and fluorescent lamps will affect the voltage recovery or delay the recovery.
- Protective relays especially thermal or over-current relays.
- Thermostatic control of loads (i.e., heaters and refrigerators).
- ULTCs on distribution transformers, voltage regulators, and voltage-controlled capacitor banks. These devices are used to mitigate a disturbance and to make the system return to the pre-disturbance levels.

Some of the load models used in industry are as follows.

WECC composite load model

The composite load model (CMPLDW) [11] developed by Western Electricity Coordinating Council (WECC) is implemented in production grade software like PSLF by GE, Siemens PTI PSS®E, Power World, PowerTech TSAT. The model is further refined and is presented in Figure 9.2.

This composite load model structure includes static loads, electronic loads, constant-torque three-phase motors (A), high inertia speed-squared load motors (B), low inertia speed-squared load motors (C), and constant-torque single-phase motors (D). A detailed modeling of each motor type is required in the CMPLDW and the total input parameters required are more than hundred. Model validation studies have been done to validate these load models. It was observed that these models can be tuned to reproduce past events like delayed voltage recovery, large under-frequency events, and inter-area power oscillation with great accuracy.

Composite load model in PSS/E

The composite load model (CLOD) used in PSS/E is an aggregation of the multiple models in the system. The structure of this model is shown in Figure 9.3.

This model consists of various sub-models consisting of following components:

1. Induction motors: Motors are characterized by typical torque-speed, current-speed, and power factor-speed curves.
2. Discharge lighting: The percentage of discharge lightning is provided in this sub-model. Above 0.75 pu of voltage, the real power is modeled as constant current and the reactive power increases with exponential factor of 4.5. As

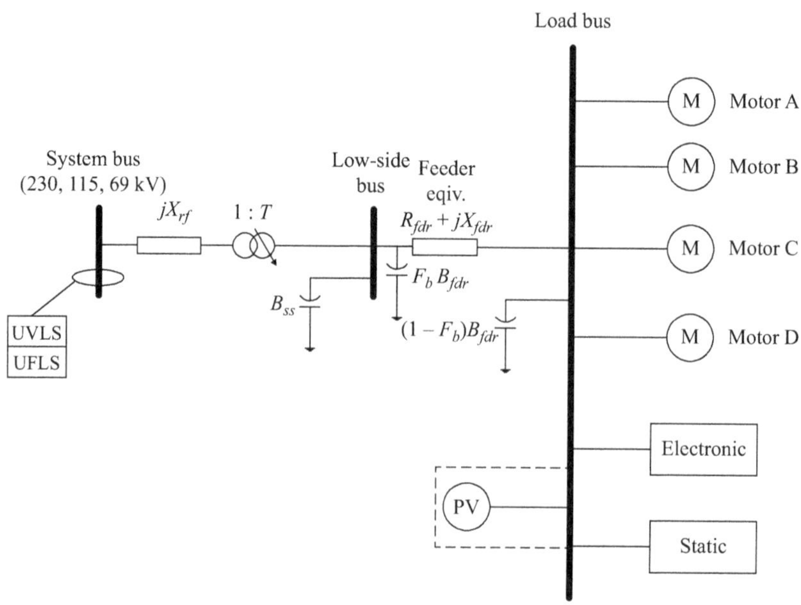

Figure 9.2 WECC CMPLDW load model

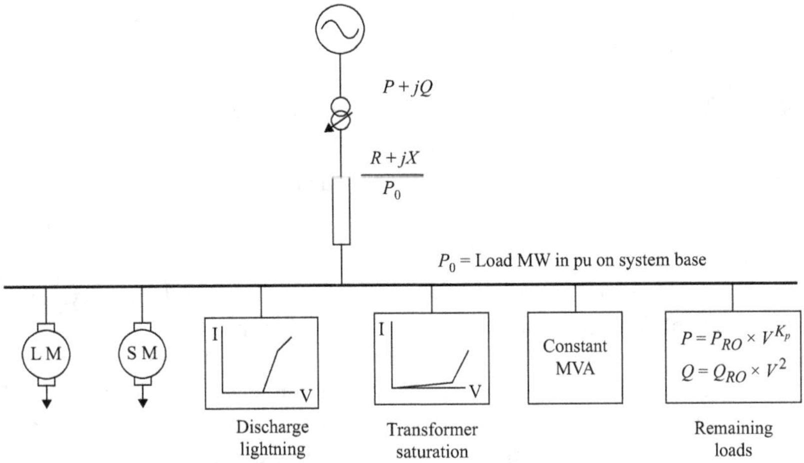

Figure 9.3 CLOD load model in PSS/E

voltage goes below 0.75 pu, both active and reactive power drop linearly with voltage until the value of voltage reaches 0.65 pu below which the lighting gets extinguished.

3. Transformer losses: The percentage of the transformer losses is provided that account for both core losses and copper losses.
4. Constant MVA: The percentage of the constant real and reactive power component of the load is provided.
5. Remaining loads: The real power is modeled as exponential function of voltage and the reactive power as constant impedance.

9.2.3 Review of estimation of load model parameters

Two basic approaches to obtain the load characteristics are: physical-component-based approach and measurement-data-based approach [8]. These approaches are usually applied for aggregate load models, but these principles can also be applied for the formulation of individual loads.

9.2.3.1 Physical-component-based approach

In this approach, the aggregate model is built from the models of individual components within the aggregated load [8,12,13]. This approach requires survey of various components in power system. This makes it a very difficult task as it involves manual survey and uncertainty. Further as the load data varies from bus to bus and is dependent on weather and time, it is necessary to often update the load data for each bus of the system.

9.2.3.2 Measurement-data-based approach

In measurement-data-based approach of load modeling, power meters and PMUs are installed at substations to measure voltage and/or frequency variations. The main advantage of using a measurement based model over the component-based approach is the availability of actual data from the system that makes it possible to track real-time variations in load behavior.

9.2.4 Advantages of using synchronized phasor measurements for load modeling

Various advantages of using PMUs for load modeling are:

- PMUs can report data at very high rates. This can improve load model development and its validation process.
- The inputs required for load modeling like voltage, current, frequency, and rate of change of frequency can be directly obtained from PMUs.
- Any further signal processing is not required since PMUs directly provide all the necessary quantities required for estimation of load parameters.
- Since PMUs can provide real-time measurements with time-stamp, they can be used for centralized load monitoring and forming a database [14].
- The measurement noise of the PMU signal is very low [15]. There may still be some offset errors caused by current transformer and/or voltage transformers. But high frequency measurement noise is not an issue with PMU data. Thus, PMUs can reliably process small variations in power systems.
- A certain level of voltage variation is required for obtaining better load responses. The normal variations in voltages are lower than 1% but since PMUs have a very low noise level, accurate modeling of the load response to small voltage changes is possible [15].

9.2.5 Parameter identification techniques

Once a load model is chosen, the next step involves the estimation of its parameters, i.e., to find a set of parameters for which the simulated results from the proposed load model best fits the measurements. There are two kinds of approaches for a nonlinear system identifications.

9.2.5.1 Analytical-based approach

In this approach, the unknown parameters of the load model are determined based on experimental test results. In 1977, Shackshaft *et al.* [16] performed a series of field tests and proposed a method for deriving the parameters of a simplified induction motor model. The load composition and characteristics are determined by conducting step response tests in which small voltage disturbances are imposed on a power station auxiliary system and consumer loads. These tests are extremely sensitive to measurement errors. The parameters obtained using small disturbances in such tests are not able to capture the parameters related to large disturbances.

9.2.5.2 Optimization-based approach

In this approach, the best fit parameters are searched that minimize the error function between the measured output variables and simulated ones. Various search algorithms have been applied in this type of approach. Search algorithms can be of two types: statistical techniques and heuristic techniques.

Some of the algorithms using statistical techniques are as follows:

- A least square approach is used in Ref. [17] to estimate the parameters of the load model. The ordinary least squares method assumes that the measurement errors are independent and normally distributed with constant standard deviation. The predicted values are such that the sum of squared difference between the actual value and predicted value is small. The basic regression equation is given as:

$$Y_k = H_k X + e_k \qquad k = 1, 2, \ldots, n \tag{9.19}$$

where

$H_k = [H_{1k}, H_{2k}, \ldots, H_{nk}]^T$ is an input column vector

Y_k is an output column vector
X is a column vector for the parameters to be estimated
e_k is the error

- In weighted least square based parameter estimation, the problem is solved by considering unequal variances of errors. The weights are given to each measurement based on its variability. The main difficulty in this method is to find proper weights, which also increases the complexity of the algorithm.
- An instrument variable (IV)-based technique was used in Ref. [18]. This method considers the co-relation between states to be estimated, measurements, and the residue error. A new variable $'I'$ is defined such that
  1. It should be highly correlated with H
  2. It should not be correlated with error e

The two staged equation solving process is as follows:

$$\widehat{H}_k = I_k Z \tag{9.20}$$

$$\widehat{Y}_k = \widehat{H}_k X + e \tag{9.21}$$

First Z is estimated from (9.20) and then X is obtained from (9.21). Hence, this is also called two-staged least squares approach.

Some of the algorithms using heuristic techniques are as follows:

- An adaptive simulated annealing (ASA) algorithm for parameter estimation is discussed in Ref. [19]. ASA performs better than the basic SA algorithm due to several substantial improvements that are discussed in Ref. [20].
- It uses a hybrid algorithm for the estimation of load parameters. The hybrid algorithm including the combination of Levenberg–Marquardt and genetic algorithm was used to find the composite load model parameters in Ref. [10]. The advantage of using Levenberg–Marquardt approach is that it has a good local search ability and the genetic algorithm has a better global search ability. The disadvantage of this hybrid approach is that since it involves genetic algorithm, it is slow and many parameters have to be chosen heuristically.

- Artificial neural networks are used for dynamic load modeling due to features like nonlinear mapping and generalization capabilities. Artificial neural network algorithms use the measurements to train and update the model structure adaptively, thereby eliminating the need to predefine the load model structure and to know the information on load composition and mix. In Refs. [21–23], the authors use artificial neural network-based technique for estimating the load parameters.

A method to reduce the number of parameters in a model based on a sensitivity method was proposed in Ref. [24]. In this study, trajectory sensitivity method is used to find the most important equivalent motor parameters for a composite model of load. This results in estimation of only eight parameters instead of 13 parameters, thus, reducing the identification time and enhancing the efficiency of the load modeling.

9.2.6 Estimation of static load parameters

In the following section, the identification and estimation techniques used for determining parameters of the static load for transmission level power system are analyzed. As the variations in the frequency during steady states are minimal in a power system, load changes due to frequency are not considered. Once the synchrophasor measurements are acquired, the identification process can start estimating the load parameters. In this study, the polynomial or the ZIP load model is chosen due to its significant physical meaning and also as it is one of the most widely accepted load models in the power system analysis software tool [25]. The dynamic loads can also be modeled as a constant ZIP model during steady state. For example, an active power load consisting of 30% motor and 70% of impedance can be modeled with a ZIP load model with Z_p= 0.3, I_p= 0, and P_p= 0.7 considering that the motor behaves as a constant power load during steady state.

The measured data should be examined for its suitability in parameter estimation. Following steps should be considered before including the measured data for estimation:

- The data obtained during transients are not included for estimation purposes. Voltage magnitude can be processed to identify transient conditions and previous ZIP model could be used as an approximate ZIP model during transient conditions.
- All measured values are taken around a steady state condition. In some cases, frequency changes continuously after a sudden change in the system while voltage, and power may return to their original operating point. These type of measurement data are discarded in the estimation procedure and previous value of ZIP could be used as an approximation.
- All measured data are checked for bad data using data mining techniques. Bad data should be eliminated or given very less weighing factor during the estimation procedure.

9.2.6.1 Recursive least square estimation

IEEE 14 bus system [26] is considered for analysis with load flow incorporating voltage dependent load models. The simulations results are obtained using MATLAB®.

To obtain the input measurements, the power flow program is modified by including a ZIP load model in the power flow equations. The ZIP load parameters are specified for each load bus in the IEEE 14 bus test system. The voltage magnitude, phase angle, real power, and reactive power are obtained as output of the power flow program that are treated as synchrophasor measurements. This voltage magnitude and real power output data are used as measurement inputs to the load parameter estimation procedure. The voltage variations at the load bus may occur due to several reasons like tap change, contingency, load change, etc. The length of the window chosen is an important factor during the parameter estimation procedure. In this study an initial window length of three data points is considered as there are three unknown parameters (*Z*, *I*, and *P*). During initialization of the estimation procedure, the window length is dynamically adjusted such that the system of equations become over determined.

The most commonly used estimation technique is the least square method [17]. The objective function for the least square optimization is to minimize the square of error between the model prediction and the observed value [27].

$$J = min|P_{measured} - P_{estimated}|^2 \tag{9.22}$$

The parameters to be estimated are *Z*, *I*, and *P*. The parameters are estimated using the following equation:

$$\hat{x} = (H^T R^{-1} H)^{-1} H^T R^{-1} b \tag{9.23}$$

where

$\hat{x}$ is the estimated variables vector
R is the co-variance matrix
b is the measurement vector
H is the Jacobian matrix of the state variable

As there are more unknowns than the number of equations, to estimate the parameters online, a sliding window approach is used. Figure 9.4 shows how the parameters are estimated at each time step. A minimum of three different voltage values are required to solve for three unknown variables (*Z*, *I*, and *P*).

In the recursive least square method, the solution from previous time step is also included in the computation of parameters for the present time step. This makes computation of parameters more accurate and efficient. The recursive least square algorithm [27] can be written as follows:

$$P_k = [\lambda P_{k-1}^{-1} + (H_k^T R^{-1} H_k)]^{-1} \tag{9.24}$$

$$K_k = P_k H_k^T R^{-1} \tag{9.25}$$

$$\hat{x}_k = \hat{x}_{k-1} + K_k(b_k - H_k \hat{x}_{k-1}) \tag{9.26}$$

where

P_k is the co-variance matrix
K_k is known as the estimator gain matrix
λ is the forgetting factor

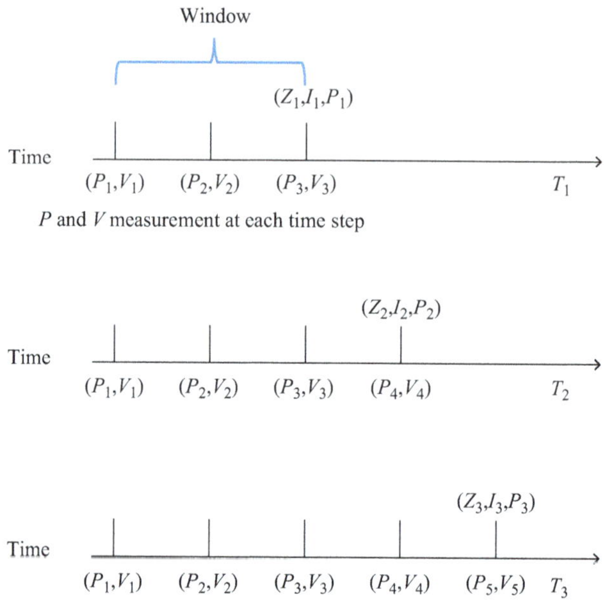

Figure 9.4 Sliding window approach

P_0 and V_0 values are chosen such that these are continuously changing at every instant. The fact that P_0 is the measured output power when a voltage V_0 is applied to the system load is used in this approach. For instance, if a window size of '3' measurements is chosen, the first measurement instant of voltage and power is chosen as the rated base values, respectively, after bad data and transients detection. Mathematically, it can be shown as follows:

The basic regression equation is given by

$$H_k X + e_k = Y_k \qquad k = 1, 2, \ldots, n \tag{9.27}$$

where

$H_k = [H_{1k}, H_{2k}, \ldots, H_{nk}]^T$ is an input column vector
Y_k is an output column vector
X is a column vector for the parameters to be estimated
e_k is the error

At time step = 3, the rated base voltage $V_0 = V_1$ and the rated base active power $P_0 = P_1$. Equation (9.27) can be written as:

$$\begin{bmatrix} 1 & 1 & 1 \\ \left|\dfrac{V_2}{V_1}\right|^2 & \left|\dfrac{V_2}{V_1}\right| & 1 \\ \left|\dfrac{V_3}{V_1}\right|^2 & \left|\dfrac{V_3}{V_1}\right| & 1 \end{bmatrix} \begin{bmatrix} Z \\ I \\ P \end{bmatrix} = \begin{bmatrix} 1 \\ \dfrac{P_2}{P_1} \\ \dfrac{P_3}{P_1} \end{bmatrix}$$

At time step = 4, the rated base voltage $V_0 = V_2$ and the rated base active power $P_0 = P_2$. Equation (9.27) becomes as follows:

$$\begin{bmatrix} 1 & 1 & 1 \\ \left|\frac{V_3}{V_2}\right|^2 & \left|\frac{V_3}{V_2}\right| & 1 \\ \left|\frac{V_4}{V_2}\right|^2 & \left|\frac{V_4}{V_2}\right| & 1 \end{bmatrix} \begin{bmatrix} Z \\ I \\ P \end{bmatrix} = \begin{bmatrix} 1 \\ \frac{P_3}{P_2} \\ \frac{P_4}{P_2} \end{bmatrix}$$

The load is monitored at the 14th bus of the system. The estimation process is then initiated and the results are summarized in Table 9.1. It is to be noted here that the current/power and voltage measurements used are obtained only for steady-state operating conditions of the system. However, this algorithm doesn't give accurate estimates for those operating scenarios where ZIP parameters are changing.

Here, an "Operating scenario" can be defined as a single snap shot of the measurement window. The different operating scenarios that may occur during the process of parameter estimation are:

1. Varying voltage, varying power, and constant ZIP parameters
2. Constant voltage, constant power, and constant ZIP parameters
3. Varying voltage, Varying power, and varying ZIP parameters
4. Constant voltage, varying power, and varying ZIP parameters

It can be seen from Table 9.1 that the estimated parameters are not estimated accurately under all operating scenarios by using the recursive least square method.

9.2.6.2 Adaptive search-based algorithm

To overcome the above-mentioned limitations, another approach is proposed here. The basic flowchart of this proposed algorithm is shown in Figure 9.5. The estimation process is started once the voltage and power measurements are obtained. The process is initialized by having the voltage and power measurements at three different voltage instances. After that, the estimation process is continued by using the recursive least square technique. After the estimation of ZIP parameters by the RLS method, the algorithm checks if the estimated ZIP parameters satisfy the following two conditions:

1. $Z + I + P = 1$
2. $Z \geq 0$, $I \geq 0$, and $P \geq 0$

The RLS technique can be applicable if and only if the ZIP parameters remain constant. Hence, if ZIP parameters have changed, the above-mentioned conditions will not hold true. If these conditions are violated, the algorithm shifts to the search-based estimation technique.

In this approach, the power is first calculated for all possible ZIP parameter values using the voltage measurements at that instant. The ZIP parameters are then selected that give the minimum square of error from the measured power values as given by

$$J = min|P_{measured} - P_{calculated}|^2 \tag{9.28}$$

Table 9.1 Summary of results using adaptive search-based algorithm v/s recursive least square approach—bus 14

Measurement instant	Voltage measurement (pu)	Active power measurement (pu)	Actual (*Z*, *I*, *P*) parameters	Estimated (*Z*, *I*, *P*) parameters using RLS method	Estimated (*Z*, *I*, *P*) parameters using adaptive search-based algorithm
1	1.0355	0.1490	(0,0,1)	–	–
2	1.0116	0.1490	(0,0,1)	–	–
3	0.9894	0.1490	(0,0,1)	(0,0,1)	(0,0,1)
4	1.0613	0.1490	(0,0,1)	(0,0,1)	(0,0,1)
5	1.0613	0.1490	(0,0,1)	(0,0,1)	(0,0,1)
6	1.0613	0.1490	(0,0,1)	(1,0,0)	(0,0,1)
7	1.0613	0.1490	(0,0,1)	(1,0,0)	(0,0,1)
8	1.0613	0.1490	(0,0,1)	(1,0,0)	(0,0,1)
9	1.0355	0.1490	(0,0,1)	(1,0,0)	(0,0,1)
10	1.0355	0.1490	(0,0,1)	(1,0,0)	(0,0,1)
11	1.0355	0.1490	(0,0,1)	(1,0,0)	(0,0,1)
12	1.0354	0.1501	(0,0,2,0,8)	(−38,0,39)	(0,0,21,0,79)
13	1.0354	0.1501	(0,0,2,0,8)	(−38,0,39)	(0,0,21,0,79)
14	1.0610	0.1508	(0,0,2,0,8)	(−38,0,39)	(0,0,2,0,8)
15	1.0610	0.1501	(0,0,2,0,8)	(−38,0,39)	(0,0,2,0,8)
16	1.0351	0.1517	(0,18,0,15,0,67)	(−18,0,19)	(0,18,0,15,0,67)
17	1.0351	0.1517	(0,18,0,15,0,67)	(−18,0,19)	(0,18,0,15,0,67)
18	1.0351	0.1517	(0,18,0,15,0,67)	(1,0,0)	(0,183,0,147,0,669)
19	1.0349	0.1517	(0,32,0,05,0,63)	(−17,0,18)	(0,33,0,04,0,63)
20	1.0349	0.1517	(0,32,0,05,0,63)	(−17,0,18)	(0,33,0,04,0,63)
21	1.0349	0.1517	(0,32,0,05,0,63)	(1,0,0)	(0,329,0,039,0,631)

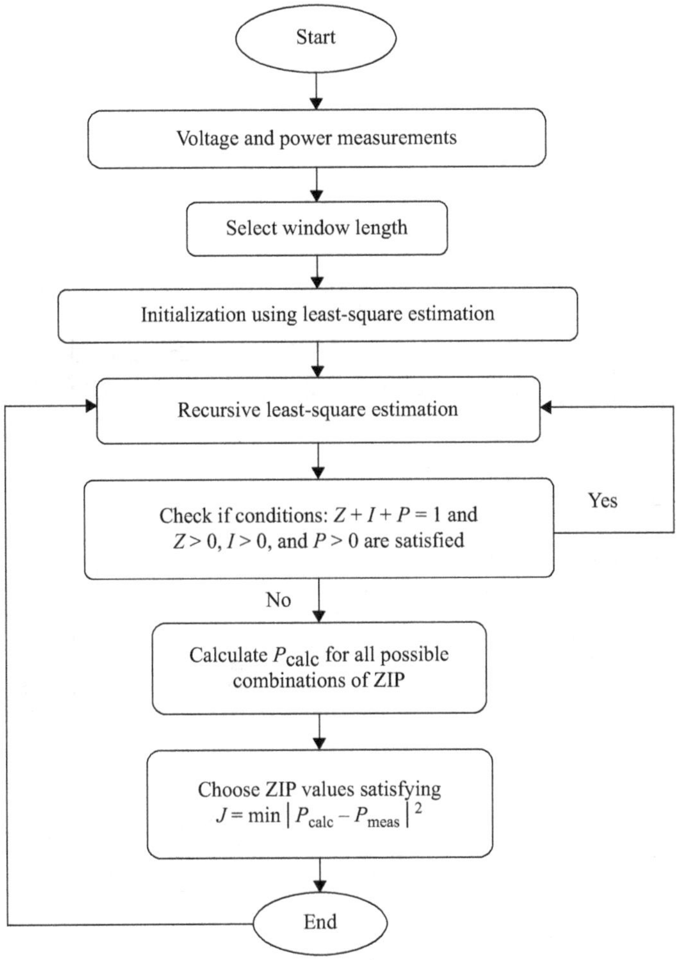

Figure 9.5 Flowchart for adaptive search-based algorithm

If multiple parameter solutions exist, then the best solution is chosen by calculating the norm from the previous set of ZIP parameters. It can be seen that the ZIP parameters are estimated accurately under all operating conditions. Table 9.1 summarizes the results of this approach for different operating scenarios. The estimation of ZIP parameters is initialized after three measurement instances using the least square approach. Until the 11th measurement instant, the recursive least square method estimates the ZIP parameters accurately since there is no change in the ZIP parameters of the load. Once the ZIP change occurs at measurement instant = 12, the RLS method fails to accurately estimate the load parameters. This occurs during operating scenarios 3 or 4. These operating scenarios can be easily identified by looking into the necessary satisfactory conditions, i.e., (1) $Z + I + P = 1$ and (2) $Z \geq 0$, $I \geq 0$, and $P \geq 0$. If these conditions are not satisfied, the algorithm switches

Table 9.2 Summary of results using adaptive search-based algorithm v/s recursive least square approach—bus 12

Measurement instant	Voltage measurement (pu)	Active power measurement (pu)	Actual (*Z*, *I*, *P*) parameters	Estimated (*Z*, *I*, *P*) parameters using RLS method	Estimated (*Z*, *I*, *P*) parameters using adaptive search-based algorithm
1	1.0548	1.0641	(0.3,0.3,0.4)	–	–
2	1.0243	1.0623	(0.3,0.3,0.4)	–	–
3	1.0884	1.0660	(0.3,0.3,0.4)	(0.3,0.3,0.4)	(0.3,0.3,0.4)
4	1.0884	1.0660	(0.3,0.3,0.4)	(0.3,0.3,0.4)	(0.3,0.3,0.4)
5	1.0884	1.0660	(0.3,0.3,0.4)	(0.3,0.3,0.4)	(0.3,0.3,0.4)
6	1.1259	1.0682	(0.5,0.1,0.4)	(−0.477,0,1,477)	(0.5,0.1,0.4)
7	1.1259	1.0682	(0.5,0.1,0.4)	(−0.477,0,1,477)	(0.5,0.1,0.4)
8	1.1259	1.0682	(0.5,0.1,0.4)	(1,0,0)	(0.5,0.1,0.4)
9	1.1259	1.0682	(0.5,0.1,0.4)	(1,0,0)	(0.5,0.1,0.4)
10	1.1257	1.0699	(0.5,0.1,0.4)	(−70.9,0,71.9)	(0.5,0.1,0.4)
11	1.1257	1.0699	(0.5,0.1,0.4)	(−70.9,0,71.9)	(0.5,0.1,0.4)
12	1.1257	1.0699	(0.5,0.1,0.4)	(1,0,0)	(0.5,0.1,0.4)
13	1.0547	1.0647	(0.5,0.1,0.4)	(1,0,0)	(0.5,0.1,0.4)
14	1.0547	1.0647	(0.5,0.1,0.4)	(1,0,0)	(0.5,0.1,0.4)
15	1.0545	1.0647	(0.5,0.1,0.1)	(−60.1,0,61.1)	(0.5,0.1,0.4)
16	1.0545	1.0647	(0.5,0.1,0.1)	(−60.1,0,61.1)	(0.5,0.1,0.4)
17	1.0548	1.0641	(0.35,0.2,0.45)	(−65.5,0,66.56)	(0.35,0.2,0.45)
18	1.0548	1.0641	(0.35,0.2,0.45)	(−65.5,0,66.56)	(0.35,0.2,0.45)
19	1.0548	1.0641	(0.35,0.2,0.45)	(1,0,0)	(0.35,0.2,0.45)
20	1.0549	1.0637	(0.15,0.5,0.35)	(−197,263,−65)	(0.15,0.5,0.35)
21	1.0549	1.0637	(0.15,0.5,0.35)	(−197,263,−65)	(0.15,0.5,0.35)

to a search-based estimation technique. It can be seen from Table 9.1 that the ZIP parameters are estimated accurately using the adaptive search-based algorithm. The estimation of load parameters is also done for bus 12 of the IEEE 14 bus system. The estimation results are summarized in Table 9.2 for all the operating scenarios.

From the above results, it can be concluded that the proposed algorithm can accurately estimate the ZIP parameters during all operating scenarios.

9.3 Generator parameter estimation using synchrophasor data

Synchrophasor measurements with high sampling rates enable cost-effective online model validation and calibration of generators. Such accurate generator model parameters are essential for realistic power system dynamic simulations. Inaccurate models have historically failed to predict large-scale events, such as the August 1996 Western Interconnection blackout, which impacted 2 million people and caused an 11,850 MW load loss. Traditional validation methods, such as stage testing, are accurate but costly and require taking generators offline. Synchrophasors offer a cost-effective, non-intrusive solution for this. This chapter illustrates how high-resolution phasor data are used for validating and calibrating a GENROU-type machine model along with an IEEE G1-type governor and an IEEE T1-type exciter connected to the power grid. The overall process is shown in Figure 9.6.

First, an unsupervised anomaly detection architecture, using a combination of regression, Chebyshev, DBSCAN, and isolation forest techniques, identifies and removes synchrophasor data quality issues before the parameter estimation process. Next, two methods are presented for the estimation process: (1) joint-extended Kalman filter technique with Ljung modifications to address the dependency between unknown parameters and states, and (2) dual-extended Kalman filter technique with recurrent derivation with two parallel filters for simultaneously estimating states and parameters. Both methods effectively address numerical stability and convergence challenges in the parameter estimation process.

9.3.1 Overview of generator parameter estimation and calibration with extended Kalman filter

The general discrete state space representation of the generator model including machine, exciter, governor, stabilizer, and turbine is described as

$$\begin{cases} x_k = f(x_{k-1}, y_{k-1}, u_{k-1}, \alpha) \\ y_k = g(x_k, u_k) \end{cases} \tag{9.29}$$

Here, x, u, and y denote the state, input, and output vectors, respectively. The input vector u includes frequency F and voltage V, while the output vector y includes active power P and reactive power Q. Also, k denotes time steps, f and g are general nonlinear functions of the state variables, and α comprises all generator model parameters. Note that not all parameters affect model outputs, so we select and tune those parameters that

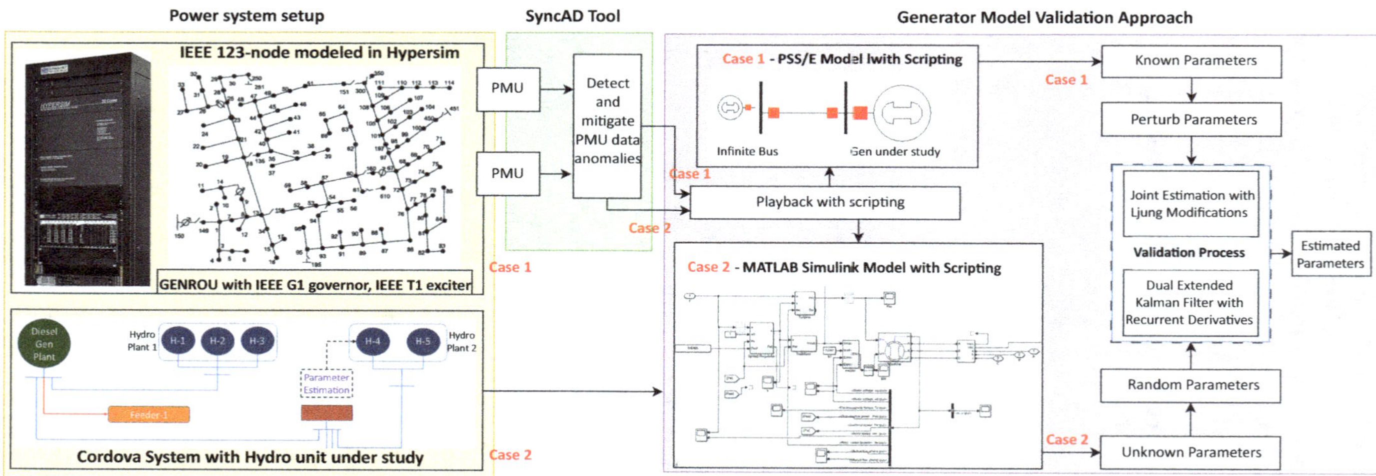

Figure 9.6 *Generator parameter estimation process using joint estimation with Ljung modification method and dual-extended Kalman filter with recurrent derivatives*

are sensitive to the event. This is done as follows. With α_0 and $\Delta\alpha_0$ as the initial parameter value and a small perturbation respectively, α_1 and α_2 are defined as

$$\begin{cases} \alpha_1 = \alpha_0 + \Delta\alpha_0 \\ \alpha_2 = \alpha_0 - \Delta\alpha_0 \end{cases} \tag{9.30}$$

With y^1 and y^2 as model outputs and T as the total simulation time steps, the sensitivity analysis is given as [28]

$$S(\alpha) = \frac{1}{T}\sum_{k=1}^{T}\left|\frac{y_k^1(\alpha_1) - y_k^2(\alpha_2)}{2(\Delta\alpha_0/\alpha_0)}\right| \tag{9.31}$$

Only those parameters with the highest sensitivity are then validated.

The generator model validation process using synchrophasor measurements employs various techniques including swarm optimization [29], frequency domain modal decomposition [30], machine learning [31], Bayesian methods [32], least squares [33], and variants of Bayesian-based filtering (BbF) approaches [32,34]. The most well-known BbF method is the Kalman filter, a recursive technique that minimizes estimation variance in linear systems with Gaussian noise. For nonlinear systems, a number of extensions have been proposed such as EKF [35,36], ensemble Kalman filter (EnKF) [37], cubature Kalman filter [38], unscented Kalman filter [36,39,40], and particle filter [41]. Among these, the EKF uses a first-order approximation to linearize the state space of the system and the Jacobian matrix to propagate the covariance and hence has been chosen here for the estimation process. EKF also has lower sensitivity to outliers and missing data, higher accuracy, and faster computation time, making it the preferred method [42]. There are two steps in the EKF method:

1. **Prediction**: A priori estimation of $\widehat{x}_k^-$ is calculated either directly by the model (9.29) or indirectly using predicted states $\widehat{x}_{k-1}$ and error covariance P_{k-1} at time step $k-1$ as follows:

$$\widehat{x}_k^- = f(\widehat{x}_{k-1}, y_{k-1}, u_{k-1}) \tag{9.32a}$$

$$\widehat{y}_k = g(\widehat{x}_k^-, u_k) \tag{9.32b}$$

$$A_{k-1} = \frac{\partial f(x, y_{k-1}, u_{k-1})}{\partial x}\Big|_{\hat{x}_{k-1}} \tag{9.32c}$$

$$P_k^- = A_{k-1}P_{k-1}A_{k-1}^T + R^v \tag{9.32d}$$

   Here, (9.32a) predicts the current state using previous estimates, inputs, outputs, and the system model; (9.32b) calculates the output from the previous estimates; (9.32c) computes the Jacobian with respect to states; and (9.32d) updates the error covariance using the Jacobian and previous covariance, with R^v as process noise.

2. **Correction**: The a priori estimation from the first step is corrected using a Kalman gain K_k and the error between estimated outputs and actual observations

$$C_k = \frac{\partial g(x, u_k)}{\partial x}\Big|_{\hat{x}_k^-} \tag{9.33a}$$

$$K_k = P_k^- C_k^T (C_k P_k^- C_k^T + R^n)^{-1} \tag{9.33b}$$

$$\widehat{x}_k = \widehat{x}_k^- + K_k(z_k - \widehat{y}_k) \tag{9.33c}$$

$$P_k = (I - K_k C_k) P_k^- \tag{9.33d}$$

$$\widehat{y}_k^+ = g(\widehat{x}_k, u_k) \tag{9.33e}$$

Here, (9.33a) is the Jacobian of the output with respect to states; (9.33b) calculates Kalman gain with the Jacobian and error covariance, where $(C_k P_k^- C_k^T + R^n)$ is the innovation matrix and R^n is observation noise covariance; (9.33c) adjusts the prediction using Kalman gain and the error between estimated and actual outputs, yielding corrected states $\widehat{x}_k$; (9.33d) updates the error covariance using Kalman gain and the Jacobian, where I is the identity matrix and P_k is the corrected covariance; and (9.33e) provides the output estimation with the corrected states. In the next section, a brief introduction to joint estimation with Ljung modifications is presented.

9.3.2 Generator parameter estimation with joint-extended Kalman filter technique with Ljung modifications

The joint estimation approach integrates inaccurate parameters into the state vector and updates both state x and parameters w using a single Kalman filter as

$$\tilde{x} = \begin{bmatrix} x_k \\ w_k \end{bmatrix} = \begin{bmatrix} f(x_{k-1}, y_{k-1}, u_{k-1}) \\ w_{k-1} \end{bmatrix} \tag{9.34}$$

The Jacobian of f w.r.t states is

$$\tilde{A} = \frac{\partial f(x, u, w)}{\partial x}\Big|_{x=\tilde{x}} = \begin{bmatrix} A(x, w) & M(x, w) \\ 0 & I \end{bmatrix} \tag{9.35}$$

The Jacobian of g w.r.t states is

$$\tilde{C} = \frac{\partial g(x, u, w)}{\partial x}\Big|_{x=\tilde{x}} = [\, C(x, w) \quad D(x, w) \,] \tag{9.36}$$

The new partitioned Kalman gain corresponds to the states and parameters,

$$\tilde{K} = \begin{bmatrix} K \\ L \end{bmatrix} \tag{9.37}$$

The error covariance is updated using the new Kalman gain and Jacobian as

$$\tilde{P} = \begin{bmatrix} P_1 & P_2 \\ P_2^T & P_3 \end{bmatrix} \tag{9.38}$$

The innovation covariance is

$$S = \left[[C \quad D]\tilde{P}[C \quad D]^T + R^n \right] \tag{9.39}$$

Although joint estimation has been used in power system contexts, its numerical stability issues are often neglected [43]. Recursive matrix operations can cause instability, leading to extreme values and potential divergence of the estimation process. The states depend on previous time-step values and parameters, but changes in parameters are usually not considered when calculating the Kalman gain, K. This lack of coupling between the gain and parameters can lead to divergence. To address this issue, matrix M is modified as follows [44] which reduces prediction error associated with model parameters w and ensures global convergence,

$$M^*(x, u, w) = M(x, u, w) + \left[\frac{\partial K(w)}{\partial w}\right](z - \widehat{y}) \tag{9.40}$$

The elements of the modification term are calculated as follows. The derivative of the Kalman gain w.r.t to parameter w is

$$\begin{aligned} \kappa_k = \left.\frac{\partial K(w)}{\partial w}\right|_{\widehat{w}_k} &= \left.\left[\frac{\partial A(w)}{\partial w} P_{1k} C_k^T + A_k \Pi_k C_K^T + A_k P_{1k} \frac{\partial C(w)^T}{\partial w}\right]\right|_{\widehat{w}_k} \\ &\cdot S_k^{-1} - K_k S_k^{-1} \sigma_k S_k^{-1} \end{aligned} \tag{9.41}$$

where the following terms are expanded as

$$\sigma_k = \left.\left[\frac{\partial C(w)}{\partial w} P_{1k} C_k^T + C_k \Pi_k C_k^T + C_k P_{1k} \frac{\partial C}{\partial w}\right]\right|_{\widehat{w}_k} \tag{9.42a}$$

$$\begin{aligned} \Pi_{k+1} = \Bigg[&\frac{\partial A(w)}{\partial w} P_{1k} A_k^T + A_k \Pi_k A_k^T + A_k P_{1k} \frac{\partial A(w)^T}{\partial w} \\ &\left. - \kappa_k S_k K^T - K_k \sigma_k K_k^T - K_k S_k \kappa_k \Bigg]\right|_{\widehat{w}_k} \end{aligned} \tag{9.42b}$$

The overall flowchart for the joint-estimation method with Ljung modifications is shown in Figure 9.7.

9.3.3 Generator parameter estimation with dual-extended Kalman filter with recurrent derivatives

The dual estimation approach uses two parallel Kalman filters: one for parameters, treating states as known, and one for states, treating parameters as known. There are three steps in the dual-EKF method: prediction step, state correction step, and parameter correction step.

The prediction for each time step k consists of the following steps. Prediction using the previous state model, current input, and previous states is given as

$$\widehat{x}_k^- = f(\widehat{x}_{k-1}, u_k, \widehat{w}_{k-1}) \tag{9.43}$$

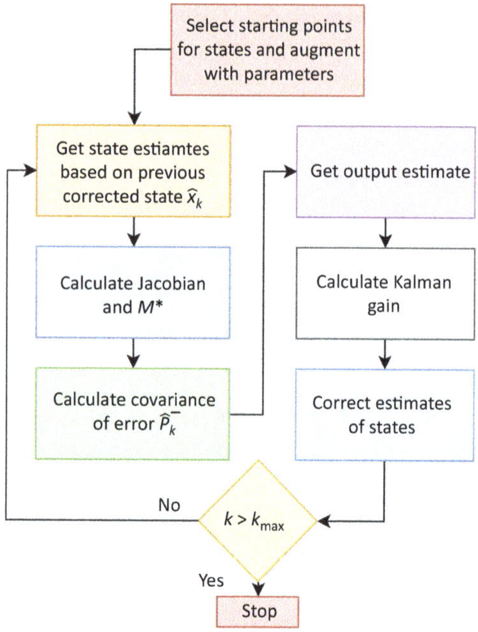

Figure 9.7 Flowchart for joint estimation with Ljung modifications

The previously estimated parameters are used for the current time step as

$$\widehat{w}_k^- = \widehat{w}_{k-1} \tag{9.44}$$

The output using the previous state estimates, model and inputs is given as

$$\widehat{y}_k = g(\widehat{x}_k^-, u_k, \widehat{w}_k^-) \tag{9.45}$$

The Jacobian of the system w.r.t states is given as

$$A_{k-1} = \frac{\partial f(x, u_{k-1}, w_{k-1})}{\partial x}\Big|_{\hat{x}_{k-1}} \tag{9.46}$$

The state error covariance is given as

$$P_{x_k}^- = A_{k-1} P_{x_{k-1}} A_{k-1}^T + R^v \tag{9.47}$$

The parameter error covariance with $\lambda = (P_{w_{k-1}} + R^r)$ is

$$P_{w_k}^- = \lambda^{-1} P_{w_{k-1}} \tag{9.48}$$

Similar to joint estimation, the state correction steps are as follows:

$$C_k = \frac{\partial g(x, u_k, \widehat{w}_k^-)}{\partial x}\Bigg|_{\hat{x}_k^-} \tag{9.49a}$$

$$K_k^x = P_{x_k}^- C_k^T (C_k P_{x_k}^- C_k^T + R^n)^{-1} \tag{9.49b}$$

$$\widehat{x}_k = \widehat{x}_k^- + K_k^x (z_k - \widehat{y}_k) \tag{9.49c}$$

$$P_{x_k} = (I - K_k^x C_k) P_{x_k}^- \tag{9.49d}$$

The last step is the parameter correction step which includes the following steps similar to the joint estimation but now w.r.t parameters

$$C_k^w = \left. \frac{\partial g(\widehat{x}_k^-, u_k, w)}{\partial w} \right|_{\widehat{w}_k^-} \tag{9.50a}$$

$$K_k^w = P_{w_k}^- (C_k^w)^T (C_k^w P_{x_k}^- (C_k^w)^T + R^e)^{-1} \tag{9.50b}$$

$$\widehat{w}_k = \widehat{w}_k^- + K_k^w (z_k - \widehat{y}_k) \tag{9.50c}$$

$$P_{w_k} = (I - K_k^w C_k^w) P_{w_k}^- \tag{9.50d}$$

The interdependencies between parameters and states are addressed by using a recurrent derivation for linearization in the parameter filter.

The Jacobian for the parameter C_k^w is computed using recurrent derivatives and the chain rule, given that $\widehat{x}_k$ depends on $\widehat{x}_{k-1}$ and both are functions of w

$$\begin{aligned} C_k^w &= \frac{\partial g(\widehat{x}_k^-, w)}{\partial \widehat{w}_k^-} + \frac{\partial g(x, \widehat{w}_k^-)}{\partial \widehat{x}_k^-} \cdot \frac{\partial \widehat{x}_k^-}{\partial \widehat{w}_k^-} \\ &= \frac{\partial g(\widehat{x}_k^-, w)}{\partial \widehat{w}_k} + C_k^x \frac{\partial \widehat{x}_k^-}{\partial \widehat{w}_k} \end{aligned} \tag{9.51a}$$

$$\begin{aligned} \frac{\partial \widehat{x}_k^-}{\partial \widehat{w}_k} &= \frac{\partial f(\widehat{x}_{k-1}, w)}{\partial \widehat{w}_k} + \frac{\partial f(x, \widehat{w}_k^-)}{\partial \widehat{x}_{k-1}} \cdot \frac{\partial \widehat{x}_{k-1}}{\partial \widehat{w}_k} \\ &= \frac{\partial f(\widehat{x}_{k-1}, w)}{\partial \widehat{w}_k} + A_{k-1} \cdot \frac{\partial \widehat{x}_{k-1}}{\partial \widehat{w}_k} \end{aligned} \tag{9.51b}$$

$$\frac{\partial \widehat{x}_{k-1}}{\partial \widehat{w}_k^-} = (I - K_k^x \cdot \frac{\partial f(x, \widehat{w}_k^-)}{\partial \widehat{w}_{k-1}^-}) \frac{\partial \widehat{x}_k^-}{\partial w_{k-1}^-} + \frac{\partial K_k^{x-}}{\partial \widehat{w}\,_k} (z_k - \widehat{y}_k) \tag{9.51c}$$

The overall flowchart for the dual-estimation method with recurrent derivatives is shown in Figure 9.8.

9.3.4 Results

The results of the validation and calibration process are described next. Two test systems were considered:

1. IEEE 123 node test feeder system: this was simulated on OPAL-RT Hypersim and synchrophasor measurements were obtained. A GENROU machine model with an IEEE G1 governor and an IEEE T1 exciter was added. Governor parameter K and exciter parameter V_{Rmax} are the top two most sensitive parameters.
2. Real-world isolated active distribution system obtained from Cordova Electric Cooperative in Alaska. The hydro generator parameters H and T'_{D0}, and exciter parameter K_{a} are the top two most sensitive parameters.

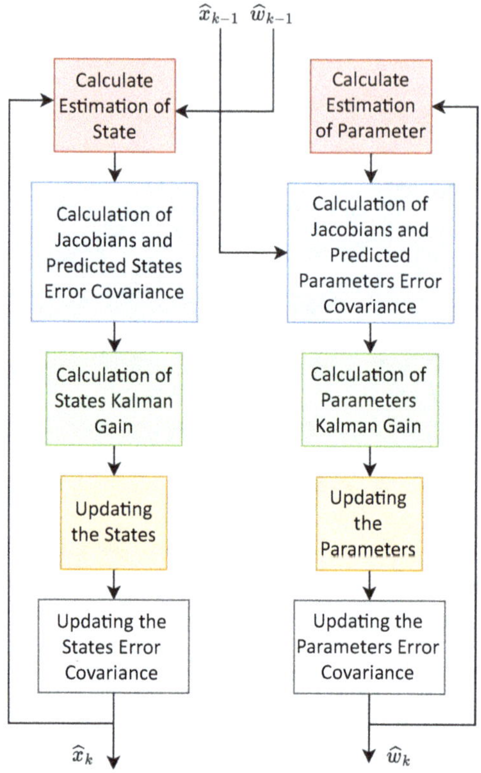

Figure 9.8 Flowchart for dual-extended Kalman filter with recurrent derivatives

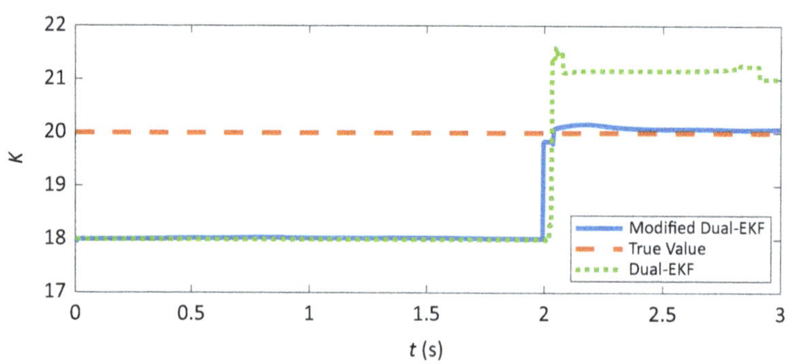

Figure 9.9 Comparison of parameter estimation of K and its true value

The original and modified dual-EKF were tested with perturbed parameters K and V_{Rmax}. For K, both methods converged, but the modified dual-EKF showed negligible error, whereas the original method was biased (Figure 9.9). For perturbed V_{Rmax}, the original dual-EKF did not converge, but the modified version converged to the true parameter value, as shown in Figure 9.10.

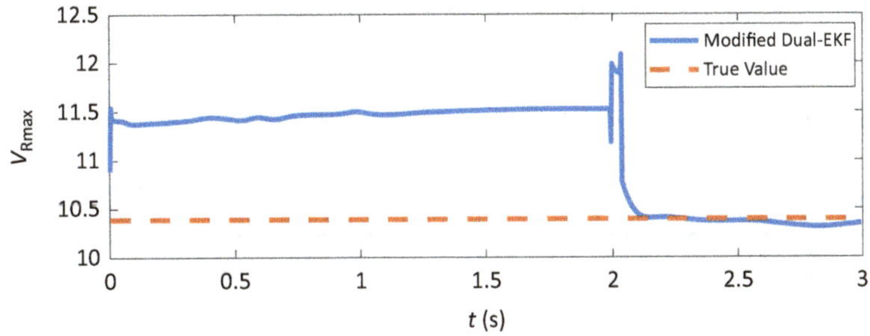

Figure 9.10 Comparison of parameter estimation of V_{Rmax}, and its true value. The original Dual-EKF did not converge.

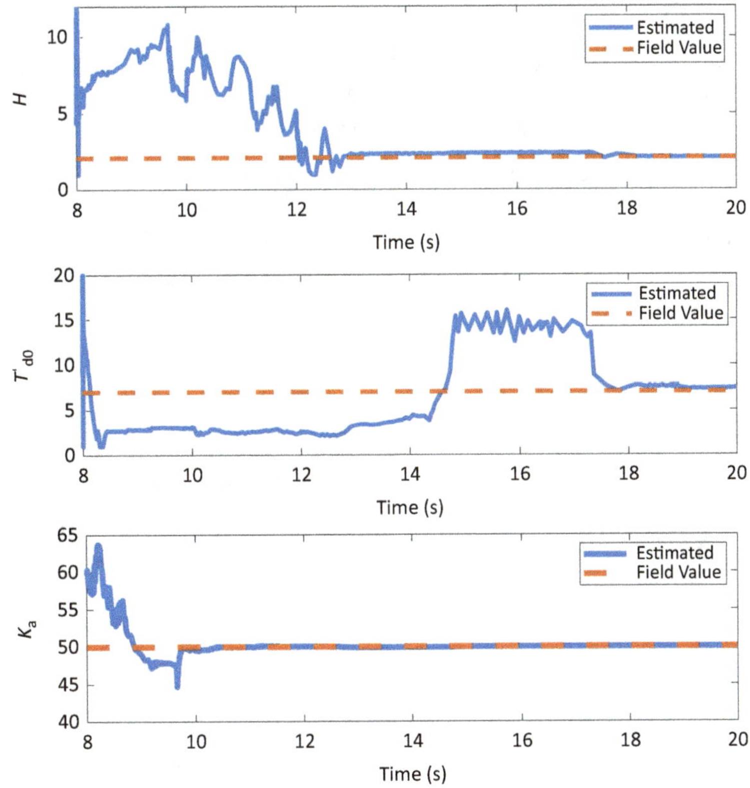

Figure 9.11 Illustration of the progress of the estimation process

The modified dual-EKF is applied to the parameters for the real-world hydro generators. Note that for this case, we only have the field measurement data and not the true values. Figure 9.11 shows the parameter estimation process, with all parameters converging to stable values. The initial and final parameter values are shown in Table 9.3.

Table 9.3 Parameter estimation results for real-world system

Parameters	Before estimation	After estimation
H	5	2.0996
T'_{D0}	10	7.4008
K_a	60	50.18

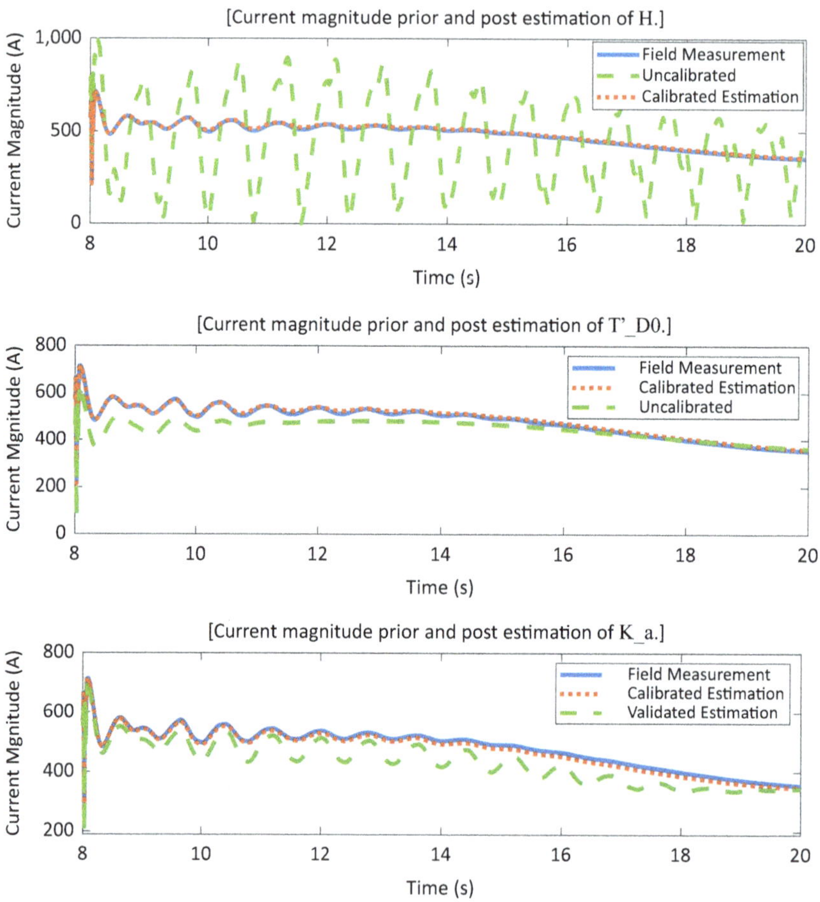

Figure 9.12 Model performance prior and post parameter estimation

Figure 9.12 shows the effect of inaccurate parameters on current magnitude and the estimation performance. Figure 9.13 shows the estimation performance on the current magnitude and angle. Finally, Table 9.4 compares the error compensation with other methods in Refs. [35,37].

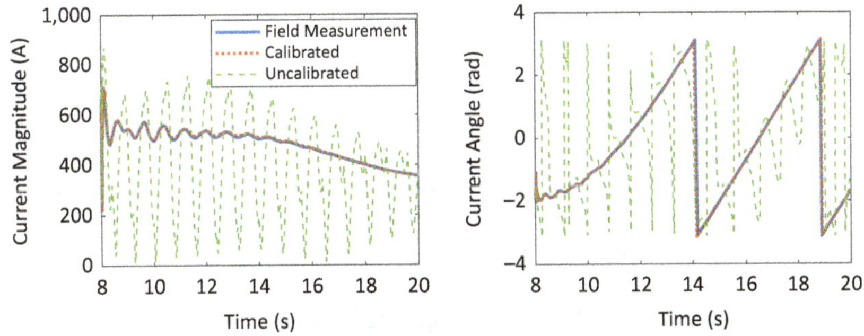

Figure 9.13 Comparison of model estimation prior and post estimation using modified-EKF versus PMU measurements

Table 9.4 Error compensation comparison of modified dual-EKF with traditional methods

Parameters	Modified dual-EKF	Joint-EKF [35]	Joint-EnKF [37]
H	138.14%	28%	23%
K_a	19.57%	16.5%	18.37%
T'_{D0}	35.12%	N/A	30.31%

9.4 Summary

The application of synchrophasor measurements for load parameter estimation and generator model validation is presented in this chapter. First, the importance of load modeling is described. There are various approaches to modeling a load, among which the measurement-based load modeling approach is discussed in detail. Various load models and parameter estimation techniques have also been discussed. The polynomial/ZIP load model, which is one of the widely used load model, is discussed with interface to the existing power system simulation software. Some of the techniques like least squares estimation and adaptive search-based algorithm have been presented in detail with simulation results. The difficulties faced during the estimation process of the load parameters have also been discussed. Lastly, the use of synchrophasor measurements for generator model validation is described. All synchrophasor anomalies are first detected and mitigated to ensure data quality before the parameter estimation process. Ljung modification for joint estimation and dual estimation with stability-based recurrent derivative modification for EKF are presented. Model inaccuracies and errors are effectively addressed by the proposed algorithm, which outperforms existing methods in stability and accuracy. The main advantage of our approach is its ability to enable more accurate model calibration through non-intrusive validation. This improvement allows for proactive identification of potential issues and timely mitigation actions.

References

[1] Y. Kokai, F. Masuda, S. Horiike, and Y. Sekine, "Recent development in open systems for EMS/SCADA," *International Journal of Electrical Power and Energy Systems*, vol. 20, no. 2, pp. 111–123, 1998.

[2] A. Bose, "Smart transmission grid applications and their supporting infrastructure," *IEEE Transactions on Smart Grid*, vol. 1, no. 1, pp. 11–19, 2010.

[3] S. Pandey, A. K. Srivastava, and B. G. Amidan, "A real time event detection, classification and localization using synchrophasor data," *IEEE Transactions on Power Systems*, vol. 35, no. 6, pp. 4421–4431, 2020.

[4] C. Taylor, *Power System Voltage Stability*. New York, NY: McGraw-Hill, 1994.

[5] IEEE Task Force, "Load representation for dynamic performance analysis (of power systems)," *IEEE Transactions on Power Systems*, vol. 8, no. 2, pp. 472–482, 1993.

[6] T. Overbye, P. Sauer, C. DeMarco, B. Lesieutre, and M. Venkatasubramanian, "Using PMU data to increase situational awareness," *PSERC Final Project Report, Tech. Rep.*, September 2010.

[7] P. Zhang, "Measurement-based load modeling," *EPRI, Tech. Rep.*, September 2006.

[8] P. Kundur, *Power System Stability and Control*. New York, NY: McGraw-Hill, 1994.

[9] S. Kalinowsky and M. Forte, "Steady state load-voltage characteristic field tests at area substations and fluorescent lighting component characteristics," *IEEE Transactions on Power Apparatus and Systems*, vol. PAS-100, no. 6, pp. 3087–3094, 1981.

[10] H. Bai, P. Zhang, and V. Ajjarapu, "A novel parameter identification approach via hybrid learning for aggregate load modeling," *IEEE Transactions on Power Systems*, vol. 24, no. 3, pp. 1145–1154, 2009.

[11] WECC Modeling Group, "Composite load model for dynamic simulations," *Tech. Rep.*, Jun. 2012.

[12] T. Ohyama, A. Watanabe, K. Nishimura, and S. Tsuruta, "Voltage dependence of composite loads in power systems," *IEEE Transactions on Power Apparatus and Systems*, vol. PAS-104, no. 11, pp. 3064–3073, 1985.

[13] D. Kosterev, A. Meklin, J. Undrill, *et al.*, "Load modeling in power system studies: WECC progress update," in *Power and Energy Society General Meeting—Conversion and Delivery of Electrical Energy in the 21st Century*, 2008 IEEE, July 2008, pp. 1–8.

[14] P. Regulski, P. Wall, Z. Rusidovic, and V. Terzija, "Estimation of load model parameters from PMU measurements," in *Innovative Smart Grid Technologies Conference Europe (ISGT-Europe)*, 2014 IEEE PES, October 2014, pp. 1–6.

[15] G. Ledwich and C. F. Moyano, "Synchrophasors for load modelling in Australia," in *IEEE Power and Energy Society General Meeting*, 2011 IEEE, July 2011, pp. 1–6.

[16] G. Shackshaft, O. Symons, and J. Hadwick, "General-purpose model of power-system loads," *Proceedings of the Institution of Electrical Engineers*, vol. 124, no. 8, pp. 715–723, 1977.

[17] Q. Liu, Y. Chen, and D. Duan, "The load modeling and parameters identification for voltage stability analysis," in *International Conference on Power System Technology, 2002. Proceedings, PowerCon 2002*, vol. 4, 2002, pp. 2030–2033.

[18] S. Zhu, J. Zheng, and G. Luo, "Effect of load modeling on voltage stability," in *IEEE Power Engineering Society Summer Meeting*, 2000, vol. 1, 2000, pp. 395–400.

[19] V. Knyazkin, C. Canizares, and L. Soder, "On the parameter estimation and modeling of aggregate power system loads," *IEEE Transactions on Power Systems*, vol. 19, no. 2, pp. 1023–1031, 2004.

[20] L. Ingber, "Adaptive simulated annealing (ASA): Lessons learned," *Control and Cybernetics*, vol. 25, pp. 33–54, 1996.

[21] A. Alves da Silva, C. Ferreira, and G. Torres, "Dynamic load modeling based on a nonparametric ANN," in *International Conference on Intelligent Systems Applications to Power Systems, 1996. Proceedings ISAP 96*, January 1996, pp. 55–59.

[22] D. Chen and R. Mohler, "Load modelling and voltage stability analysis by neural networks," in *Proceedings of the 1997 American Control Conference*, vol. 2, June 1997, pp. 1086–1090.

[23] T. Hiyama, M. Tokieda, W. Hubbi, and H. Andou, "Artificial neural network based dynamic load modeling," *IEEE Transactions on Power Systems*, vol. 12, no. 4, pp. 1576–1583, 1997.

[24] J. Ma, D. Han, R. M. He, Z. Y. Dong, and D. J. Hill, "Reducing identified parameters of measurement-based composite load model," *IEEE Transactions on Power Systems*, vol. 23, no. 1, pp. 76–83, 2008.

[25] Powertech Labs Inc., *DSA Tools*, Surrey, British Columbia, Canada.

[26] R. Christie, *IEEE Power System Test Cases Archive*. August 1999. [Online]. https://labs.ece.uw.edu/pstca/ [Accessed May 22, 2022].

[27] D. Simon, *Optimal State Estimation: Kalman, H Infinity, and Nonlinear Approaches*. Hoboken, NJ: Wiley-Interscience, 2006.

[28] R. Huang, R. Diao, Y. Li, *et al.*, "Calibrating parameters of power system stability models using advanced ensemble Kalman filter," *IEEE Transactions on Power Systems*, vol. 33, no. 3, pp. 2895–2905, 2017.

[29] C.-M. Huang, Y.-C. Huang, S.-J. Chen, and S.-P. Yang, "A hierarchical optimization method for parameter estimation of diesel generators," *IEEE Access*, vol. 8, pp. 176467–176479, 2020.

[30] A. Gorbunov, A. Dymarsky, and J. Bialek, "Estimation of parameters of a dynamic generator model from modal PMU measurements," *IEEE Transactions on Power Systems*, vol. 35, no. 1, pp. 53–62, 2019.

[31] S. R. Khazeiynasab, J. Qi, and I. Batarseh, "Generator parameter estimation by q-learning based on PMU measurements," in *2021 IEEE Power & Energy Society Innovative Smart Grid Technologies Conference (ISGT)*. IEEE, 2021, pp. 01–05.

[32] Y. Xu, L. Mili, X. Chen, M. Korkali, and L. Min, "A Bayesian approach to real-time dynamic parameter estimation using phasor measurement unit measurement," *IEEE Transactions on Power Systems*, vol. 35, no. 2, pp. 1109–1119, 2019.

[33] Y. Wehbe, L. Fan, and Z. Miao, "Least squares based estimation of synchronous generator states and parameters with phasor measurement units," in *2012 North American Power Symposium (NAPS)*. IEEE, 2012, pp. 1–6.

[34] Y. Xu, L. Mili, M. Korkali, and X. Chen, "An adaptive Bayesian parameter estimation of a synchronous generator under gross errors," *IEEE Transactions on Industrial Informatics*, vol. 16, no. 8, pp. 5088–5098, 2019.

[35] Z. Huang, P. Du, D. Kosterev, and S. Yang, "Generator dynamic model validation and parameter calibration using phasor measurements at the point of connection," *IEEE Transactions on Power Systems*, vol. 28, no. 2, pp. 1939–1949, 2013.

[36] L. Fan, "Least squares estimation and Kalman filter based dynamic state and parameter estimation," in *2015 IEEE Power & Energy Society General Meeting*. IEEE, 2015, pp. 1–5.

[37] R. Huang, R. Diao, Y. Li, *et al.*, "Calibrating parameters of power system stability models using advanced ensemble Kalman filter," *IEEE Transactions on Power Systems*, vol. 33, no. 3, pp. 2895–2905, 2018.

[38] M. González-Cagigal, J. A. Rosendo-Macías, and A. Gómez-Expósito, "Parameter estimation of wind turbines with PMSM using cubature Kalman filters," *IEEE Transactions on Power Systems*, vol. 35, no. 3, pp. 1796–1804, 2019.

[39] A. Rouhani and A. Abur, "Constrained iterated unscented Kalman filter for dynamic state and parameter estimation," *IEEE Transactions on Power Systems*, vol. 33, no. 3, pp. 2404–2414, 2017.

[40] J. Qi, K. Sun, J. Wang, and H. Liu, "Dynamic state estimation for multi-machine power system by unscented Kalman filter with enhanced numerical stability," *IEEE Transactions on Smart Grid*, vol. 9, no. 2, pp. 1184–1196, 2016.

[41] K. Emami, T. Fernando, H. H.-C. Iu, H. Trinh, and K. P. Wong, "Particle filter approach to dynamic state estimation of generators in power systems," *IEEE Transactions on Power Systems*, vol. 30, no. 5, pp. 2665–2675, 2014.

[42] N. Zhou, D. Meng, Z. Huang, and G. Welch, "Dynamic state estimation of a synchronous machine using PMU data: A comparative study," *IEEE Transactions on Smart Grid*, vol. 6, no. 1, pp. 450–460, 2014.

[43] L. Ljung, "Asymptotic behavior of the extended Kalman filter as a parameter estimator for linear systems," *IEEE Transactions on Automatic Control*, vol. 24, no. 1, pp. 36–50, 1979.

[44] L. Ljung, *System Identification: Theory for the User*. Upper Saddle River, NJ: Prentice Hall PTR, 1999.

Chapter 10

State estimation in the presence of synchronized measurement

Sanjeev Kumar Mallik[1], Saikat Chakrabarti[2] and Sri Niwas Singh[2]

State estimation plays a critical role for the wide-area monitoring system (WAMS) at the control center. The power system network, geographically spread over a large area, is monitored and controlled using the measurements collected at regular intervals. A number of application functions are performed based on the estimated states of the system to ensure secure and reliable operation of the power system. The amplitude and angle of the voltage phasor at each bus constitute the states of the system. The measurements are collected from the field and telemetered to the control center, where they are processed, typically using standard weighted least squares (WLS) static state estimator (SE), to estimate the states of the system.

10.1 State estimation

Since the first introduction of static state estimation as a data processing algorithm [1–3] in 1970s, for converting meter readings and other available information into estimates of the states, the SE algorithms have traversed a long path of development. This chapter discusses the inclusion of synchronized measurements from phasor measurement units (PMUs) into the static SE.

The states of a power system are usually estimated using equivalent single-phase asynchronous measurements, assuming the system to be balanced. In the SE, typically the positive sequence model is considered. The measurements collected from the field are telemetered to the control center and then processed through the SE to provide the estimate of the states. The measurements from the metering devices constitute the core inputs for the estimation of the power system states, and hence, there is a need to pay attention to the measuring device technology. Power system measurements may be classified into two categories based on the measuring device technologies, namely, supervisory control and data acquisition (SCADA) system measurements, and the measurements from PMUs.

[1]Department of Electrical Engineering, National Institute of Technology (NIT) Patna, India
[2]Department of Electrical Engineering, Indian Institute of Technology (IIT) Kanpur, India

10.1.1 SCADA measurements

A SCADA system involves collection of data through monitoring devices, and transferring them to the control center via a communication system. The first SCADA systems used panel of meters, lights, and strip chart recorders for data acquisition. With the technology advancement, remote terminal units (RTUs) and intelligent electronic devices (IEDs) have become integral parts of the modern SCADA system. IEDs are the sensors designed with the intelligence of programmable logic controllers (PLCs) and distributed control systems (DCS). An RTU is a microprocessor-based data acquisition and control unit, used to control and acquire data from the measuring devices [4]. This acquired data is telemetered to the control center.

SCADA-based energy management system (EMS) has been widely deployed in power system network. It utilizes RTUs and IEDs in the field to acquire measurement data through the measuring devices installed at the substations. The acquired measurements are transmitted to the control center through the communication network. The conventional SCADA measurements consist of real and reactive power injection measurements at the bus, real and reactive power flow measurements through the branch, magnitude of the current flowing through the branch, as well as the voltage magnitude at the bus. These measurements are scanned typically at a rate of 2–10 s. The breaker and switch statuses are also acquired and transmitted to the control center in order to construct the network topology.

10.1.2 Phasor measurement unit

The synchrophasor technology provides time-stamped voltage and current phasors from different locations in a power system network. The time stamps are derived from the global positioning system (GPS) clock. The PMU is a measuring device based on synchronized measurement technology [5,6]. The concept of the PMU came from the symmetrical component distance relay and the synchronization of sampling clock [7]. The first prototype of the PMU was developed at Virginia Tech, USA. Based on this prototype, Macrodyne Co. had started the production of the PMU commercially. Now there are a large number of PMU manufacturers in the market.

The analog voltage and the current from the field through the instrument transformers are the input to the PMU. The PMU then samples the waveforms and computes phasors using discrete Fourier transform (DFT) type of algorithms. One pulse per second timing signal obtained from the GPS is used to time-stamp these computed phasors. These phasors may be stored locally, or transmitted to remote locations using a communication network. Phasor data concentrators (PDCs) collect these measurements for further applications or transmission to other locations. The PMU data standard is specified in IEEE Std. C37.118 [8,9].

Measurement accuracy of a PMU is expressed in terms of total vector error (TVE), defined as [8,10],

$$\%\mathrm{TVE} = \frac{\left|\overline{\mathbf{X}}_{\mathrm{est}} - \overline{\mathbf{X}}_{\mathrm{ideal}}\right|}{\left|\overline{\mathbf{X}}_{\mathrm{ideal}}\right|} \times 100 \tag{10.1}$$

where $\overline{\mathbf{X}}_{est}$ and $\overline{\mathbf{X}}_{ideal}$ are the estimated phasor and the true phasor, respectively, for the measurements at a given instant. According to IEEE C37.118 Std., all the PMUs installed in the power system network should have less than 1% TVE under steady state. The TVE in the PMU measurements may be due to the error in the phasor magnitude, angle, and inaccuracy in time synchronization.

Due to the ability to provide time-synchronized phasor measurements at high refresh rates, PMUs are being increasingly deployed in power systems. It is envisaged that there will be increasing use of the PMUs in state estimation and its corresponding functions [6].

10.1.3 State estimation functions

State estimation is an integral part of the EMS. Based on the estimated states, subsequent actions are taken at the control center. There are a number of functions performed by the SE.

- **Topology processing**: It collects the status data from the circuit breakers and switches placed in the whole power system network to configure one-line diagram of the system.
- **Observability analysis**: It ensures whether the unique SE solution can be achieved or not, with the available set of measurements. If the system is not observable, it identifies the unobservable branches and the observable islands in the systems. A number of techniques have been evolved by the researchers for the observability analysis [11–20].
- **SE solution**: It provides an estimate for the system states by processing the given set of measurements.
- **Bad data processing**: It detects the bad data in the measurements, identifies the bad data, and processes or eliminates the bad data.
- **Parameter estimation**: It identifies the most likely parameters in the network.

These functions are integrated in the SE software of the EMS at the control center. This chapter focuses mainly on the state estimation solution techniques in the presence of PMU measurements.

10.1.4 Measurement model for static state estimation

The measurements for the power system network can be modeled as

$$\mathbf{z} = \mathbf{h}(\mathbf{x}) + \mathbf{e}_m \tag{10.2}$$

where $\mathbf{z}$ is the vector of m measurements consisting of SCADA measurements as well as PMU measurements; $\mathbf{h}(\mathbf{x})$ is the relationship between the measurement vector $\mathbf{z}$ and the state vector $\mathbf{x}$ that consists of voltage magnitudes and phase angles at the buses; $\mathbf{e}_m$ is the error vector for the measurements.

The state estimation problem in the presence of equality constraints can be formulated by minimizing the following objective function:

$$J(\mathbf{x}) = \frac{1}{2}[\mathbf{z} - \mathbf{h}(\mathbf{x})]^{\mathrm{T}}\mathbf{W}[\mathbf{z} - \mathbf{h}(\mathbf{x})] \tag{10.3}$$

Subject to

$$\mathbf{c}(\mathbf{x}) = 0 \tag{10.4}$$

where $\mathbf{W}$ is the diagonal weight matrix (assuming the measurements to be independent); and $\mathbf{c}(\mathbf{x})$ represents the zero injection power measurements (both real and reactive power injection).

The state estimation solution is achieved by iteratively solving the following set of equations:

$$\begin{bmatrix} \mathbf{G} & \mathbf{C}^{\mathrm{T}} \\ \mathbf{C} & \mathbf{0} \end{bmatrix} \begin{bmatrix} \Delta\mathbf{x} \\ \lambda_{\mathrm{L}} \end{bmatrix} = \begin{bmatrix} \mathbf{H}^{\mathrm{T}}\mathbf{W}(\mathbf{z} - \mathbf{h}(\mathbf{x})) \\ -\mathbf{c}(\mathbf{x}) \end{bmatrix} \tag{10.5}$$

where at kth iteration, $\Delta\mathbf{x} = \mathbf{x}^{k+1} - \mathbf{x}^{k}$ is the change in the state vector; λ_{L} is the vector of Lagrangian multiplier corresponding to zero injection measurements; $\mathbf{H} = \frac{\partial \mathbf{h}(\mathbf{x})}{\partial \mathbf{x}}$ and $\mathbf{C} = \frac{\partial \mathbf{c}(\mathbf{x})}{\partial \mathbf{x}}$ are the measurement Jacobian matrices; $\mathbf{G} = \mathbf{H}^{\mathrm{T}}\mathbf{W}\mathbf{H}$ is the gain matrix corresponding to WLS-SE. Above equation can be written as

$$\mathbf{E}\mathbf{y} = \mathbf{b} \tag{10.6}$$

where $\mathbf{E}$ is the so-called information matrix, $\mathbf{y}$ is the augmented state vector; the vector $\mathbf{b}$ is defined as $\mathbf{b} = [[\mathbf{H}^{\mathrm{T}}\mathbf{W}(\mathbf{z} - \mathbf{h}(\mathbf{x}))]^{T}; -\mathbf{c}(\mathbf{x})^{T}]^{T}$.

10.1.5 Assigning weights to the measurements

Weights are usually assigned to the measurements in proportion to their accuracy. Quantitatively speaking, a widely used approach is to assign weights inversely proportional to the variation in the measurements. The manufacturer usually specifies the maximum error in the measurements obtained from the measurement devices. In the absence of any measurement error distribution specified by the manufacturer, the guide to the uncertainty in measurements (GUM) prescribes adoption of uniform measurement error distribution [21]. An alternative error distribution, viz., Gaussian distribution, is also widely used for the measurements. Following subsections briefly discuss the assignment of measurement weights for these two types of error distributions.

10.1.5.1 Gaussian error distribution

Gaussian error distribution is specified by the mean and the standard deviation (SD) in the measurements. The mean value of any error in the measurements is assumed to be zero. Typical values of the standard deviation that are widely used in the literature for different types of measurements are listed in Table 10.1 [11,22]. However, these values may differ for different metering devices, and may be specified by the manufacturer.

The variance in the ith measurement, R_i, for the Gaussian error distribution can be expressed as

$$R_i = SD_i^{\,2} \tag{10.7}$$

where SD_i is the standard deviation in the ith measurement.

Table 10.1 Typical standard deviation in various measurements

Conventional measurements		PMU measurements	
Power injection 0.01 pu	Power flow 0.008 pu	|V| or |I| 0.002 pu	Phase angle 0.0017 radian

Table 10.2 Typical values of maximum measurement uncertainties

Conventional measurements		PMU measurements		
Power injection 3%	Power flow 3%	|V| 0.02%	|I| 0.03%	Phase angle 0.01 degree

10.1.5.2 Uniform error distribution

The maximum measurement uncertainty of a measuring device is usually specified by the manufacturer as a percentage of the meter reading, typical value of which are listed in Table 10.2. However, these values may differ for different metering devices, depending on the manufacturer specifications.

In the absence of any specified probability distribution for the measurement uncertainty, the standard uncertainty in the direct measurement can be expressed in terms of maximum specified uncertainty, by assuming a uniform probability distribution for the measurement uncertainty over the entire range, as prescribed in [21,23]. For the ith measurement, for example, the standard uncertainty or standard deviation in measurement can be expressed as

$$u_i = \frac{\Delta u_i}{\sqrt{3}} \tag{10.8}$$

where Δu_i is the specified maximum uncertainty in the ith measurement.

Example 10.1 The voltage magnitude measured by a PMU is 1.04 pu. Using Table 10.2, determine the standard uncertainty in this magnitude measurement.

Solution: The standard uncertainty for this measurement is

$$u_{V_i} = \frac{1.04 \times 0.02/100}{\sqrt{3}} = 1.2 \times 10^{-4}$$

Similarly, the standard uncertainty in the phase angle measurement will be

$$u_{\theta_i} = \frac{0.01 \times \pi}{180 \times \sqrt{3}} = 1.0077 \times 10^{-4}$$

The variance in the ith measurement, R_i, can then be expressed as

$$R_i = u_i^2 \tag{10.9}$$

The weight matrix, **W**, in (10.3) is the inverse of the error co-variance matrix, **R**, which is a diagonal square matrix, with R_i as (i,i)th element. During state estimation process, direct measurements may need to be converted into different form. For example, the voltage phasor computed at a bus by using voltage phasor directly measured at the other end and current phasor through the line is a derived or pseudo-measurement. The combined standard uncertainty in the derived measurements may be evaluated using the classical theory of propagation of measurement uncertainty [21].

Calculation of combined uncertainty

Let k independent measurement variables be $v_1, v_2, \ldots, v_k$ in the system. A measurement y is derived from the available k independent measurement variables. The combined standard uncertainty, u_y, for the measurement y can be obtained by combining the standard uncertainties of the independent variables, as shown below.

$$u_y = \sqrt{\sum_{j=1}^{k} \left(\frac{\partial y}{\partial v_j}\right)^2 (u_{v_j})^2} \tag{10.10}$$

where u_{v_j} is the standard uncertainty in the measurement variable v_j.

10.2 Conventional state estimation

After collecting all the measurements, observability analysis is carried out to determine whether the system is observable with the given set of measurements or not. After ensuring the observability, the measurements are processed through an SE algorithm to provide an estimate of the power system states. Depending on the type of measurements, the SE can be classified into three categories, namely, conventional SE, linear SE, and hybrid SE. The SE may include both conventional and synchronized measurements, and may be classified as pre-processed hybrid SE and post-processed sequential SE. Conventional SE is discussed in this section.

Conventional SE processes the conventional SCADA measurements to estimate the voltage phasor angle and magnitude at all the buses. Bus power injection and branch power flow (both real and reactive), as well as the magnitude of the bus voltage and the branch current, form the conventional SCADA measurement set. The measurement vector and its corresponding Jacobian (assuming state vector in polar form, voltage phasor angle and magnitude vector, δ and $\mathbf{v}$) for the SE process can be written as

$$\mathbf{z}_{\text{conventional}} = \begin{bmatrix} \mathbf{zP}_{\text{inj}} \\ \mathbf{zQ}_{\text{inj}} \\ \mathbf{zP}_{\text{flow}} \\ \mathbf{zQ}_{\text{flow}} \\ \mathbf{z}_{|\mathbf{V}|} \\ \mathbf{z}_{|\mathbf{I}|} \end{bmatrix} \tag{10.11}$$

$$\mathbf{H}_{\text{conventional}} = \begin{bmatrix} \frac{\partial \mathbf{z}_{\mathbf{P}_{\text{inj}}}}{\partial \delta} & \frac{\partial \mathbf{z}_{\mathbf{P}_{\text{inj}}}}{\partial \mathbf{V}} \\ \frac{\partial \mathbf{z}_{\mathbf{Q}_{\text{inj}}}}{\partial \delta} & \frac{\partial \mathbf{z}_{\mathbf{Q}_{\text{inj}}}}{\partial \mathbf{V}} \\ \frac{\partial \mathbf{z}_{\mathbf{P}_{\text{flow}}}}{\partial \delta} & \frac{\partial \mathbf{z}_{\mathbf{P}_{\text{flow}}}}{\partial \mathbf{V}} \\ \frac{\partial \mathbf{z}_{\mathbf{Q}_{\text{flow}}}}{\partial \delta} & \frac{\partial \mathbf{z}_{\mathbf{Q}_{\text{flow}}}}{\partial \mathbf{V}} \\ \frac{\partial \mathbf{z}_{|\mathbf{V}|}}{\partial \delta} & \frac{\partial \mathbf{z}_{|\mathbf{V}|}}{\partial \mathbf{V}} \\ \frac{\partial \mathbf{z}_{|\mathbf{I}|}}{\partial \delta} & \frac{\partial \mathbf{z}_{|\mathbf{I}|}}{\partial \mathbf{V}} \end{bmatrix} \tag{10.12}$$

Once the measurement vector and its corresponding Jacobian is formed, an iterative method is used to estimate the states for the conventional SE by solving (10.5) in presence of equality constraints, since the relationship between the measurement vector and the state vector is non-linear. The solution of the conventional SE involves matrix inversion of **E**. The convergence and the accuracy must be ensured while solving the iterative equation of the WLS-SE. The convergence of the SE is mathematically represented in terms of the condition number of the gain matrix. The high condition number of the gain matrix may deteriorate the accuracy and convergence. Therefore, it is essential to address the numerical conditioning issues associated with the SE problems, which is indicated by the condition number. The condition number of the matrix is defined as

$$\text{cond}(\mathbf{E}) = \|\mathbf{E}\|\|\mathbf{E}^{-1}\| \tag{10.13}$$

where $\|\mathbf{E}\|$ represents the matrix norm of **E**. The second-order norm is typically used for the SE problems. Higher-order norms are not considered mainly due to large computational effort. However, in some of the literature, the first-order norm is used to simplify the analysis of the condition number [24,25]. A high value of the condition number usually signifies numerical ill-conditioning. Some of the commonly encountered factors causing numerical ill-conditioning of the state estimation problem, as identified by the researchers, are existence of negative line reactance, existence of very small series impedance branch, presence of numerous power injection measurements, low measurement redundancy, size of the system, position of the slack bus, differential weight assigned for the measurements, inclusion of zero bus power injection as an equality constraints, and many more. To address these numerical conditioning issues, a number of techniques [24,26–29] have been developed to find the solution of (10.5) by the researchers. The majority of the existing SE methods are based on the inversion and decomposition of the gain matrix, **G** or information matrix, **E**. A new regularization method to solve the numerical instability of an equality constrained SE using L-curve method is discussed later [29]. The complete flowchart for the conventional SE is shown in Figure 10.1.

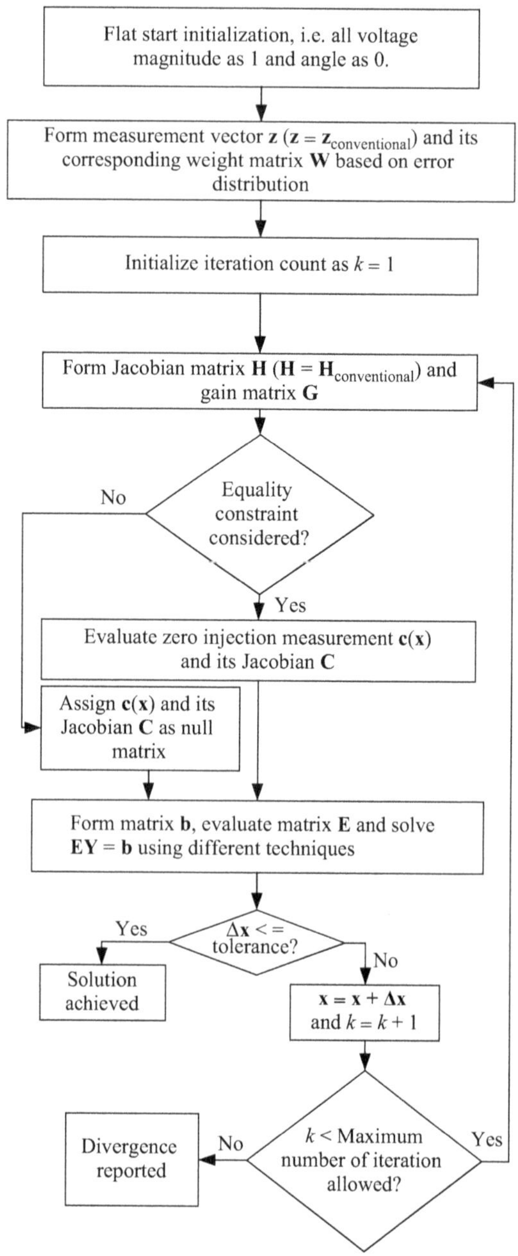

Figure 10.1 Conventional state estimation flowchart

10.3 Linear state estimation

The synchrophasor technology has brought a revolution in the SE due to its capability to measure time-stamped and synchronized phasor data at a high refreshing rate (as high as one phasor per cycle). The PMUs also provide direct measurements of the voltage and current phase angle with respect to a global reference obtained from the GPS clock. In short, the time-stamped synchronized measurement of voltage and current phasors with high accuracy and high reporting rate makes the PMUs preferable over the conventional SCADA measurements. In the presence of sufficient number of PMUs, the WLS-SE becomes linear and non-iterative [30].

In linear estimator, the power system network is observed through PMU measurements only. In this case, the measurement vector, $\mathbf{z}$, and its corresponding Jacobian can be expressed as

$$\mathbf{z}_{\mathrm{PMU}} = \begin{bmatrix} \mathbf{z}_{\mathbf{V}_{\mathrm{real}}} \\ \mathbf{z}_{\mathbf{V}_{\mathrm{imaginary}}} \\ \mathbf{z}_{\mathbf{I}_{\mathrm{real}}} \\ \mathbf{z}_{\mathbf{I}_{\mathrm{imaginary}}} \end{bmatrix}_{\mathrm{PMU}} \tag{10.14}$$

$$\mathbf{H}_{\mathrm{PMU}} = \begin{bmatrix} \mathbf{H}_{11} & \mathbf{0} \\ \mathbf{0} & \mathbf{H}_{22} \\ \mathbf{H}_{31} & \mathbf{H}_{32} \\ \mathbf{H}_{41} & \mathbf{H}_{42} \end{bmatrix} \tag{10.15}$$

In general, the voltage and current phasor measurements from the PMU are available in polar form, which needs to be converted into rectangular form (as expressed in (10.14)) for the linear estimator process, and the corresponding weight of the converted measurement is evaluated using classical theory of propagation of uncertainty, as discussed earlier. In case of linear estimator, the real and reactive parts of the voltage at all buses constitute the state vector, $\mathbf{x}$ ([$\mathbf{e}$;$\mathbf{f}$]). With the given measurement vector, $\mathbf{z}_{\mathrm{PMU}}$, the measurement Jacobian with respect to the state vector (represented by (10.15)) is independent of the state vector, and hence, the state vector can be directly evaluated in a non-iterative manner.

$$\mathbf{x}_{\mathrm{PMU}}^{\mathrm{est}} = \begin{bmatrix} \mathbf{e} \\ \mathbf{f} \end{bmatrix} = \left(\mathbf{H}_{\mathrm{PMU}}^{\mathrm{T}} \mathbf{W}_{\mathrm{PMU}} \mathbf{H}_{\mathrm{PMU}}\right)^{-1} \mathbf{H}_{\mathrm{PMU}}^{\mathrm{T}} \mathbf{W}_{\mathrm{PMU}} \mathbf{z}_{\mathrm{PMU}} \tag{10.16}$$

where $\mathbf{W}_{\mathrm{PMU}}$ is the weight matrix corresponding to the PMU measurements.

For the linear estimator, sufficient number of PMUs should be employed in the power system network. Due to the communication as well as financial constraints, limited numbers of the PMUs are being installed in the network that may not make the system completely observable. Therefore, the measurements in terms of voltage and current phasors data from the PMUs should be embedded with the conventional SCADA measurements in the SE to exploit the advantages of the PMUs. In the fusion of the PMU current and voltage phasor data with the conventional SCADA measurements, two approaches are reported in the literature, namely, post-processing sequential estimator and pre-processing hybrid estimator.

10.4 Post-processing sequential state estimation

The post-processing algorithm for the SE combines the conventional and the linear SE in a two stage process in a sequential manner, and hence may be called as sequential SE. The first stage is exactly similar to the conventional SE processing the conventional SCADA measurements to estimate the voltage magnitude and angle at all buses of the system, as described in Section 10.3. The estimated states, $\mathbf{x}_{\text{stageI,polar}}$ from the conventional SE are in polar form. The error covariance matrix (nxn matrix) for the estimated states can be expressed as [30]

$$\text{covariance}(\mathbf{x}_{\text{stageI,polar}}) = \left(\mathbf{H}^{\text{T}}_{\text{conventional}}\mathbf{W}_{\text{conventional}}\mathbf{H}_{\text{conventional}}\right)^{-1} \tag{10.17}$$

where $\mathbf{W}_{\text{conventional}}$ is the weight matrix corresponding to the conventional SCADA measurements. The measurement Jacobian for the conventional SCADA measurements, $\mathbf{H}_{\text{conventional}}$, is calculated at the estimated states from stage I, $\mathbf{x}_{\text{stageI,polar}}$. The variance matrix, $\mathbf{R}_{\text{stageI,polar}}$ of the estimated voltage phasors from the conventional SE can be obtained by the corresponding diagonal element of the matrix expressed in (10.17), by ignoring the off-diagonal elements.

In a sequential SE, the estimates from the existing SCADA software are combined with PMU measurements in stage II, to obtain improved estimates of the states. The complete measurement vector and its Jacobian for stage II estimator may be expressed as

$$\mathbf{z}_{\text{II}} = \begin{bmatrix} \begin{pmatrix} \mathbf{V}_{\text{real}} \\ \mathbf{V}_{\text{imaginary}} \end{pmatrix}_{\text{stageI}} \\ \begin{pmatrix} \mathbf{V}_{\text{real}} \\ \mathbf{V}_{\text{imaginary}} \end{pmatrix}_{\text{PMU}} \\ \begin{pmatrix} \mathbf{I}_{\text{real}} \\ \mathbf{I}_{\text{imaginary}} \end{pmatrix}_{\text{PMU}} \end{bmatrix} \tag{10.18}$$

$$\mathbf{H}_{\text{II}} = \begin{bmatrix} \mathbf{H}_{11} & \mathbf{0} \\ \mathbf{0} & \mathbf{H}_{22} \\ \mathbf{H}_{31} & \mathbf{0} \\ \mathbf{0} & \mathbf{H}_{42} \\ \mathbf{H}_{51} & \mathbf{H}_{52} \\ \mathbf{H}_{61} & \mathbf{H}_{62} \end{bmatrix} \tag{10.19}$$

For the linear SE in stage II, all the voltage phasors are either estimated from stage I or collected from the PMUs. The current phasors from the PMU must be in rectangular form. Since the states are estimated from conventional SE (stage I) in polar form, they need to be converted into rectangular form for the inclusion in the measurement vector of stage II. The variance, $\mathbf{R}_{\text{stageI,rect}}$ corresponding to the converted states from stage I in rectangular form can be evaluated using classical theory of propagation of uncertainty.

The measurement vector from stage II includes the real and reactive parts of the voltage phasor obtained from stage I, and the real and reactive parts of the voltage and the current phasor from the PMU, as expressed in (10.18). The complete variance matrix for the stage II can be written as

$$\mathbf{R}_{\mathrm{II}} = \begin{bmatrix} \mathbf{R}_{\mathrm{stageI,rect}} & \mathbf{0} & \mathbf{0} \\ \mathbf{0} & \mathbf{R}_{\mathrm{V,PMU}} & \mathbf{0} \\ \mathbf{0} & \mathbf{0} & \mathbf{R}_{\mathrm{I,PMU}} \end{bmatrix} \tag{10.20}$$

where $\mathbf{R}_{\mathrm{V,PMU}}$, $\mathbf{R}_{\mathrm{I,PMU}}$ are the variance matrix corresponding to the voltage and the current phasor measurements, respectively, obtained from the PMUs placed in the network. The weight matrix for the stage II is the inverse of $\mathbf{R}_{\mathrm{II}}$. The measurement Jacobian expressed by (10.19) is independent of the states of the system. Therefore, the final states of the power system can be estimated directly without involving any iteration.

$$\mathbf{x}_{\mathrm{II}} = \begin{bmatrix} \mathbf{e} \\ \mathbf{f} \end{bmatrix} = \left(\mathbf{H}_{\mathrm{II}}\mathbf{R}_{\mathrm{II}}^{-1}\mathbf{H}_{\mathrm{II}}\right)^{-1}\mathbf{H}_{\mathrm{II}}\mathbf{R}_{\mathrm{II}}^{-1}\mathbf{z}_{\mathrm{II}} \tag{10.21}$$

The complete algorithm for the sequential SE is shown in Figure 10.2. In the sequential SE, the states obtained from the conventional SE are in polar form, which are converted into rectangular form along with their variance or weight. The converted estimated states are combined with the PMU measurements (voltage and current phasor in rectangular form) and processed through linear SE to estimate the final states of the system, i.e., the real and reactive parts of the voltage phasor at all buses. In this sequential SE, the stage I estimator is a conventional, iterative, and nonlinear process, whereas the stage II estimator is a non-iterative and linear process. The sequential approach is preferred due to the

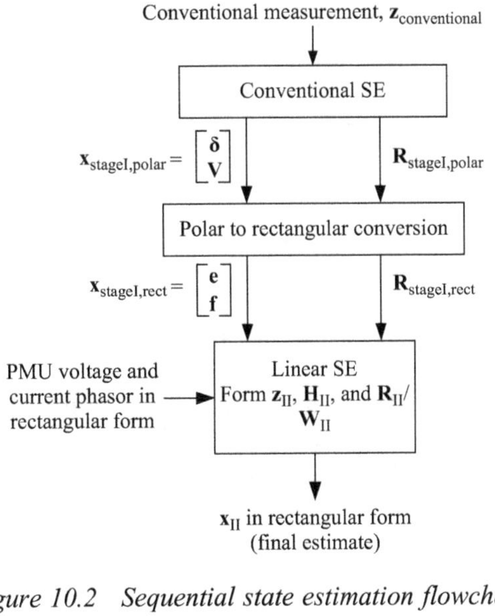

Figure 10.2 Sequential state estimation flowchart

reusability of the existing SE software [30]. However, this method is relevant when the number of PMUs is small in the system. As the number of PMUs will increase, the natural approach will be to combine the conventional and PMU measurements in a single stage. This type of hybrid estimators are referred here as the pre-processing estimator.

10.5 Pre-processing hybrid state estimation

In Sections 10.3 and 10.4, the voltage and current phasor from the PMUs were considered for the estimator in rectangular form. In hybrid SE, the voltage and current phasor from the PMU may be considered either in polar or rectangular form. In the hybrid SE, the PMU measurements are combined with the conventional SCADA measurements in a pre-processing algorithm, where the estimation process is similar to the conventional SE. The hybrid SE differs with the conventional SE in the formation of measurement vector and its corresponding Jacobian. The hybrid SE is iterative and nonlinear with respect to the states of the system. The states are estimated by solving the set of linear equations (10.5) or (10.6) in each iteration. For estimating the states, the states, the measurement vector and its corresponding Jacobian must be defined.

For the hybrid SE, the voltage magnitude and angle at all buses constitute the states of the system, and can be expressed as

$$\mathbf{x} = \begin{bmatrix} \delta \\ \mathbf{V} \end{bmatrix} \tag{10.22}$$

The conventional measurement vector consists of the real and reactive power injection at the buses as well as the real and reactive power flow through the line, and expressed as

$$\mathbf{z}_{\text{conventional}} = \begin{bmatrix} \mathbf{zP}_{\text{inj}} \\ \mathbf{zQ}_{\text{inj}} \\ \mathbf{zP}_{\text{flow}} \\ \mathbf{zQ}_{\text{flow}} \end{bmatrix} \tag{10.23}$$

The voltage and the current phasor are the measurements available from the PMU. Due to the iterative nature of the solution of hybrid SE, the inclusion of PMU in the hybrid SE has become important for ensuring the convergence and the accuracy of the estimated states. Since the states of the system are in polar form, so the voltage phasor measurements from the PMU are included directly in polar form in the measurement vector. The inclusion of the PMU currents in the hybrid SE has been a major challenge for the researchers due to ill-conditioned gain matrix, which may result in divergence problems. There are three possible ways to include the PMU current measurements in the state estimation problem [31]:

- *Current phasor magnitude and angle*: In this case, certain elements (partial derivatives) of the Jacobian matrix may become undefined (divided by zero) or extremely large at flat start or specific loading conditions. Hence, the inclusion of the current phasor in polar form is not preferred for the hybrid SE.

- *Real and imaginary part of the current phasor*: The problem associated with the polar form of current phasor is alleviated if the current phasor is included in rectangular form. In general, the rectangular form of PMU current phasor is considered for the hybrid SE.
- *Pseudo-voltage form derived from the PMU current and voltage phasor*: For a line between bus i and j, if the PMU is located at bus i, then the voltage phasor at bus j can be calculated using the voltage phasor at bus i, current phasor flowing from bus i and j, and the line parameters.

Since the PMU current is a major concern for power system engineers in terms of the SE convergence, the researchers have attempted to use the PMU current in pseudo-voltage form. The pseudo-voltage calculation involves the voltage and current measurements along with the network parameter information. Due to the uncertainty in the network parameter, which is very difficult to evaluate, there is a need to convert the PMU current in other pseudo-form depending only on the measurements. Pseudo-flow measurement ($\mathbf{V}_i\mathbf{I}_{ij}^*$) derived from the PMU current and voltage phasor is one such form to be considered by the researchers. For the pseudo-flow form, the accuracy of the pseudo-measurement depends on the PMU measurement only. With the direct inclusion of the PMU current in rectangular form, the gain matrix may become ill-conditioned due to high weight associated with the PMU current measurements. The convergence issues associated with the PMU currents can be resolved by inclusion of PMU current as branch flow pseudo-measurements, which are reflected by less number of iteration [32].

For the hybrid measurement configurations, the PMU current measurement is being used in the SE either in polar form or rectangular form. The rectangular form of PMU current is frequently used for the state estimation. By considering the voltage phasor magnitude and current, as well as real and imaginary parts of the current phasor collected from PMUs, the measurement vector corresponding to the PMU measurements can be expressed as

$$\mathbf{z}_{\text{PMU}} = \begin{bmatrix} \mathbf{z}_{\delta} \\ \mathbf{z}_{\mathbf{V}} \\ \mathbf{z}_{\mathbf{I}_{\text{real}}} \\ \mathbf{z}_{\mathbf{I}_{\text{imaginary}}} \end{bmatrix} \tag{10.24}$$

The complete measurement vector consisting of conventional measurements expressed by (10.23) and the PMU measurements expressed by (10.24), can be written as

$$\mathbf{z}_{\text{Hybrid}} = \begin{bmatrix} \mathbf{z}_{\text{conventional}} \\ \mathbf{z}_{\text{PMU}} \end{bmatrix} \tag{10.25}$$

The measurement Jacobian and the weight matrix corresponding to the measurement vector expressed by (10.25) can be written as

$$\mathbf{H}_{\text{Hybrid}} = \begin{bmatrix} \mathbf{H}_{\text{conventional}} \\ \mathbf{H}_{\text{PMU}} \end{bmatrix} \tag{10.26}$$

$$\mathbf{W}_{\text{Hybrid}} = \begin{bmatrix} \mathbf{W}_{\text{conventional}} & \mathbf{0} \\ \mathbf{0} & \mathbf{W}_{\text{PMU}} \end{bmatrix} \tag{10.27}$$

where $\mathbf{H}_{\text{conventional}}$ and $\mathbf{H}_{\text{PMU}}$ are the measurement Jacobian corresponding to the conventional measurements and the PMU measurements, respectively; $\mathbf{W}_{\text{conventional}}$ and $\mathbf{W}_{\text{PMU}}$ are the measurement weights corresponding to the conventional measurements and the PMU measurements, respectively.

The algorithm for the hybrid SE is similar to the algorithm shown in Figure 10.1. The difference lies in the formation of the measurement vector ($\mathbf{z} = \mathbf{z}_{\text{Hybrid}}$), the corresponding Jacobian ($\mathbf{H} = \mathbf{H}_{\text{Hybrid}}$), and the weight matrix ($\mathbf{W} = \mathbf{W}_{\text{Hybrid}}$). Due to the poor conditioning of the gain matrix, a number of techniques have been developed for the SE to process the conventional measurements using inversion/decomposition/factorization methods and these methods are also extended to the hybrid measurement set. In the next section, a regularization based method for solving the linear set of equations, $\mathbf{Ey} = \mathbf{b}$ expressed in (10.6) during each iteration, is discussed. This method is better equipped in handling numerical ill-conditioning problems in the SE execution.

10.6 Regularization method

The inversion of a matrix having large condition number is prone to numerical problems. A numerically robust technique is, therefore, needed to handle the ill-conditioning issues. In this section, a methodology based on Tikhonov regularization is presented to handle the ill-conditioning of the state estimation problem.

A substantial amount of research work has been done in the field of regularization and optimization of inverse ill-posed problems. The least squares method is one of the methods to solve the problem of minimization of $||\mathbf{Ey} - \mathbf{b}||^2$ (refer (10.6)). The least square solution for (10.6) is also known as pseudo-inverse solution. The output of least squares optimization is affected by the data error and rounding error [33]. Regularization methods can be used to reduce these errors in solving the inverse problems. Tikhonov regularization is a widely used method for solving ill-conditioned problems, and is described below in the context of solving (10.6) [34].

$$\min_{\mathbf{y}} \left\{ ||\mathbf{Ey} - \mathbf{b}||^2 + \lambda^2 ||\mathbf{L}(\mathbf{y} - \mathbf{y}_0)||^2 \right\} \tag{10.28}$$

where $\mathbf{L}$ is referred to as a regularization operator and λ is a regularization parameter. The common choice of $\mathbf{L}$ is the identity matrix $\mathbf{I}$. The vector $\mathbf{y}_0$ is an apriori estimate of $\mathbf{y}$. In the absence of apriori information, $\mathbf{y}_0$ is assumed to be zero. With these assumptions applied, (10.28) is simplified as

$$\min_{\mathbf{y}} \left\{ ||\mathbf{Ey} - \mathbf{b}||^2 + \lambda^2 ||\mathbf{y}||^2 \right\} \tag{10.29}$$

The first term $\|\mathbf{Ey} - \mathbf{b}\|^2$ is known as the least square or residual norm and $\|\mathbf{y}\|^2$ is the regularized or solution norm. From (10.29), it is clear that the estimate of $\mathbf{y}$ depends on the regularization parameter. The unregularized solution ($\lambda = 0$) of (10.29) corresponds to the pseudo-inverse solution. Pseudo-inverse solution only considers the idea of the curve fitting, i.e., minimization of the least square norm. It does not consider the solution norm, i.e., the accuracy of the estimate of $\mathbf{y}$. Tikhonov regularization provides the solution by minimizing the least square norm as well as the solution norm, and the solution is controlled by the regularization parameter. The next step is to analyze the effect of regularization on the solution. If too much regularization is imposed on the solution, then it will not fit the given data $\mathbf{b}$. However, if very small regularization is imposed, the fit will be good, but the solution will be dominated by the contribution from the data errors. Hence, there is a tradeoff between the size of the regularized solution and the quantity of the fit. This tradeoff is controlled by selection of proper regularization parameter. Some of the well-known methods for selecting the regularization parameter are listed below:

- Discrepancy principle
- Generalized cross validation (GCV) method
- L-curve method

According to the discrepancy principle, the regularization parameter λ is chosen so that the discrepancy is satisfied assuming accurate estimation in ε [35]:

$$\|\mathbf{b} - \mathbf{Ey}\| = \varepsilon \tag{10.30}$$

where ε is the norm of the error in $\mathbf{b}$. The discrepancy principle faces difficulty in the absence of the information on ε. The GCV method is based on cross validation of each data point as reported in [36]. The GCV method is not satisfactory when the errors are highly correlated. The L-curve method is robust and has the ability to handle perturbations caused by correlation noise. Hence, the L-curve method for the selection of regularization parameter has gained attention in recent years, and is considered for estimating the regularization parameter.

10.6.1 Selection of regularization parameter by L-curve method

L-curve is a log-log plot between the squared norm of the regularized solution and the squared norm of the residual for a range of values of the regularization parameter. The name is derived from the fact that this plot resembles letter "L" closely. Taking the following notations:

$$\eta = \|\mathbf{y}\|^2,\ \rho = \|\mathbf{Ey} - \mathbf{b}\|^2 \tag{10.31}$$

$$\widehat{\eta} = \log \eta,\ \widehat{\rho} = \log \rho \tag{10.32}$$

L-curve is the plot of $\widehat{\eta}/2$ versus $\widehat{\rho}/2$. With the above notations expressed in (10.31) and (10.32), the solution norm as well as least square norm depends on $\mathbf{y}$, which is the solution of (10.29), expressed as

$$\mathbf{y} = \left(\mathbf{E}^{\mathrm{T}}\mathbf{E} + \lambda^2\mathbf{I}\right)^{-1}\mathbf{E}^{\mathrm{T}}\mathbf{b} \tag{10.33}$$

The sparsity of the matrix $(\mathbf{E}^{\mathrm{T}}\mathbf{E} + \lambda^2\mathbf{I})$ used in (10.33) is reduced, and the evaluation of **y** is computationally expensive. Hence, a computationally efficient tool is required to simplify the calculation of two norms associated with **y**, and singular value decomposition (SVD) is one such tool. The SVD component of the matrix **E** can be expressed as

$$\mathbf{E} = \sum_{i=1}^{ns} u_i \sigma_i {v_i}^{\mathrm{T}} \tag{10.34}$$

$$\sigma_1 \geq \sigma_2 \geq \sigma_3 \geq \cdots \geq \sigma_{ns} \geq 0 \tag{10.35}$$

where u_i and v_i are the left and right singular vectors for the corresponding singular value σ_i; ns is the number of non-zero singular values of **E**. After inserting the value of SVD component of **E** into (10.33), Tikhonov solution is given as

$$\mathbf{y} = \sum_{i=1}^{ns} f_i \frac{u_i^{\mathrm{T}} b}{\sigma_i} v_i \tag{10.36}$$

$$f_i = \frac{\sigma_i^2}{\sigma_i^2 + \lambda^2} \tag{10.37}$$

where $f_1, f_2, \ldots, f_{ns}$ are the Tikhonov filter factors. With this SVD formulation, the solution and residual norm can be expressed as

$$\eta = \|\mathbf{y}\|^2 = \sum_{i=1}^{ns} \left(f_i \frac{u_i^{\mathrm{T}} b}{\sigma_i}\right)^2 \tag{10.38}$$

$$\rho = \|\mathbf{E}\mathbf{y} - \mathbf{b}\|^2 = \sum_{i=1}^{ns} \left((1 - f_i) u_i^{\mathrm{T}} b\right)^2 \tag{10.39}$$

For different values of λ, the solution norm as well as least square norm based on (10.38) and (10.39) may be plotted on a logarithmic scale to depict L-curve. L-curve for the unperturbed data is mostly flat at the value $\|\mathbf{y}\|$, except for larger values of the residual norm, where the curve approaches the axis corresponding to the residual norm. L-curve corresponding to the error in **b** is overall very steep, located slightly left of ε except for small value of λ, where it approaches the axis corresponding to the regularized norm [34]. Combining both the component of the data, it is concluded that L-curve is dominated by pure noise in case of small λ and by the unperturbed data for large value of λ. The above conclusion is valid only when L-curve is plotted on a logarithmic scale. In linear scale, L-curve is always convex irrespective of **b**.

A graphical strategy to select the regularization parameter uses the location of the corner of this curve. The point corresponding to the suitable λ lies on the corner of the L-curve. The logic behind this choice is that the corner separates the flat and vertical parts of the curve, where the solution is dominated by the regularization errors and perturbation errors, respectively [34]. Graphically, it is very difficult to locate the corner point of the L-curve with sufficient accuracy. Hence, there is a need to convert the graphical approach into a numerical technique for locating the corner point. Hansen [34] suggested a numerical technique for finding the corner of the L-curve by tracing the point of maximum curvature. The curvature $k(\lambda)$can be found by

$$k(\lambda) = \frac{\widehat{\rho}'\widehat{\eta}'' - \widehat{\rho}''\widehat{\eta}'}{\left((\widehat{\rho}')^2 + (\widehat{\eta}')^2\right)^{3/2}} \tag{10.40}$$

where the prime denotes the differentiation with respect to λ. The point of maximum curvature corresponds to the corner of the L-curve, where the values of both the regularized norm and residual norm are low simultaneously. The formulation of the curvature involves the evaluation of first- and second-order derivative of $\widehat{\eta}$ and $\widehat{\rho}$ with respect to λ [29].

With this formulation, the curvature is evaluated for different values of λ, depending upon the singular values of matrix **E,** to make sure that the maximum curvature point is included. Out of these values of the curvature, the value of λ corresponding to the maximum curvature is evaluated. Let this value be λ_c, where c is the index for λ. Further, the maximum curvature is evaluated by using Golden section search between two points λ_{c-1} and λ_{c+1}.

10.6.2 Regularized state estimation problem

The iterative equation for the state estimation problem (10.5) or (10.6) is solved through Tikhonov regularization, and the regularization parameter is estimated using L-curve criterion. A corrective step is taken at the post-estimation stage to include zero injection. Without this corrective step, the SE results for the bus having zero injection measurement will become erroneous.

The flowchart for finding the solution for **Ey** = **b** is shown below.

[STEP 1] Formulate SVD component of **E**, as shown in (10.34).

[STEP 2] Generate a number of values of λ. Let this number be 200. Then,

$$\lambda_{200} = \sigma_{\min} \tag{10.41}$$

$$\frac{\lambda_i}{\lambda_{i+1}} = \left(\frac{\sigma_{\max}}{\sigma_{\min}}\right)^{\frac{1}{200-1}} \forall i = 199 : -1 : 1 \tag{10.42}$$

[STEP 3] Evaluate the least square norm as well as the solution norm for all 200 values of λ.

[STEP 4] Plot L-curve for all λ.

[STEP 5] For each λ, the value of the curvature is evaluated.

[STEP 6] Find the maximum curvature and its corresponding λ (say, λ_c).

[STEP 7] Search λ_{opt} between λ_{c-1} and λ_{c+1} for finding out the absolute maximum curvature point using Golden section search method.

[STEP 8] Evaluate the solution **y** for the corresponding λ_{opt}. From **y**, separate out $\Delta\mathbf{x}$ required for SE algorithm.

From Step 6, the regularization parameter estimated may not correspond to the maximum curvature. However, it is certain to have maximum curvature between λ_{c-1} and λ_{c+1}. Hence, there is a need to refine the search for regularization parameter between λ_{c-1} and λ_{c+1}. For this purpose, Golden section search method is attempted while considering the parabolic nature of the curvature.

After completing the regularization-based algorithm for the conventional or hybrid state estimation problem, the estimated value of injection at zero injection bus may not satisfy the equality constraint. To satisfy the equality constraints, the zero injection bus voltage is modified after completing the regularization algorithm. The voltage phasor at *iz*th bus having zero injection can be expressed in terms of non-zero injection bus voltage phasors as

$$\overline{\mathbf{V}}_{iz} = -\frac{1}{\overline{\mathrm{Y}}_{iz,iz}} \sum_{j \neq iz}^{j=1^{n}} \overline{\mathrm{Y}}_{iz,j} * \overline{\mathbf{V}}_{j} \tag{10.43}$$

where $\overline{\mathrm{Y}}_{iz,j}$ is the (iz, j)th element of the complex bus admittance matrix. $\overline{\mathrm{V}}_j$ is the complex voltage at bus j.

10.7 Case study

In this section, the regularized state estimation is implemented and tested on IEEE 14-bus test system [37]. The system is used for investigating the performance of the regularized method with some of the other existing methods. In this chapter, the regularization-based SE is compared with LU factorization-based SE (referred as LU-SE) [38]. In LU factorization-based SE, iterative equation (10.6) is solved using LU factorization of the information matrix, **E**.

PMU measurements include voltage phasor and current phasor in polar form at the buses where it is deployed. The PMU current phasor measurements, thus collected, are converted into rectangular form and the corresponding weights are evaluated using the principle of combined measurement uncertainty [31]. The PMU inputs for the SE methods are PMU current phasor in rectangular form and PMU voltage phasor in polar form. Zero injection measurements are treated as equality constraints in this chapter. A total of 1,000 Monte Carlo simulations are carried out for testing the accuracy of the SE for IEEE 14-bus test system. A flat start is assumed and the convergence tolerance is set to 10^{-6}. All simulations are carried out on an Intel Core processor i7-2600 CPU at 3.4 GHz and 4 GB of RAM.

The true value of the state vector is acquired from load flow solutions. The true measurements, such as real and reactive power injections as well as power flows, and line currents in polar form, are evaluated from the true state vector as well as line parameter information. The measurements are generated by adding Gaussian errors to the true value of the measurements. One thousand Monte Carlo trials are performed by randomly selecting the measurements following a Gaussian distribution from a range defined by the mean and the standard deviation of the distribution. The SD of the error in the conventional measurements (real and reactive power injection as well as power flow) and the PMU measurements (voltage as well as current phasor magnitude and angle) is considered as listed in Table 10.1. All the measurement errors are assumed to have zero mean.

The performance of the regularized SE is evaluated by its absolute error (average and maximum) in voltage phasor (magnitude in pu as well as angle in radian). The behavior of the SE is evaluated by the performance indices, namely, the filter effect (FE) and the total statistical error variance (TSEV) [22]. The FE of the SE algorithm is defined as

$$\mathrm{FE} = \frac{\sum_{i=1}^{m}\left(\mathrm{S}_i^{\mathrm{E}} - \mathrm{S}_i^{\mathrm{T}}\right)^2}{\sum_{i=1}^{m}\left(\mathrm{S}_i^{\mathrm{M}} - \mathrm{S}_i^{\mathrm{T}}\right)^2} \tag{10.44}$$

where $\mathrm{S}_i^{\mathrm{E}}$, $\mathrm{S}_i^{\mathrm{T}}$, and $\mathrm{S}_i^{\mathrm{M}}$ are the final estimated measurements, true value of measurements, and the measurements at each Monte Carlo run, respectively. In this chapter, the average filter effect over the entire Monte Carlo run is reported for the performance analysis.

TSEV of the estimated state variables of size n averaging over the T Monte Carlo run is expressed as

$$\mathrm{TSEV} = \sum_{i=1}^{n}\left(\frac{1}{T}\sum_{mc=1}^{T}\left(\mathbf{x}(i)_{mc} - \mathbf{x}(i)_t\right)^2\right) \tag{10.45}$$

where $\mathbf{x}(i)_{mc}$ is the vector of the estimates of the ith state variable in the *(mc)*th Monte Carlo trial. $\mathbf{x}(i)_t$ is the true value of ith state.

The regularized SE is implemented and tested on the IEEE 14-bus system for two cases. In Case I, a set of measurements consists of the injection meters, the flow meters, and some PMUs. In Case II, the network is covered by few injection and flow meters without any PMU.

10.7.1 Case I for IEEE 14-bus test system

The real and reactive power injection measurements at all buses except the zero injection buses, and flow measurements (real as well as reactive) at all lines are part of the conventional measurements for this case. Two PMUs at buses 1 and 6 are placed for the validation of the regularized SE.

The regularized SE is demonstrated for the second iteration of 550th Monte Carlo sample (randomly selected). The singular value of **E** obtained for this particular iteration and sample is found to vary from 2.5569×10^{-5} ($\sigma_{\min}$) to 1.1355×10^8 ($\sigma_{\max}$). The least square norm and the solution norm are evaluated for 200 points of λ such that $\lambda_1 \geq \lambda_2 \geq \lambda_3 \geq \cdots \geq \lambda_{200}$. The ratio of two consecutive value of λ is evaluated using (10.42).

The values of both the norms are plotted on logarithmic scale for the second iteration of a 550th Monte Carlo trial and shown in Figure 10.3 (only few values of λ are indicated in this figure). The indicated values in the plot correspond to different values of the regularization parameter. Its corresponding curvature represented by (10.40) is plotted against different values of λ, as shown in Figure 10.4.

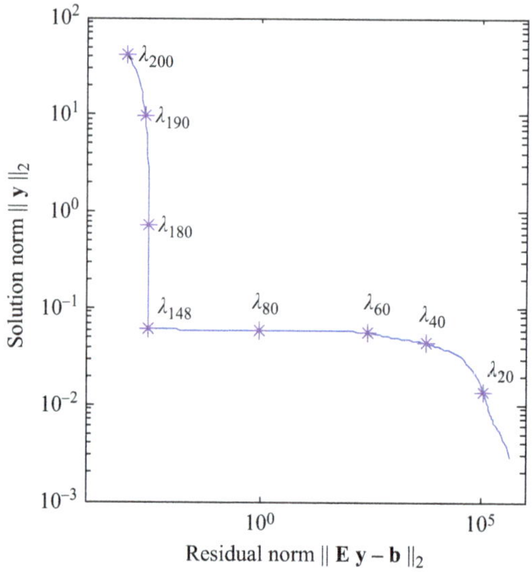

Figure 10.3 L-curve during second iteration of 550th Monte Carlo sample of IEEE 14-bus system for different values of λ

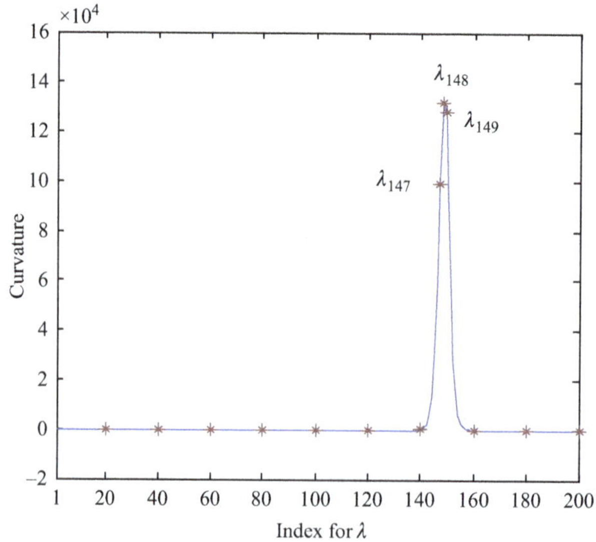

Figure 10.4 Curvature corresponding to Figure 10.6 for IEEE-14 bus system

From Figure 10.4, the maximum point of $k(\lambda)$ is captured to be 1.3173 × 10^5 at λ_{148} = 0.0516. This maximum value of $k(\lambda)$ may not be the exact maximum point and hence further processing is required to evaluate the exact maximum point within the narrow search between λ_{147} = 0.0597 and λ_{149} = 0.0446. The golden section search method is employed to find the maximum value of $k(\lambda)$ in the interval [0.0446 0.0597] of λ. The regularization parameter selected corresponding to Figure 10.4 is taken as 0.0485, with the maximum curvature 1.3542 × 10^5. For this particular 550th Monte Carlo trial, the values of regularization parameter selected in all iterations are listed in Table 10.3.

With two PMUs at buses 1 and 6, the performance of the regularized estimator is compared with the LU-SE and listed in Table 10.4. The accuracy of the estimator is better as compared to the LU-SE, as is evident from Table 10.4. The average FE over the entire Monte Carlo trials from the regularized method is comparable to the LU-SE. The small value of variance shows the accuracy of the solutions, whereas the FE indicates the fit of the estimated states for the measurements. The estimated values of the state variables through the Tikhonov regularization method by processing the erroneous measurements are accurate up to four decimal digits compared with the true values of the states. The average value of objective function represented by (10.3) is minimal for the regularized SE in comparison to LU-SE.

Table 10.3 Regularization parameter for 550th Monte Carlo Sample of IEEE 14-bus system

Iteration	Condition number		2Λ
	E	**G**	
1st	4.0512×10^{12}	1.711×10^{17}	0.0572
2nd	4.4410×10^{12}	5.837×10^{8}	0.0485
3rd	4.3716×10^{12}	2.097×10^{9}	0.1718
4th	4.3710×10^{12}	2.291×10^{9}	0.1768

Table 10.4 The SE performance for Case I of IEEE 14-bus system

Performance index	LU-SE	Regularized SE
Maximum absolute error in $\|V\|$ (in pu, $\times 10^{-5}$)	5.0037	4.9287
Maximum absolute error in $\angle V$ (in radian, $\times 10^{-5}$)	3.9713	3.9023
Average max absolute error in $\|V\|$ (in pu, $\times 10^{-5}$)	2.1434	2.1259
Average absolute error in $\angle V$ (in radian, $\times 10^{-5}$)	1.5626	1.6320
Average filter effects	0.15605	0.159193
TSEV ($\times 10^{-5}$)	3.103274	3.124207
Average objective function J	92.1093	90.1511
Average condition number of **E** ($\times 10^{12}$)	4.3822	4.3822
Average time (in ms)	15.4185	53.7973
Average iteration counts	4.091	4.319

The accuracy of the SE is ensured by the proper selection of the regularization parameter. The condition number of the coefficient matrix is of the order of 10^{12} for IEEE 14-bus system, which indicates ill-conditioning. The regularization parameter selected to overcome the effect of ill-conditioning for providing the best estimate of the state change vector ranges from 0.0211 to 7.987×10^6 at the solution point for the Monte Carlo trials.

10.7.2 Case II for IEEE 14-bus test system

In an effort to generate a case of SE divergence, a lot of measurement meter placement strategies are tested for IEEE 14-bus system and one of those placements is listed for Case II in Table 10.5. The SE problem for this case was solvable by using the regularized method only.

Non-convergence issues are reported for 757 samples of measurement sets out of 1,000 Monte Carlo trials for LU-SE methods. The regularized SE provides the converged solution for all Monte Carlo trials in Case II. The SE output for this case with regularized SE is listed in Table 10.6.

Regularized method is able to solve the SE problem in case of numerically ill-conditioned problem, i.e., Case II. This divergence is due to the measurement meter placement. This issue is resolved by considering the regularized norm along with the residual norm using Tikhonov regularization. The tradeoff between the both norms is controlled by proper selection of λ using the L-curve method. For Case II, the voltage magnitude and angle estimated by Tikhonov regularization-based SE are plotted against the true values obtained from load flow studies and shown in Figures 10.5 and 10.6.

In case of non-convergence, the value of regularization parameter plays a major role to provide the converged solution by trading off between the solution norm and the residual norm. From Figures 10.5 and 10.6 for Case II, it is evident that the solution may not be very accurate, but a converged and reasonable one.

Table 10.5 Measurement meter placement of Case II for IEEE 14-bus system

Meter Type	Reference	Location
Injection	Bus	1, 2, 6, 7, 9, and 14
Flow	Line	1–2, 2–4, 4–5, 5–6, 6–12, 7–8, 10–11, 13–14

Table 10.6 The regularized SE performance for Case II of IEEE 14-bus system

Average filter effect	0.9227
TSEV	0.0383
Average objective function J	0.9577
Average time (in ms)	44
Average iteration counts	4.089
Average λ at solution point $\times 10^6$	6.6119

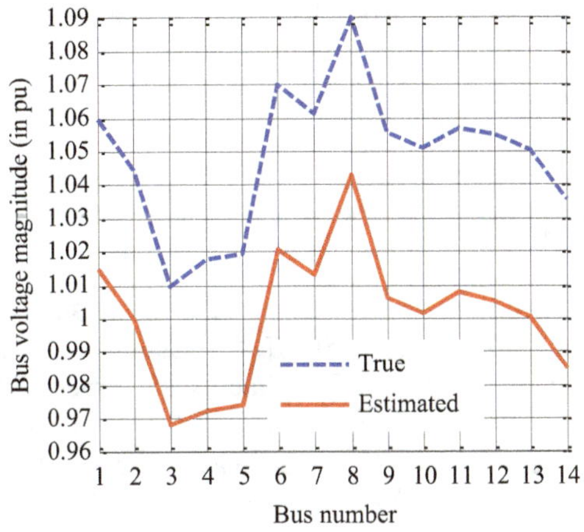

Figure 10.5 Bus voltage magnitude (in pu) for Case II of IEEE 14-bus system by the regularized SE

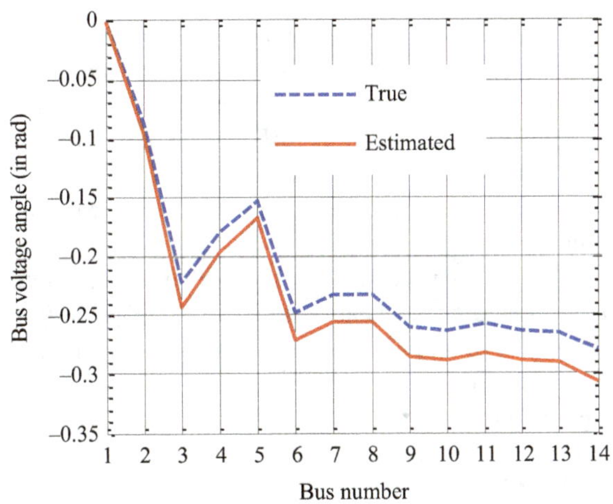

Figure 10.6 Bus voltage angle (in radian) for Case II of IEEE 14-bus system by the regularized SE

The regularized method is also tested for an ill-conditioned 13-bus test system [39]. The injection meters are placed at buses 1, 6, 9, and 12; flow meters are placed at the branch connecting buses 10 and 11 near bus 10. Zero injection measurements are treated as equality constraints. With the given measurement placement, divergence issues are reported for the LU-SE. Out of 1,000 Monte Carlo trials, 375 sets of measurement samples failed to converge for the LU-SE method. The Tikhonov regularization-based SE converged successfully for the same 1,000 trials used for the LU-SE. One important advantage of the regularization method is its ability to solve ill-conditioned problems. The regularized method has focused mainly on the numerical stability aspect via Tikhonov regularization, while keeping the solution time low and maintaining high accuracy by minimizing the SE objective function.

10.8 Conclusion

In this chapter, a single-phase, single-area, static SE is presented in presence of the PMUs along with the conventional SCADA measurements. In absence of the PMUs, the conventional SCADA measurements are processed through conventional SE to estimate the states of the system. With sufficient number of PMUs being placed in the network, the linear estimator is a good choice ignoring the conventional measurements. In the presence of the limited number of PMUs in the network, the sequential SE is preferred due to reuse of the existing SCADA software running at the control center. The hybrid SE has been extensively used by the researchers to exploit the advantages of the PMUs. However, the presence of the PMUs may cause ill-conditioning, which results in inaccuracy in the estimates and in extreme cases, it may diverge too. This chapter presents an analysis for the inclusion of PMU current phasor in different form with uniform distribution of measurement errors. From the results, it is concluded that the pseudo branch flow measurements derived from the PMU current measurements eliminates the high weights corresponding to the low values of current measurements, resulting in improvement of the convergence as well as the accuracy.

A number of methods are available to solve the numerical problem associated with the hybrid state estimation. The numerical problem mainly appears from an ill-conditioned information matrix, which may further escalate into divergence. A regularized SE is presented in this chapter to solve the ill-conditioned hybrid state estimation problems. The computational effort with the regularized SE is on the higher side, compared to the conventional methods. However, conventional methods often fail to converge due to numerical ill-conditioning. The SE algorithm in the control center may be equipped with this regularized method along with existing SE methods. Whenever some convergence issues are encountered, the SE program can switch to the regularized method to obtain a reasonable solution.

The SE algorithm may also be employed for transmission system as well as distribution system. Also, for a large power system network, the computational effort for this type of single-area SE may be high. To resolve this issue, the control area is usually divided into a number of sub-areas and each area has its own SE using local measurements. The estimates from these sub-areas are combined at the unified control center. This type of estimator is known as the multi-area state estimator (MASE). Many researchers have shown interest in analyzing the different aspects of the MASE [40–42].

To simplify the static SE process, the different types of measurements are processed at different levels. At the substation level, a local distributed linear estimation is performed and then the output of the first-level estimator, comprising state variables and associated variances, is passed to a centralized nonlinear estimator at the control center [43]. This two-level SE has the advantage in computation. Also, all measurements need not be transmitted to the control center and thus relieving the congestion in communication bandwidth.

References

[1] F. C. Schweppe and J. Wildes, "Power system static state estimation, Part I: exact model," *IEEE Transactions on Power Apparatus and Systems*, vol. PAS-89, no. 1, pp. 120–125, 1970.

[2] F. C. Schweppe and D. B. Rom, "Power system static state estimation, Part II: approximate model," *IEEE Transactions on Power Apparatus and Systems*, vol. PAS-89, no. 1, pp. 125–130, 1970.

[3] F. C. Schweppe, "Power system static state estimation, Part III: implementation," *IEEE Transactions on Power Apparatus and Systems*, vol. PAS-89, no. 1, pp. 130–135, 1970.

[4] D. Bailey and E. Wright, *Practical SCADA for Industry*, Elsevier, Amsterdam, 2003.

[5] A. G. Phadke and J. S. Thorp, *Synchronized Phasor Measurements and Their Applications*, Springer, Berlin, 2008.

[6] S. Chakrabarti, E. Kyriakides, T. Bi, D. Cai and V. Terzija, "Measurements get together," *IEEE Power and Energy Magazine*, vol. 7, no. 1, pp. 41–49, 2009.

[7] A. G. Phadke, "Synchronized phasor measurements a historical overview," *Proceedings of IEEE PES Asia Pacific Transmission and Distribution Conference and Exhibition*, vol. 1, pp. 476–479, 2002.

[8] IEEE standard for synchophasors for power systems. *IEEE Std. C37.118-2005*, 2005.

[9] IEEE standard for synchophasor measurements for power systems. *IEEE Std. C37.118.2-2011*, 2011.

[10] K. E. Martin, D. Hamai, M. G. Adamiak, *et al.*, "Exploring the IEEE Standard C37.118 2005 synchrophasors for power systems," *IEEE Transactions on Power Delivery*, vol. 23, no. 4, pp. 1805–1811, 2008.

[11] A. Abur and A. G. Exposito, *Power System State Estimation: Theory and Implementation*, Marcel Dekker, New York, 2004.

[12] A. Monticelli, *State Estimation in Electric Power Systems: Generalized Approach*, Kluwer: London, 1999.

[13] G. R. Krumpholz, K. A. Clements and P. W. Davis, "Power system observability—a practical algorithm using network topology," *IEEE Transactions on Power Apparatus and Systems*, vol. PAS-99, no. 4, pp. 1534–1542, 1980.

[14] R. R. Nucera and M. L. Gilles, "Observability analysis: a topological algorithm," *IEEE Transactions on Power Systems*, vol. 6, no. 2, pp. 466–475, 1991.

[15] B. Gao and A. Abur, "A direct numerical method for observability analysis," *IEEE Transactions on Power Systems*, vol. 15, no. 2, pp. 625–630, 2000.

[16] B. Gou, "Jacobian matrix based observability analysis for state estimation," *IEEE Transactions on Power Systems*, vol. 21, no. 1, pp. 348–356, 2006.

[17] Q. Ding and V. A. Emesih, "A simple factorization based observability analysis and meter placement method," *International Journal of Electrical Power & Energy Systems*, vol. 29, pp. 729–737, 2007.

[18] M. C. de Almeida, E. N. Asada and A. V. Garcia, "Power system observability analysis based on gram matrix and minimum norm solution," *IEEE Transactions on Power Systems*, vol. 23, no. 4, pp. 1611–1618, 2008.

[19] C. Solares, A. J. Conejo, E. Castillo and R. E. Pruneda, "Binary arithmetic approach to observability checking in state estimation," *IET Generation, Transmission and Distribution*, vol. 3, no. 4, pp. 1336–1345, 2009.

[20] R. E. Pruneda, C. Solares, A. J. Conejo and E. Castillo, "An efficient algebraic approach to observability analysis in state estimation," *Electric Power Systems Research*, vol. 80, pp. 277–286, 2010.

[21] ISO-IEC-OIML-BIPM: Guide to the Expression of Uncertainty in the Measurement, 1992.

[22] T. S. Bi, X. H. Qin and Q. X. Yang, "A novel hybrid state estimator for including synchronized measurements," *Electric Power Systems Research*, vol. 78, no. 8, pp. 1343–1352, 2008.

[23] JCGM, "Evaluation of measurement data—Supplement 1 to the 'Guide to the expression of uncertainty in measurement'—Propagation of distributions using a Monte Carlo method," *Joint Committee for Guides in Metrology*, 2006.

[24] J. W. Gu, K. A. Clements, G. R. Krumpholz and P. W. Davis, "The solution of ill-conditioned power system state estimation problems via the method of Peters and Wilkinson," *IEEE Transactions on Power Apparatus and Systems*, vol. PAS-102, no. 10, pp. 3473–3480, 1983.

[25] R. Ebrahimian and R. Baldick, "State estimator condition number analysis," *IEEE Transactions on Power Systems*, vol. 16, no. 2, pp. 273–279, 2001.

[26] A. Simoes-Costa and V. H. Quintana, "A robust numerical technique for power system state estimation," *IEEE Transactions on Power Apparatus and Systems*, vol. PAS-100, no. 2, pp. 691–698, 1981.

[27] C. E. Lin and S. J. Huang, "Integral state estimation for well-conditioned and ill-conditioned power systems," *Electric Power Systems Research*, vol. 12, pp. 219–226, 1987.

[28] G. N. Korres, "A robust algorithm for power system state estimation with equality constraints," *IEEE Transactions on Power Systems*, vol. 25, no. 3, pp. 1531–1541, 2010.

[29] S. K. Mallik, S. Chakrabarti and S. N. Singh, "A robust regularized hybrid state estimator for power systems," *Electric Power Components and Systems*, vol. 42, no. 7, pp. 671–681, 2014.

[30] M. Zhou, V. Centeno, J. S. Thorp and A. G. Phadke, "An alternative for including phasor measurements in state estimators," *IEEE Transactions on Power Systems*, vol. 21, no. 4, pp. 1930–1937, 2006.

[31] S. Chakrabarti, E. Kyriakides, G. Ledwich and A. Ghosh, "Inclusion of PMU current phasor measurements in a power system state estimator," *IET Generation, Transmission and Distribution*, vol. 4, no. 10, pp. 1104–1115, 2010.

[32] S. K. Mallik, S. Chakrabarti and S. N. Singh, "Improving the convergence characteristic of hybrid state estimation using pseudo measurement," *17th Power Systems Computation Conference (PSCC)*, Sweden, August 2011.

[33] V. Agarwal, "Total variation regularization and L-curve method for the selection of regularization parameter," ECE 599, University of Tennessee, summer 2003.

[34] P. C. Hansen, "The L-curve and its use in the numerical treatment of inverse problems," *in Computational inverse problems in electro cardiology*, P Johnston ed. WIT Press: Moscow, 2001, pp 119–142.

[35] B. Lewis and L. Reichel, "Arnoldi-Tikhonov regularization methods," *Journal of Computational and Applied Mathematics*, vol. 226, pp. 92–102, 2009.

[36] P. C. Hansen and D. P. O'Leary, "The use of the L-curve in the regularization of discrete ill-posed problems," *SIAM Journal of Scientific & Statistical Computing*, vol. 14, no. 6, pp. 1487–1503, 1993.

[37] http://labs.ece.uw.edu/pstca/pf14/pg_tca14bus.htm

[38] S. K. Mallik, S. Handa, S. Chakrabarti and S. N. Singh, "Performance study of a regularized method for solving non-converging power system state estimation problems," in *Proceedings of 5th International Conference on Power and Energy Systems*, Kathmandu, Nepal, October 2013.

[39] N. D. Rao and S. C. Tripathy, "Power system static state estimation by the Levenberg-Marquardt algorithm," *IEEE Transactions on Power Apparatus and Systems*, vol. PAS-99, no. 2, pp. 695–702, 1980.

[40] T. Van Cutsem, J. L. Horward and M. Ribbens-Pavella, "A two-level static state estimator for electric power systems," *IEEE Transactions on Power Apparatus and Systems*, vol. PAS-100, no. 8, pp. 3722–3732, 1981.

[41] L. Zhao and A. Abur, "Multiarea state estimation using synchronized phasor measurements," *IEEE Transactions on Power Systems*, vol. 20, no. 2, pp. 611–617, 2005.

[42] G. N. Korres, "A distributed multiarea state estimation," *IEEE Transactions on Power Systems*, vol. 26, no. 1, pp. 73–84, 2011.

[43] A. G. Exposito and A. de la Villa Jaen, "Two-level state estimation with local measurement pre-processing," *IEEE Transactions on Power Delivery*, vol. 24, no. 2, pp. 676–684, 2009.

Chapter 11
PMU-based wide-area security assessment

Subhransu Ranjan Samantaray[1], *Innocent Kamwa*[2] *and Geza Joos*[3]

11.1 Introduction

Dynamic security assessment (DSA) of a power system usually refers to the degree to which a particular system condition can withstand all credible contingencies considering the detailed dynamic characteristic of the system. Due to ever-increasing power demand exceeding the transmission and generation facilities, power systems are forced to operate with narrower margins of security. Security is defined as the capability of guaranteeing the continuous operation of a power system under normal operation even following some significant perturbations. As security is a major concern in power system, a fast and reliable security assessment is necessary. DSA can deal with transient stability problems and/or voltage stability problems that, respectively, require transient stability assessment and voltage stability assessment. Thus, predicting early warning of the deteriorating system conditions is one of the important requirements of the security assessment, which enables the system operator to take corrective actions. The early warning systems should provide more diagnostic tools than the existing one which can allow effective use of automatic controls for self-correction including automatic switching or power flow control.

The defense plan against these rare contingencies is based on event detection [1] using breaker status and fault signals from relays in combination. This is because the more appealing response-based approach [2] is not fast enough for effective remedial actions. However, with recent developments in phasor measurement units (PMUs)-based wide-area measurement (WAM), fast-response-based stability assessment of extreme contingencies finds better prospects. This includes ranking the contingency (severity/stability) using online dynamic information which is potentially generalized and robust compared to the existing event-detection based schemes alone relying heavily on off-line simulations of system conditions. The lack of transmission system expansion means that, in any case,

[1]School of Electrical and Computer Sciences, Indian Institute of Technology Bhubaneswar, India
[2]Faculty of Science and Engineering, Department of Electrical and Computer Engineering, Laval University, Canada
[3]Department of Electrical and Computer Engineering, McGill University, Canada

current defense plans will reach their limit at a time when it may be appropriate to supplement them with more refined and context-sensitive wide-area response-based remedial actions or special protection systems.

Recently, a comprehensive time-frequency-based approach for contingency severity ranking and rapid stability assessment [3] has been proposed and the aim was to assist the classical off-line simulations-based DSA task [4–12] by correctly classifying all single or multiple contingencies those may result in a loss of stability in the first 20 s following the fault-clearing. Here, the number of strategic monitoring buses is selected where the PMUs are placed to capture the representative wide-area voltage magnitudes and angles during real-time operation [13,14]. The short-time Fourier transform (STFT) [15,16] is then dynamically applied to the responses for extracting selected decision features as the post-disturbance time frame undergoes significant changes. It is observed that the frequency-domain features such as the peak spectral density of the angle, the frequency, and their dot product evaluated over the grid areas (referred to the system center of inertia) are reliable time-varying stability indicators that can form the basis of an entirely reliable classifier.

In the wide-area framework, there is an abundance of historical and real-time quality data generally collected through the PMU infrastructure [17]. Thus, in a data-centric smart-grid, converting data to information and corresponding controls is a fundamental step triggering the development of new predictive analytical tools to predict the status of catastrophes following disturbances, or to quantify the relative status of the system from the edge of a safe operating boundary [12]. However, developing predictive models from historical or simulated PMU records is no small task because power system responses are highly nonlinear, and therefore very context-dependent. Although data mining technique is one of the emerging tools to perform the task of synthesizing such massive data bases into predictive models, however they need thousands of configurations and instances to minimally capture the intrinsic characteristics of the underlying power grid dynamics [18].

Among the data-mining tools, decision tree (DT) has gained lots of momentum in power engineering applications. Most high-accuracy DTs are opaque (they contain a large number of deep terminal nodes) and it is not possible for a human to understand the logic behind the prediction which is mandatory for stability control and power system protection. This is to note that if simpler and transparent models are built, the accuracy is often sacrificed, resulting in the accuracy vs. transparency tradeoff. In this context, two competing data-mining trends are emerging: (1) the fuzzy-logic rules-based approach [19–22], which has the advantage of transparency and interpretability [23], and (2) the machine learning-based black-box approach which relies on several complementary tools from statistical learning such as neural network [3,10,24], Support Vector Machine (SVM) [25,26], DT [2,12,27,28], and Ensemble DTs (random forests) [18,29]. All of the aforementioned data-mining tools have strengths and weaknesses in terms of computational burden, training time, accuracy, robustness to noise, etc. Normally, they are selected purely on the basis of their immediate availability to the user or prior familiarity with them.

Thus, tradeoff between the choice of a powerful model with high performance (i.e., a model with high accuracy) and a transparent model (i.e., a model with high comprehensibility) needs to be discussed. To address this tradeoff in the context of PMU data analysis for wide-area stability assessment and control, we will consider three black box models [30] (NNET, SVM, RF) and two transparent (comprehensible models) [31] (DT, DT_Fuzzy) for catastrophe predictor. For the high-performance black-box models, the RF has been found to be the most accurate model, while SVM is the second [30]. Regarding the transparent model family, since the DT_ Fuzzy combines the strength of DT and Fuzzy-logic to form a systematic Fuzzy logic and is emerging as the most transparent predictive model.

11.2 System studied using wide-area monitoring

The Hydro-Québec 783-bus system using PMU base wide-area monitoring shown in Figure 11.1 is considered for operations planning [32]. For this study, winter and summer operation planning models are used with about 1,000 load flow patterns generated by the transfer limit search and critical clearing time search processes based on 30 carefully chosen 735-kV contingencies.

The Hydro-Québec network is divided into a number of electrically coherent areas [33] associated with weak inter-tie lines or systematically identified stability boundaries [34]. In the shown representation, these inter-tie lines are the primary clusters of the system, defining the finite number of possible ways to split the system into unconnected islands with embedded generation. Further, selecting an optimum PMU set allocated for monitoring using a sequential addition algorithm to expand their number, while maximizing the amount of information added by each new PMU [13]. In this study, the buses with PMU are only considered for monitoring the wide-area post-disturbance records. The system is configured in such a way that the simulation will be smoothly replaced by actual measurements in real-time situation considering the same PMU-based wide-area monitoring system.

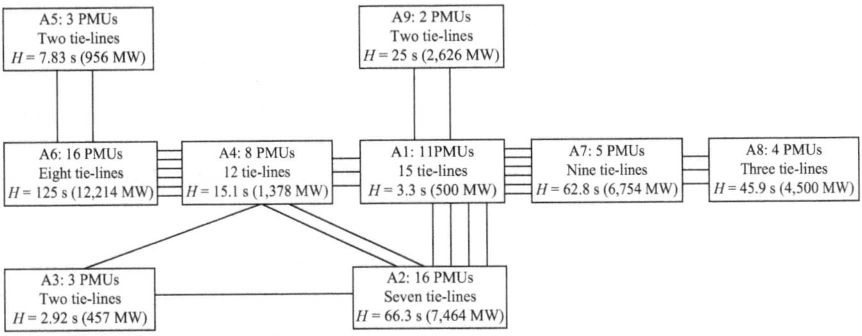

Figure 11.1 Hydro-Québec system (783-bus) with intra-area and inter-tie PMUs distributed over nine electrically coherent areas (H-Inertia and MW-generation)

11.3 Wide-area severity indices

11.3.1 WASI features

The wide-area severity indices (WASI) are defined by an equivalent inertia associated with each area, representing the total inertia of the generation located in that area. It is reasonable to consider its behavior to that of a single large machine with the same inertia and generation assuming each area is coherent following a disturbance. Even though this assumption is not perfect, it offers a straightforward means of deriving the center of inertia (COI), which is very useful information for tracking the stability of interconnected areas. In real time, a defense plan or an SPS [6,32,35] could readily derive these inertia constants through low-speed communication with the control center state estimator, which holds the actual load and generation dispatch.

The voltage and COI angle deviation are the time-domain features considered in this study which are the key descriptors of the power grid stability state. To determine the level of system stress induced by a contingency, the most obvious criterion is the transient stability after a single or multiple swings of the generator angles. The proposed study considers the system as unstable if there is a loss of synchronism between any of the areas monitored earlier in the prescribed observational time frame, i.e., the angle shift exceeds 180°. When an early discovery of impeding instability is sought, this criterion is useless because it offers little pre-emption time. The COI angle deviation with respect to the pre-fault value is computed for each area and the maximum value of these differences is found to be a good measure of the topological stress induced by the contingency.

It is observed that the frequency-domain features such as the peak spectral density of the product of angle and frequency evaluated over the grid areas and referred to the system COI are reliable time-varying stability indicators that can form the basis for assessing contingency severity [3]. Here, the WASI concept is extended to rapid stability assessment by using simultaneously a short (16-cycle) and long (192-cycle) STFT analysis window [15,16], which is respond fast enough for the first swing instability as well as the post-inertial and mid-term dynamic instability [10]. Further, multi-signal state-space based modal analysis [36] of the output of the long filters is shown to be highly useful for assessing the post-contingency damping condition of the grid [37].

Following a fault (which is automatically detected by the PMUs), the computation sequence is as follows:

1. Compute the pilot phasor and frequency of each area by averaging the within-area measurements: $\bar{V}_i$, $\bar{\omega}_i$
2. Compute the system COI of the angle and frequency variables using the available area inertias (M_i)
3. Project the pilot angle and frequency of step 1) into the COI reference of step 2): $\bar{\theta}_i^{COI}$, $\bar{\omega}_i^{COI}$

4. Compute the shift from the pre-fault to post-fault COI-angle for each area: *PostFltAngle*
5. Compute the power spectrum of the dot-product of energy and angle to obtain a tracking severity index by area.

From the above algorithm, wide-area severity indices in time and frequency domains are defined as follows [22]:

V min *Ts* System-wise minimum voltage over the time span of $T = 1$ s or $T = 2$ s after fault-clearing
V min *Ts J* Area-wise minimum voltage over the time span $T = 1$ s or $T = 2$ s following fault-clearing, considering only the buses in area *j*
V min *TsR* Area-wise minimum voltage over the time span of $T = 1$ s or $T = 2$ s after fault-clearing, considering only the buses in the faulted area
FastWASI(*Ts*) System-wise frequency-domain severity index defined over the time span of $T = 1$ s or $T = 2$ s after fault-clearing
FastWASI(*Ts*) *j* Area-wide frequency-domain severity index for area *j*, defined over the time span of $T = 1$ s or $T = 2$ s after fault-clearing
FastWASITsR Area-wise frequency-domain severity index for the faulted area, defined over the time span of $T = 1$ s or $T = 2$ s after fault-clearing
VLowPass2s j, *VCriterion2s j* Filtered area-wise minimum voltage
VLowPass2s, *VCriterion2s* Filtered system-wise minimum voltage
TDEF Fault duration
PostFltAngle System-wise maximum COI angle deviation from steady-state to fault-clearing time
PostFltAngle j Area-wide maximum COI angle deviation from steady-state to fault-clearing time, for area *j*
*TsimSt*600 Duration of simulation up to normal end or loss of synchronism, whichever occurs first

The schematic diagram for computing frequency-domain severity indices is shown in Figure 11.2. It is observed that for transient-stability assessment, short window lengths are the most suitable. The 16-point Hamming window is used in this study.

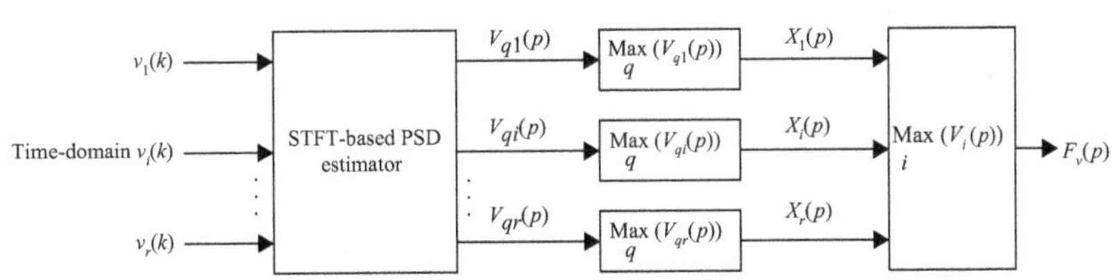

Figure 11.2 Schematic of computing frequency-domain-based WASI feature (number of areas $r = 1, \ldots, 9$)

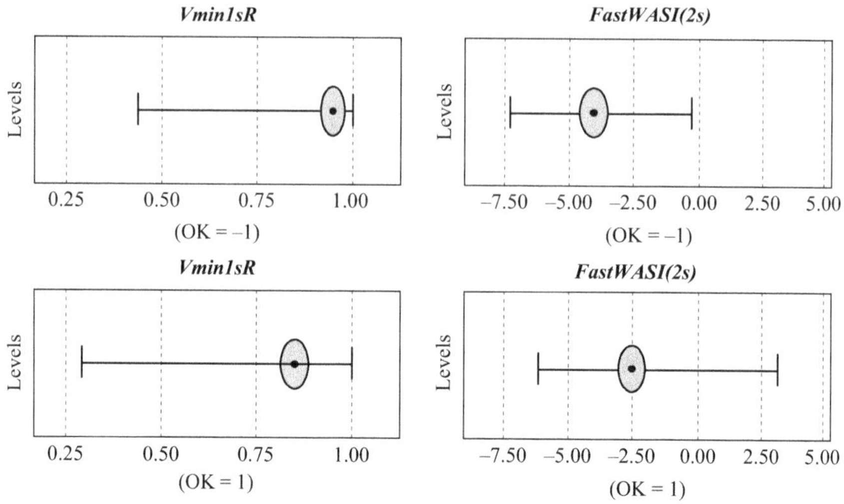

Figure 11.3 Vmin1sR and FastWASI (2s) for stability limits

11.3.2 Wide-area severity indices and stability condition

The relationship between WSAI features and stability conditions is depicted in Figure 11.3. It shows that the stability limits with respect to the voltage-based *Vmin1sR* (which is the 1s voltage minimum over the faulted area) and the energy-based *FastWASI (2s)* (the system-wise maximum value of the dot-product PSD). These plots are categorized according to the stability condition of the case as: OK = –1 (stable) and OK = 1 (unstable) and defines 50% of the whole sample. The left limit of the box defines the first quartile of the data, while the right limit corresponds to the third quartile. It is observed that $Vmin1sR > 0.91$ includes 75% of stable cases and $Vmin1sR < 0.90$ includes 75% of unstable cases. Further, the relationship $FastWASI2s < -3.2$ includes 75% of stable cases, while $FastWASI2s > -3$ includes 75% unstable cases. Thus, *FastWASI2s* is able to split the stability domain into two largely disjointed sets.

11.4 Data-mining models

11.4.1 Decision tree

DT [38,39] is an effective supervised data-mining tool for solving the classification problem in high dimension. Each internal node in the tree tests the value of a predictor, while each branch of the tree represents the outcome of a test. The terminating nodes, also referred to as leaf nodes, represent a classification. The number of predictors used in the classification problem indicates the dimension of the problem. Associated with each decision (leaf) of the tree is the confidence of the decision. This is simply a measure of the ratio of the particular class to all the classes present in the dataset for that node.

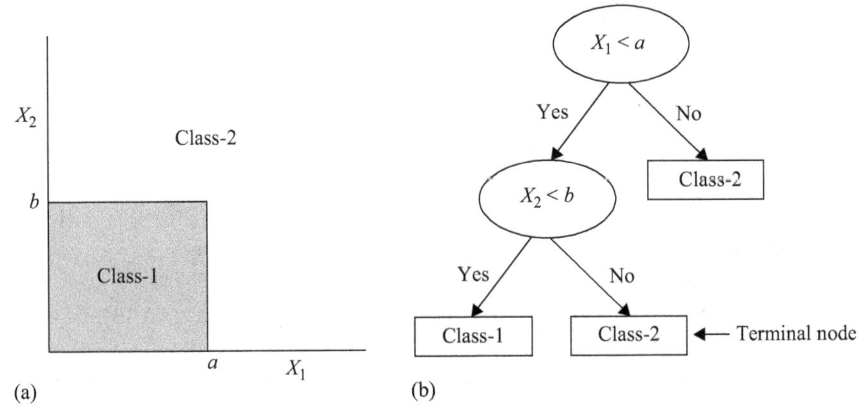

Figure 11.4 (a) Two-dimensional binary classification and (b) DT for binary classification

Any binary DT T is a collection of nested binary partitions, denoted here by a quintuple (u, v, s, p^L, q^R), represented in the following recursive form [47]:

$$T = \{(\mathrm{u}, \ \mathrm{v}, \ \mathrm{s}, \ \mathrm{p}^{\mathrm{X}}, \ \mathrm{p}^{\mathrm{Y}}), \ T^X, \ T^Y\} \tag{11.1}$$

where u denotes a decision node label for the partition, v is the feature axis, s is the threshold value used for the partition, and p^X and p^Y are the node labels for the partition of the left and right sets, respectively. T^X and T^Y denote the subtrees defined on the left and right sets of a partition, respectively. In summary, (11.1) defines the DT T in terms of the pattern lattice L created by partitioning the features plane. The equation states that the lattice L can be binary-partitioned on the feature axis v into mutually exclusive left and right sets. The left set includes lattice elements with feature v values smaller than the threshold value while the right set includes lattice elements with feature v values exceeding the threshold value. The example of two-dimensional binary classification is shown in Figure 11.4(a) and the corresponding DT is shown in Figure 11.4(b).

A common measure of impurity of *the predictive model T is the entropy, which is defined* for the subset T_k of all possible states of the prediction model T by the following equation [22,40]:

$$i(k) = -\sum_S p(S|T_k)\log_2 p(S|T_k) \tag{11.2}$$

The information carried by the prediction model T with K subset data models is then the weighted average of the entropies given by (11.2). Thus, the impurity of the prediction model $T$is given by:

$$I(T) = \sum_{k=1}^{K} i(T_k)p(T_k) \tag{11.3}$$

Assessment of the entropy reduction ascribed to each subset model T_k provides the importance ranking of independent features toward the model total impurity.

11.4.2 DT-induced Fuzzy approach (DT_Fuzzy)

The sharp boundaries used by the crisp DT will cease to exist [41,42] when the Fuzzy theory is applied to the individual tree branches and the partitions become a series of Fuzzy regions. The objective is to relax these sharp decision boundaries by introducing fuzzification onto the DT. This involves in introducing a pair of membership functions at each decision node and the decision space will then become fragmented with a number of overlapping Fuzzy regions [43,44]. In order to classify a particular case, all branches will fire to some degree. The degree of membership at each branch for a given attribute value depends upon the margin of the Fuzzy set for that particular branch. A Fuzzy region around any decision node in a crisp DT is defined using a pair of complementary membership functions C_1 and C_2 around a decision threshold x (Figure 11.5). For example, a five-node tree would be represented by ten membership functions. Figure 11.5 illustrates two complementary membership functions over the domain $p \dots q$ of attribute i, where x_i represents the branching threshold of attribute i. A specific value m of attribute i, passing through the tree would then be assigned a membership grade based on its proximity to x_i. Each membership function will have an associated domain (p_i, q_i) whose scope is determined by the attribute at each specific branch, which is defined as [45]

$$p_i = x_i - n_j\sigma_i \tag{11.4}$$

$$q_i = x_i + n_j\sigma_i \tag{11.5}$$

where σ_i is the standard deviation of attribute i, n is a real number, $n \rightarrow [0.0, \infty]$ used to determine the effect of σ_i on the membership function domain, and p, q are the lower and upper bounds of membership function, respectively. In practice, "n"

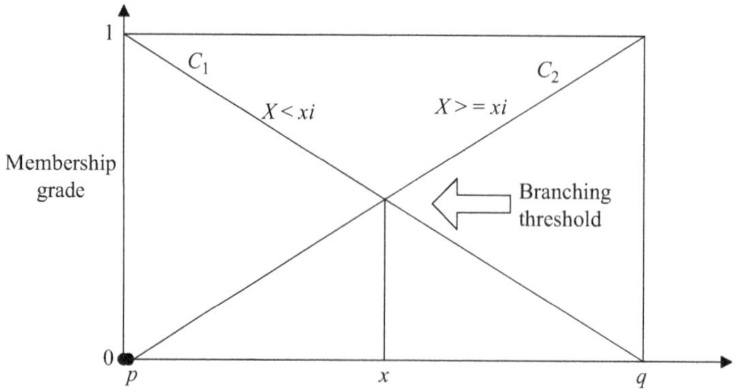

Figure 11.5 DT transformation to Fuzzy MFs

will remain typically small, $n \to [0.0, 5]$. This is because the large value of n would introduce too much fuzzification onto the tree and the process of making the decision would become too vague.

The DT is transformed into a Fuzzy rule base by developing the Fuzzy membership functions from the partition boundaries of the DT. After a pair of complementary fuzzy MFs has been developed at each node, the fuzzy IF-THEN rules are framed considering each node and leaf of the DT for final decision-making. The crisp output is obtained after a defuzzyfication process [23].

11.4.3 Ensemble decision trees

Ensemble DTs [38] (also known as Random Forests: RF) are a large combination of de-correlated tree predictors such that each tree depends on the values of a random vector sampled independently. Individual trees are noisy and unstable compared to the ensemble DTs as it grows to sufficiently deep level having relatively low bias. Thus, Ensemble DTs are the ideal candidates for ensemble growing as they can capture complex interactions, while fully benefit from aggregation-based variance reduction. Using a random selection of features to split each node and re-sampling (with replacement), the training set to grow each tree yields error rates that are de-correlated and more robust with respect to noise. The generalization error for forests converges as to a limit as the number of trees in the forest becomes large.

The ensemble tree growing procedure [18] is that for the kth tree ($k \le n_{tree}$, the number of trees in the ensemble) a random vector Φ_k is generated, independent of the past random vectors $\Phi_k, \ldots, \Phi_{k-1}$ but with the same distribution. A single tree is grown using the training set $\mathbb{S}$ and the set of attributes in Φ_k, resulting in a classifier $T_k(x, \Phi_k)$ where x is an input vector. In random split selection, Φ consists of a number n_{try} of independent random integers, where $n_{try} < n_a$, the number of attributes in $\mathbb{S}$. A random forest (RF) consists of a collection of tree-structured classifiers $\{T_k(x, \Phi_k),\ k = 1, \ldots, n_{tree}\}$, where $\{\Phi_k\}$ are independent identically distributed random vectors and each tree casts a unit vote for the most popular class at input x.

The algorithmic view of the RF growing process is summarized below:

1. For k=1 to n_{tree}:
   (a) Draw a bootstrap sample $\mathbb{S}^*$ of size N from the training data $\mathbb{S}$(which contains $M > N$ samples)
   (b) Grow a random forest tree $T_k(x, \Phi_k)$ to the bootstrapped data, by recursively repeating the steps below for each terminal noted of the tree, until the no other split is possible (unpruned tree of maximal depth):
      (i) Select n_{try} variables from the n_aWASI features
      (ii) Pick the best variable/split-point among the n_{try}
      (iii) Split the node into two daughter nodes
2. Output the ensemble of trees $\{T_k(x, \Phi_k),\ k = 1, \ldots, n_{tree}\}$.

Although RF is a relatively young data mining tool, scholars [31,46] have started recognizing its strengths: (i) it is simple and easy to use, (ii) very high accuracy, (iii) its relatively robust to outliers and noise, (iv) it gives useful internal estimates of error, strength, correlation, (v) not over fitting if selecting large number of trees, and (vi) insensitive to choice of split.

11.5 Data-mining model-based catastrophe predictors

11.5.1 Scenarios and data count generation

The models studied in the proposed approach include combined Hydro-Québec (HQ) and test as shown in Figure 11.6. The data sets studied with different scenarios are trained with extended data files and tested with performance data files. The extended data files are generated after extending the actual data sets triplicating the unstable cases in each scenario. The complete statistics of the scenarios studied are summarized in Table 11.1.

11.5.2 Data-mining models for catastrophe predictor

The catastrophe predictors proposed in this study use open-source software R [47] which includes the implementation of conventional DTs, RFs, Neural Net (NNET) and SVM. The training is carried out using the extended data files separately for each scenario and the testing is performed using the performance files (flat files). The data-mining models were configured to randomly select 70% only of the assumed data file to build the model during training. DT_Fuzzy is developed form the decision nodes of the DT. The NNET-based predictor has a feed-forward structure with ten hidden layers. Similarly, the SVM is set with a radial basis kernel. The DT is trained with the "R" package, which is called with the following settings: convergence error of 10^{-4}, with a minimum bucket, before and after a node split, equal to 600 and 300 samples, respectively.

11.5.3 Performance assessment

The performance assessment is carried out using various statistical indices defined as follows [3,8].

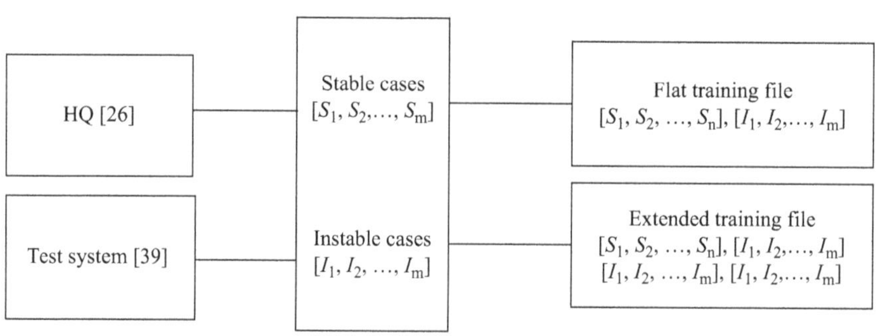

Figure 11.6 Schematic of the data-count generation process

Table 11.1 Summary count of the data sets used in the study

Model	Extended data file	Flat data file
	System Wise	
HQ_1s	93425	55196
Stable	42453	42453
HQ_2s	93425	55196
Stable	42453	42453
	Area_wise_1s	
BJN_1s	16434	10647
Stable	8718	8718
BJS_1s	24725	16766
Stable	14113	14113
CHU_1s	12440	8684
Stable	7432	7432
MQ_1s	39826	19099
Stable	12190	12190
	Area_Wise_2s	
BJN_2s	16434	10647
Stable	4680	8718
BJS_2s	24725	16766
Stable	14113	14113
CHU_2s	12440	8684
Stable	7432	7432
MQ_2s	39826	19099
Stable	12190	12190

(i) Reliability: (Total number of unstable cases – total number of cases converted to stable cases)/total number of unstable cases

(ii) Security: (Total number of stable cases – total number of cases converted to unstable cases)/total number of stable cases

(iii) Accuracy: (Total numbers of cases – number of mis-classification)/total number of cases

11.5.3.1 Decision tree

The study considers 70% of the data for training, the rest unseen data being reserved for testing. The simulation carried out on MATLAB®/SIMULINK® platform and the training took less than 22 s on a 2-GHz Centrino duo Laptop. The DT [48–51] generated during the training process for 1s decision time is shown in Figure 11.7. While observing the relative importance of the candidate features which played a role in the classification process, it is found that the V min 1sR is the most important feature based on total entropy reduction. Similarly, for other features, the importance of candidate features follows the descending order. Similar observations are made for the DT generated for 2-s decision time with same training strategy, where *FastWASI2s* turned out to be the most relevant feature. An attempt was made to improve these results by developing separate classifiers according to the geographic area where the fault occurred. This led to four different

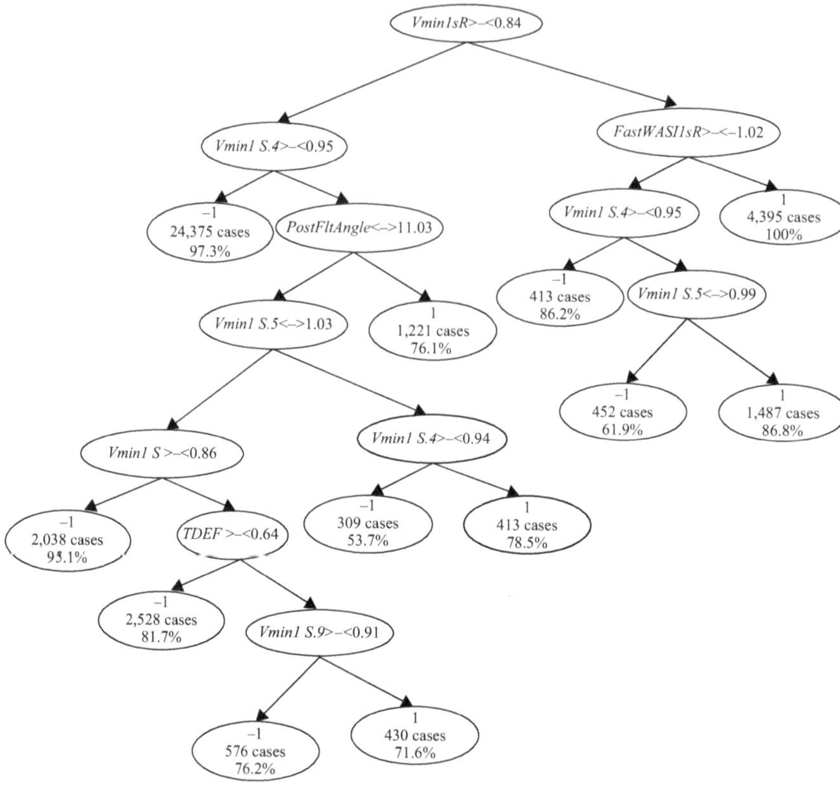

Figure 11.7 DT generated with training sample of 70% the complete dataset for 1-s decision time

area-wise DTs as follows, each for $T = 1$ s and $T = 2$ s:BJN: faults in the northern James Bay area (areas 5 and 6 in Figure 11.1), BJS: faults in the southern James Bay (area 4), CHU: faults in the Churchill Falls area (area 8), MQ: faults in the Manic-Québec area (area 7).

The performance results for flat data file with single DT are shown in Figure 11.8. It is observed that the reliability for HQ_1s and HQ_2s is 78.46% and 76.70%, respectively, while the accuracy and security are above 92.0% and 96.0%, respectively as shown in Figure 11.8(a). However, the reliability is more than 85 % in BJN, BJS, and CHU in cases of 1-s and 2-s early termination for area-wise assessment as shown in Figure 11.8(b) and (c), respectively. It is found that reliability for MQ is less than 80% considering 1-s and 2-s early termination. Thus, the reliability stays at low compared to security and accuracy for flat data sets. To observe the improvement further, the single DT has been trained and tested with extended data file. Figure 11.9(a) shows the results for extended data sets with single DT. It is observed that the reliability improves to 89.90% and 93.78% for HQ_1s and HQ_2s, respectively. Though there is a little sacrifice in security

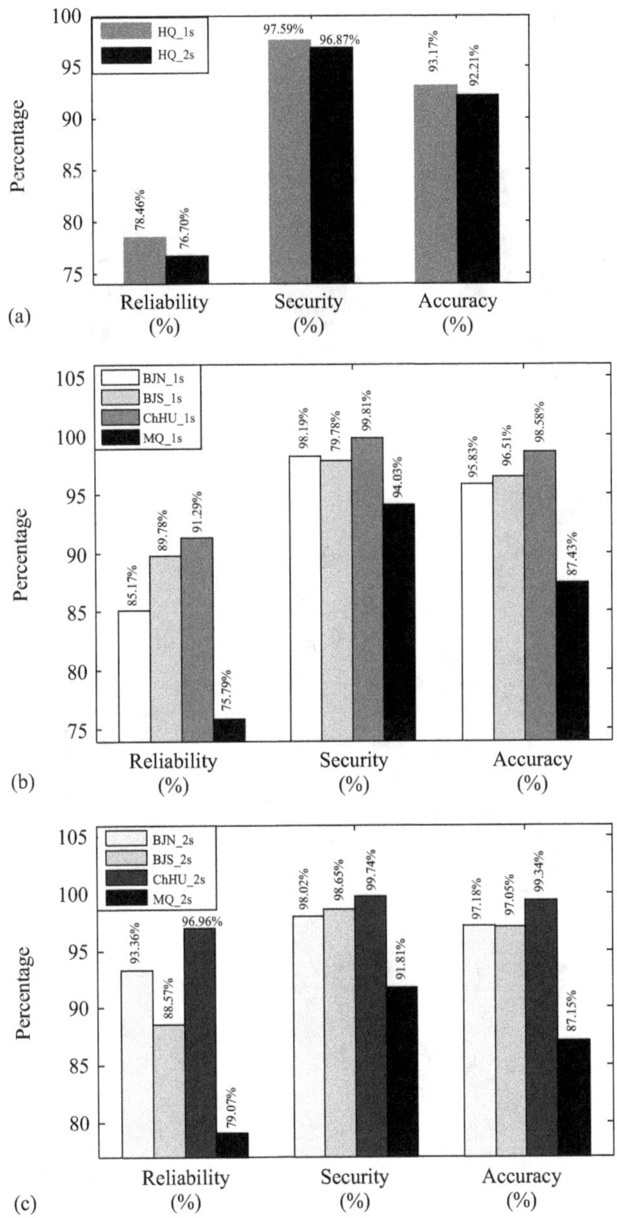

Figure 11.8 *(a) Performance of DT for HQ_1s and HQ_scenarios, (b) performance of DT for area-wise scenarios for 1 s (flat data), (c) performance of DT for area-wise scenarios for 2 s (flat data)*

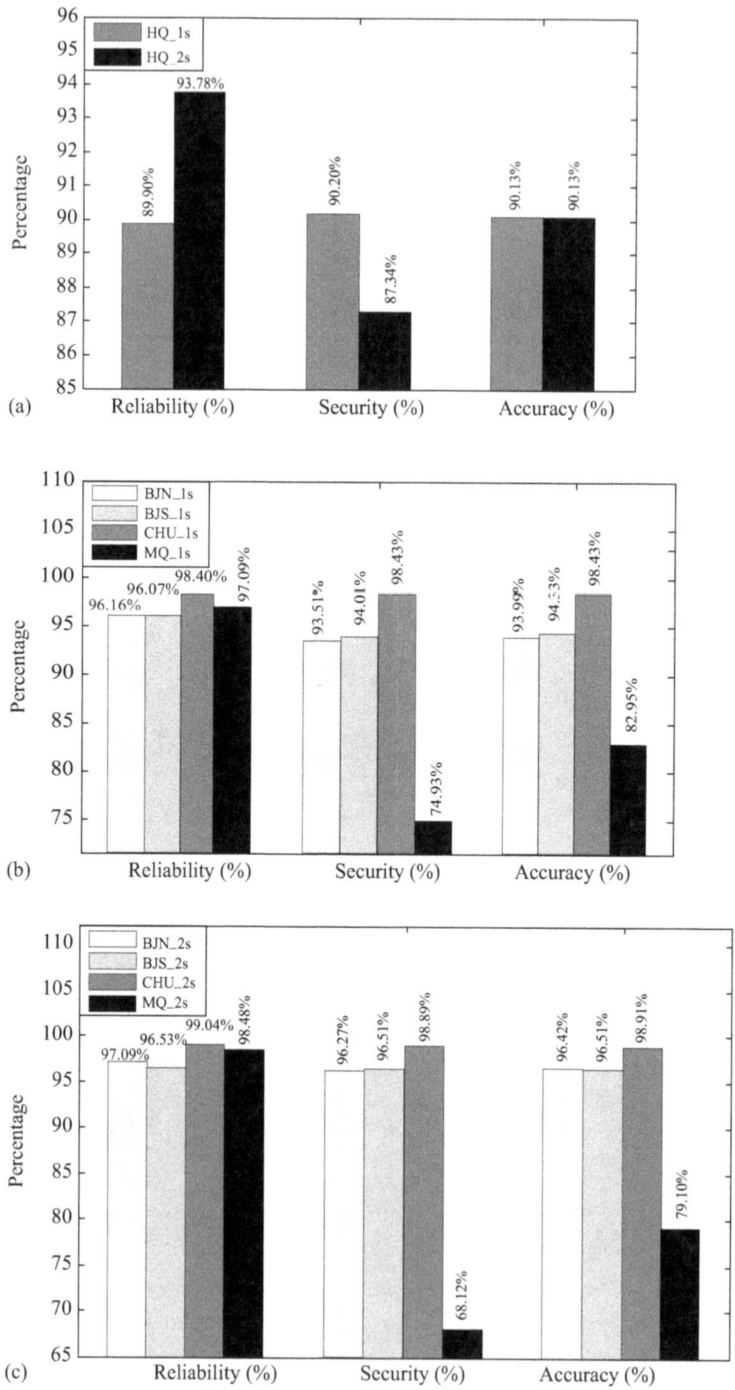

Figure 11.9 *(a) Performance of DT for HQ_1s and HQ_scenarios, (b) performance of DT for area-wise scenarios for 1 s, (c) performance of DT for area-wise scenarios for 2 s (extended data)*

(90% from 97%) and accuracy (90% from 93%), reliability is the important measure compared to security and accuracy in stability assessment. There is also a substantial jump in reliability in case of BJN, BJS, CHU, and MQ crossing 96% considering both 1-s and 2-s early termination as shown in Figure 11.9(b) and (c).

11.5.3.2 Ensemble decision trees (random forests)

The classification accuracy mostly depends on the quality of attributes describing the security concept. The correlation between the decision (OK) and the attributes for HQ_1s resulted from RF during training is shown in Figure 11.10. Highly correlated variables are close together and presented in the same color: frequency domain features on one side and voltage features on the other side, with *TDEF* in the middle. We interpret the degree of any correlation by both the shape and color of the graphic elements [52]. Any variable is perfectly correlated with itself, and this is reflected as the diagonal lies on the graphic. The graphic element is a perfect circle for the cases where there is no correlation between the variables (as is the

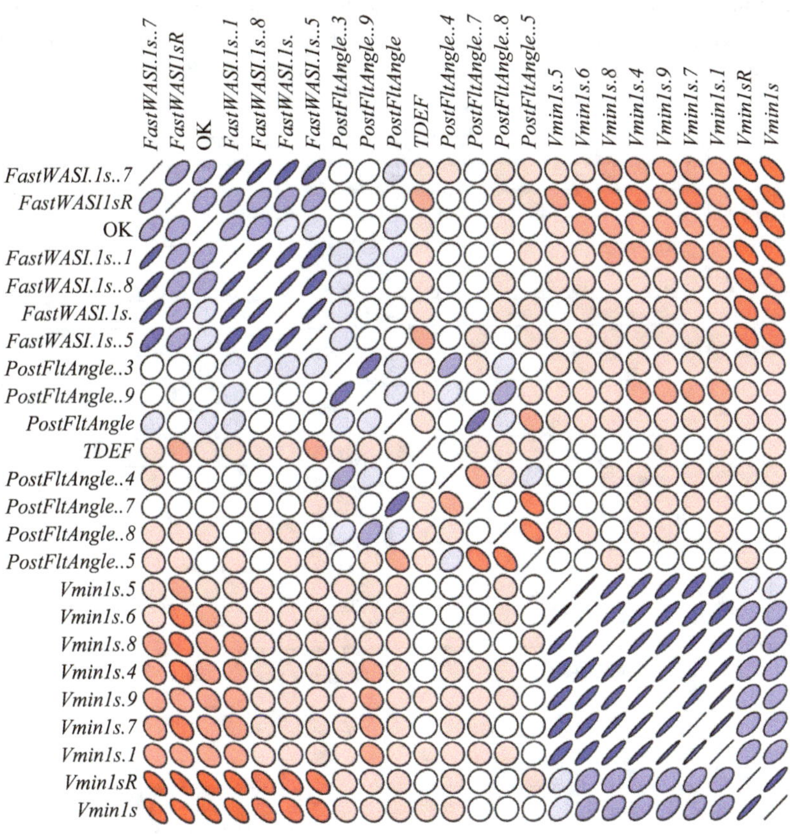

Figure 11.10 Correlation between candidate features. Reprinted, with permission, from [54].

case in the correlation between *PostFaultAngle.3*, *PostFaultAngle.9*, and *PostFaultAngle.7*).

The colors used to shade the circles give another clue to the strength of the correlation. The intensity of the color is maximal for a perfect correlation and minimal (white) if there is no correlation. Shades of red are used for negative correlations and blue for positive correlations. This provides the information on the high relevance of the energy-based features like WASI for explaining the security concept based on PMU measurements. *FastWsai1s.7* is the variable most positively correlated to OK, while *Vmin1s* is the most negatively correlated to OK (largest ellipses in both cases). This complementary behavior is highlighted by the variable importance analysis results from the RF learning in Figure 11.11.

The performance test results for both system wise and area wise for rapid stability assessment is shown in Table 11.2, which is really exciting [52–54]. Here, the extended data file is used to train the RF, where the training occurs for larger number of tree generations instead of single DT. It is observed that the reliability becomes 99.93% and 99.96% in case of HQ_1s and HQ_2s, respectively. Area-wise study provides similar performance as depicted in Table 11.2. It is observed that the reliability is more than 99% (except BJS, it is 99.9% in all cases), considering both 1s and 2s early termination. It is also found that there is substantial

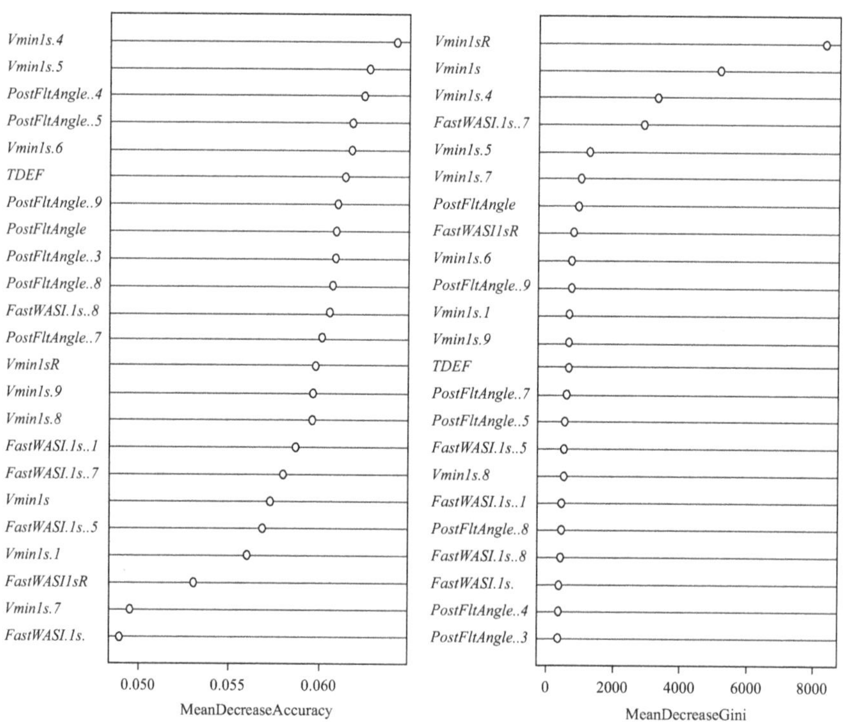

Figure 11.11 Top-down importance of the variables

Table 11.2 Reliability, security, and accuracy for extended data sets with Random Forest. Reprinted, with permission, from [54].

Early termination	Reliability (%)	Security (%)	Accuracy (%)
	System-wise		
HQ_1s	99.93	98.92	99.16
HQ_2s	99.96	99.01	99.23
	Area wise_1s		
BJN_1s	99.84	99.32	99.47
BJS_1s	99.92	99.44	99.51
CHU_1s	100	99.91	99.93
MQ_1s	99.94	97.62	98.46
	Area wise_2s		
BJN_2s	99.89	99.57	99.63
BJS_2s	99.92	99.76	99.79
CHU_2s	100	99.94	99.95
MQ_2s	99.97	97.75	98.55

increase in security and accuracy in system-wise and area-wise study. The security and accuracy are around 99% for HQ_1s and HQ_2s system wise as well as area wise (BJN_1s and 2s, BJS_1s and 2s). Only in case of MQ, the security and accuracy are around 97.5% and 98%, respectively.

From the aforementioned result analysis, it is observed that the reliability achieved is 99.9% (3 nine) using RF compared to single DT. Our analysis of recent DT applications to DSA results 80%–85% reliability [55] on system-wise study, which is definitely unsatisfactory, giving that modern electric power systems are normally designed and operated to meet a "3 nines" reliability standard [56]. Even if a DT predictor is easily interpretable and transparent for the user compared to Random Forests, however, the reliability and security measures are substantially low to be accepted for the critical issues such as DSA. The RF not only provides high jump in reliability but also the security and accuracies are around 99%, considering system-wise and area-wise security assessment for 1-s and 2-s early termination.

11.5.3.3 DT_Fuzzy

The DT_Fuzzy [57,58] is derived from the partitioning boundaries of the decision node of the DT. From the boundaries of the system-wise and area-wise classification, the fuzzy MFs of each and every feature involved in both the 1 s and 2 s response time DTs are developed. The corresponding fuzzy rule base is also developed considering the respective fuzzy MFs and looking at the original DT structure. The Mamdani Model with Centroid defuzzification is used for implementing the developed DT-induced fuzzy rule base. The fuzzy rule base developed for 1-s early termination is as follows:

Rule base for 1-s early termination decision:

R-1: If X_1 is P_{11} and X_2 is Q_1, then "−1" (stable).

R-2: If X_1 is P_{11} and X_2 is Q_{11} and X_3 is R_1, then "1" (unstable).

R-3: If X_1 is P_{11} and X_2 is Q_{11} and X_3 is R_2 and X_4 is S_{11} and X_5 is T_2, then "−1" (stable).

R-4: If X_1 is P_1 and X_2 is Q_{11} and X_3 is R_2 and X_4 is S_{11} and X_5 is T_2 and X_6 is U_1, then "−1" (stable).

R-5: If X_1 is P_5 and X_2 is Q_{11} and X_3 is R_2 and X_4 is S_{11} and X_5 is T_2 and X_6 is U_1, then "−1" (stable).

R-6: If X_1 is P_1 and X_2 is Q_9 and X_3 is R_2 and X_4 is S_{11} and X_5 is T_2 and X_6 is U_2, then "−1" (stable).

R-7: If X_1 is P_1 and X_2 is Q_{11} and X_3 is R_2 and X_4 is S_{11} and X_5 is T_2 and X_6 is U_2, then "−1" (stable).

R-8: If X_1 is P_1 and X_2 is Q_{11} and X_3 is R_2 and X_4 is S_7 and X_5 is T_2 and X_6 is U_2 and X_7 is V_1, then "−1" (stable).

R-9: If X_1 is P_1 and X_2 is Q_{11} and X_3 is R_2 and X_4 is S_7 and X_5 is T_2 and X_6 is U_2 and X_7 is V_2, then "1" (unstable).

R-10: If X_1 is P_1 and X_2 is Q_{11} and X_3 is R_2 and X_4 is S_4, then "1" (unstable).

R-11: If X_1 is P_1 and X_2 is Q_{10} and X_3 is R_2 and X_4 is S_4, then "−1" (stable).

R-12: If X_1 is P_4 and X_8 is W_2, then "1" (unstable).

R-13: If X_1 is P_4 and X_2 is Q_1 and X_8 is W_1, then "−1" (stable).

R-14: If X_1 is P_4 and X_2 is Q_{11} and X_4 is S_{11} and X_8 is W_1, then "−1" (stable).

R-15: If X_1 is P_4 and X_2 is Q_{11} and X_4 is S_6 and X_8 is W_1, then "1" (unstable).

where X_1 = Vmin1sR, X_2 = Vmin1s.4, X_3 = PostFaultAngle, X_4 = Vmin1s.5, X_5 = Vmin1s, X_6 = TDEF, X_7 = PostltAngle.8, X_8 = FastWasi1sR. Similar framework for building fuzzy rule base for 2-s early termination is also carried out for classifying stable cases from instable ones.

The performance of the proposed DT_fuzzy predictor is shown in Figure 11.12 (a) for 1-s and 2-s early termination. It is observed that the reliability improves with fuzzy-rule base formulation, which reflects the robustness of the proposed fuzzy rule base. Similar performance results for the area-wise study are shown in Figure 11.12(b) and (c) for 1-s and 2-s early termination time decision-making, respectively. As expected, the overall accuracy of the area-wise classifiers is better compared to the system-wise classifiers in all scenarios except for the MQ area, which connects directly to the major load center of Québec City. Therefore, most contingencies with a line outage under heavy load resulted in voltage collapse or in an inadmissible transient voltage profile. While the latter situation is well covered by the present severity indices-based method, the former is definitely not in its scope and should be managed by other complementary means. Further, it is observed that the reliability improves even after fuzzy-rule base simplification reflecting the robustness of the proposed fuzzy rule base approach.

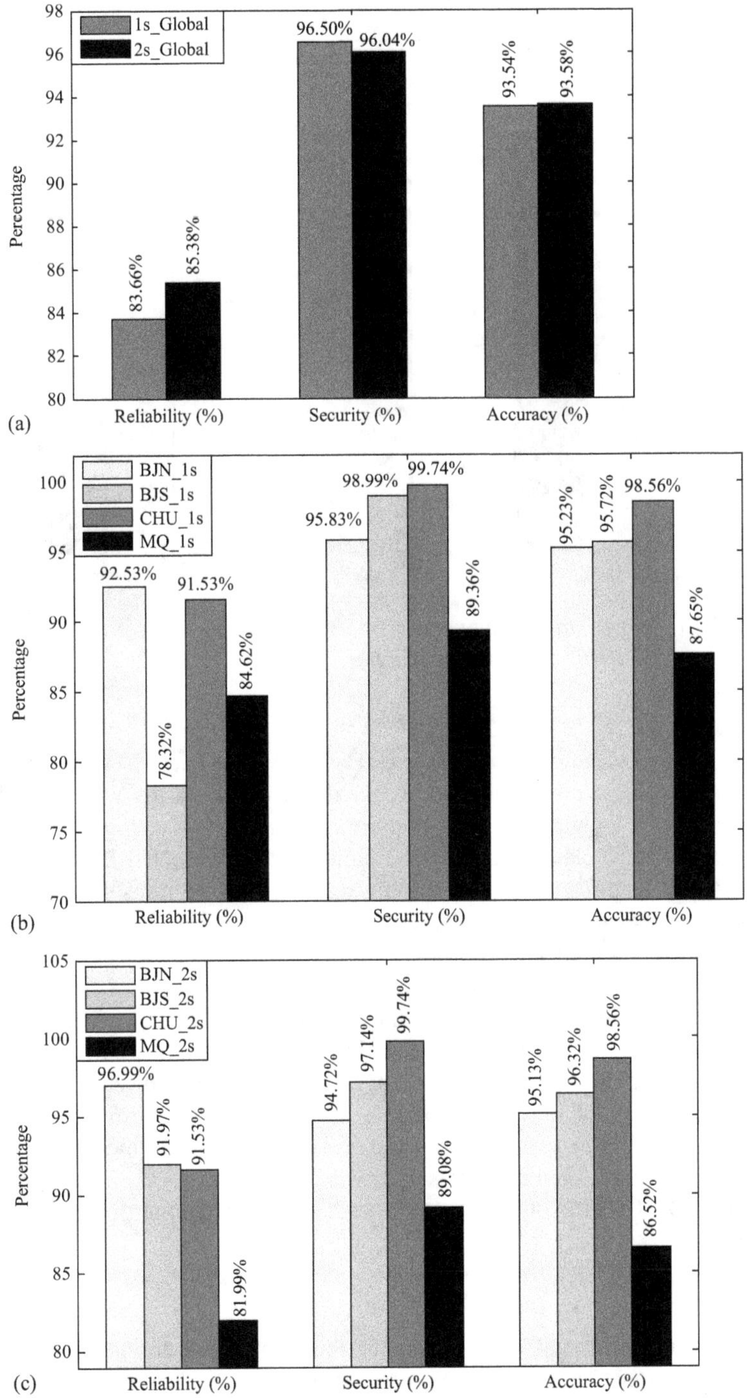

Figure 11.12 *(a) Performance of DT_Fuzzy for 1s_Global and 2s_Global scenarios, (b) performance of DT_Fuzzy for area-wise scenarios (1s), and (c) performance of DT_Fuzzy for area-wise scenarios (2s)*

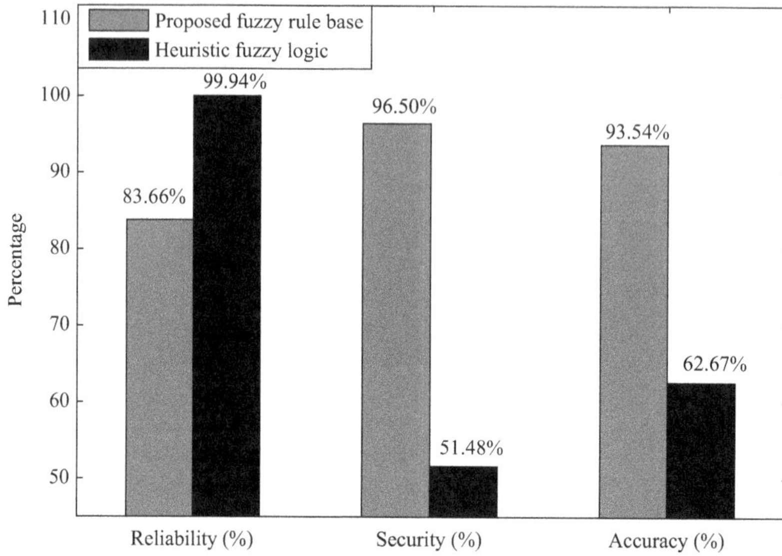

Figure 11.13 Performance comparisons between proposed DT_Fuzzy and heuristic fuzzy for 1s_Global

The performance of the proposed DT_Fuzzy approach is compared with heuristic fuzzy logic classifiers for two system-wise scenarios which are hand-developed after a lengthy and tedious process [3] as shown in Figure 11.13. The heuristic fuzzy logic classifiers share the same 15 inputs selected from the pool of best candidate features. They are combined in 16 and 10 rules for 1-s and 2-s response-time decision-making, respectively. It is found that the fuzzy-rule classifiers developed automatically from the DTs provide a substantially better security and overall accuracy compared to the heuristic fuzzy classifiers. It is found that the yield of the proposed automated fuzzy rule bases exceeds 93% compared to 51.48% and 64.21%, respectively, with the heuristic fuzzy logic for 1 s and 2 s response times decision-making. However, a key point is that the reliability of the heuristic method is close to 100%. The responsible factor for low performance indices in case of heuristic fuzzy classifier is the classifier design process which is heuristic in nature. This example demonstrates that when reliability is a prime factor, the automated DT-based solution may need additional fine-tuning to meet requirements.

11.5.4 Accuracy vs transparency trade-off

The performance of the different data-mining tools has been discussed extensively in previous sections. The complete process for building the data mining models for catastrophe predictor is summarized in Figure 11.14. Data mining for predictive analytics starts invariably with the adoption of a process of data collection and

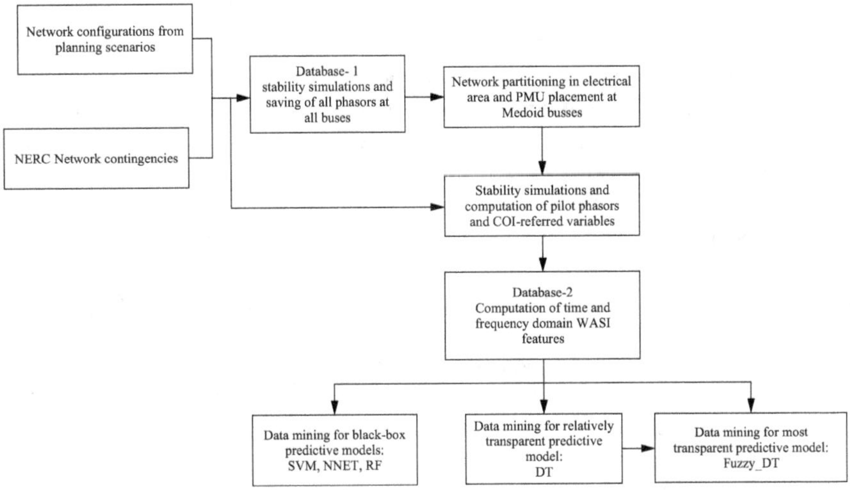

Figure 11.14 The complete framework for building data-mining model for catastrophe predictors

Table 11.3 Performance comparisons between black-box and transparent predictive models

Data-mining tools	**Performance (%)**		
	Reliability	**Security**	**Accuracy**
NNET	82.0	87.9	85.7
SVM	83.9	92.5	87.7
RF	99.9	99.0	99.2
DT	80.1	86.0	84.3
DT_Fuzzy	86.4	82.9	83.7

preparation. To further summarize the first stage includes: (i) collecting information about network configurations and contingencies, (ii) obtaining a reduced-size database of stability results through a high-performance computation center, (iii) partitioning into medoid buses which are the most suitable locations for PMUs. The second stage includes extended set of simulations and collection of the bus phasor signals aggregated by area and refereed to the COI reference frame. The data-mining features in the time and frequency domains (WASI) are finally derived according to [3,22], and saved in a large database for subsequent data-mining investigations. However, while moving from black-box data mining tools to transparent one, there is a tradeoff [59] that exists between accuracy and transparency in classification which needs to be addressed.

Table 11.3 provides the complete statistics of the catastrophe predictors developed using different data-mining tools including NNET, SVM, RF, DT,

DT_Fuzzy. It is observed that NNET provides a classification accuracy of 87.0% for the combined HQ and test system considered in the proposed study. SVM stays at 87.7%, while DT and DT_Fuzzy stay below NNET and SVM for accuracy measure. However, RF provides most accurate results with more than 99.0% accuracy. While comparing the reliability and security measures, it is observed that the reliability is more than 99.9% with the RF-based predictor compared to other data-mining models for the same system configuration. The reliability is above 80% for NNET, SVM, and DT. However, the reliability index of DT_ Fuzzy stands at 86.0% which is improved compared to NNET, SVM, and DT. In fact, the security performance of all data-mining models is better than 86% except for the DT_ Fuzzy.

While analyzing the complete statistics of accuracy, reliability, and security indices, the RF-based predictor emerges as the only tool with a more than 99% level of performance on all three indices. It is observed that while switching from black box solutions such as NNET, SVM and RF to semi-transparent solutions such as DT, and transparent solutions such as DT_Fuzzy, there is a tradeoff between accuracy, reliability, and security for the combined HQ and test, HQ and test systems.

Further validation is carried out to test the robustness of the algorithms. The black-box RF, semi-transparent algorithm (DT), and transparent algorithm (DT_Fuzzy) are further tested under noisy conditions with signal-to-noise ratio (SNR) 30 dB, and the performance is depicted in Table 11.4. It is observed that the performance of the DT is more degraded compared to that of RF and DT_Fuzzy with noise. Thus, RF and DT_Fuzzy are found to be more robust algorithms than DT for the proposed application. Table 11.5 provides a more synthetic view of the same

Table 11.4 Robustness testing under noisy condition (SNR: 30 dB)

Data-mining tools	**Performance (%)**		
	Performance accuracy (%)	**Reliability (%)**	**Security (%)**
RF	98.18	95.33	99.02
DT	81.49	84.36	80.66
DT_Fuzzy	84.46	64.62	91.07

Table 11.5 Performance loss of data-mining models

Data-mining tools	**Performance loss (in %)**		
	Reliability	**Security**	**Accuracy**
RF	4.50	−0.25	0.75
DT	−4.25	5.25	2.75
DT_Fuzzy	−0.5	1.00	1.25

findings. It is observed that in the presence of noise, crisp models (DT and RF) tend to suffer a more significant loss of reliability than the fuzzy logic-based model. This observation is in line with the previous investigations [60] and represents another benefit of fuzzy logic in addition to transparency and comprehensibility [23,60]. Indeed, a fuzzy rule-based system which includes symbolic rules enables ease of understanding and transfer of high-level knowledge. Further, the fuzzy sets along with fuzzy logic and approximate reasoning provide the ability to model the fine knowledge base which can handle uncertainties in application such as power systems which is highly uncertain. Thus, fuzzy representation is becoming increasingly popular which is more transparent, leading to easy implementation and ability to deal with problems of uncertainty, noise, and inexact data [31].

11.6 Conclusions

This chapter discusses data-mining-based catastrophe predictors using PMU-based WASI features. The study validates the performance of the black-box models such as SVM and RF through semi-transparent data-mining models such as DT and transparent models such as DT_Fuzzy. It is observed from the extensive studies of the developed data-mining-based catastrophe predictors that while switching from the black box solutions to transparent and interpretable solutions, there is an unavoidable tradeoff between accuracy, reliability, and security measures. The more transparent the predictor, the easier the implementation and maintenance by human actors. Overall, the fuzzy logic-based transparent solutions are preferred over black box solutions to ease the implementation with improved robustness and enhance their suitability for auditing process, even sacrificing the predictive performance indices.

References

[1] G. Trudel, S. Bernard, and G. Scott, "Hydro-Québec's defense plan against extreme contingencies," *IEEE Transactions on Power Systems*, PWRS-8(2), 1993, pp. 445–451.

[2] K. Mei and S. M. Rovnyak, "Response-based decision trees to trigger one-shot stabilizing control," *IEEE Transactions on Power Systems*, PWRS-19 (1), 2004, pp. 531–537.

[3] I. Kamwa, R. Grondin, and L. Loud, "Time-varying contingency screening for dynamic security assessment using intelligent-systems techniques," *IEEE Transactions on Power Systems*, PWRS-16(3), 2001, pp. 526–536.

[4] D. Ernst, D. Ruiz-Vega, M. Pavella, P. Hirsch, and D. Sobajic, "A unified approach to transient stability contingency filtering, ranking and assessment," *IEEE Transactions on Power Systems*, PWRS-16(3), 2001, pp. 435–443.

[5] Y. Mansour, E. Vaahedi, and A. El-Sharkawi, "Dynamic security contingency screening and ranking using neural networks," *IEEE Transactions on Neural Networks*, 8(15), 1997, pp. 942–950.

[6] *CIGRE Technical Brochure on "Review of On-line Power System Security Assessment Tools &Techniques,"* 2007 (K. Morison, Convener).

[7] G.C. Ejebe, C. Jing, J.G. Waight, *et al.*, "On-line dynamic security assessment: Transient energy based screening and monitoring for stability limits," *Presented at the 1997 IEEE/PES Summer Meeting*, Berlin, Germany.

[8] H.D. Chiang, C.S. Wang, and H. Li, "Development of BCU classifiers for on-line dynamic contingency screening of electric power systems," *IEEE Transactions on Power Systems*, PWRS-14(2), 1999, pp. 660–666.

[9] C. Fu, and A. Bose, "Contingency ranking based on severity indices in dynamic security analysis," *IEEE Transactions on Power Systems*, PWRS-14 (3), 1999, pp. 980–986.

[10] V. Brandwajn, A.B.R. Kumar, A. Ipakchi, A. Bose, and S.D. Kuo, "Severity indices for contingency screening in dynamic security assessment," *IEEE Transactions on Power Systems*, PWRS-12(3), 1997, pp. 1136–1142.

[11] R. Schainker, P. Miller, W. Doubleday, P. Hirsch, and Z. Guorui, "Real-time dynamic security assessment: Fast simulation and modeling applied to emergency outage security of the electric grid," *IEEE Power and Energy Magazine*, 4(2), 2006, pp. 51–58.

[12] K. Sun, S. Likhate, V. Vittal, V.S. Kolluri, and S. Mandal, "An online dynamic security assessment scheme using phasor measurements and decision trees," *IEEE Transactions on Power Systems*, PWRS-22(4), 2007, pp. 1935–1943.

[13] I. Kamwa, and R. Grondin, "PMU configuration for system dynamic performance measurement in large multi-area power systems," *IEEE Transactions on Power Systems*, PWRS-17(2), 2002, pp. 285–394.

[14] I. Kamwa, J. Béland, G. Trudel, R. Grondin, C. Lafond, and D. McNabb, "Wide-area monitoring and control at Hydro-Québec: Past, present and future," *Panel Session on PMU Prospective Applications, 2006 IEEE/PES General Meeting*, Montreal, QC, Canada, 18–22 June 2006.

[15] S.H. Nawab, and T.E. Quatieri, "Short-time Fourier transform," in *Advanced Topics in Signal Processing* (J.S. Lim, and A.V. Oppenheim, eds.), Englewood Cliffs, NJ: Prentice Hall, 1988.

[16] D.R. Ostojic, "Spectral monitoring of power system dynamic performances," *IEEE Transactions on Power Systems*, PWRS-8(2), 1993, pp. 445–451.

[17] C.W. Taylor, D.C. Erickson, K.E. Martin, R.E. Wilson, and V.V. Venkatasubramanian, "WACS—Wide-area stability and voltage control system: R&D and on-line demonstration," *Proceedings of the IEEE*, 93(5), 2005, pp. 892–906.

[18] I. Kamwa, S.R. Samantaray, and G. Joos, "Catastrophe predictors from ensemble decision-tree learning of wide-area severity indices," *IEEE Transactions on Smart Grid*, 1(2), 2010, pp. 144–158.

[19] B. Kirby, S. Richards, J. Attas, R. Grondin, I. Kamwa, and M. Rousseau, "Fast topology detection based on open line detection," *Eighth IEE International Conference on Developments in Power System Protection*, Amsterdam, Netherlands, 5–8 April 2004, pp. 1–4.

[20] M.-C. Shin, C.-W. Park, and J.-H. Kim, "Fuzzy logic-based relaying for large power transformer protection," *IEEE Transactions on Power Delivery*, 18(3), 2003, pp. 718–724.

[21] W. Rebizant, and K. Feser, "Fuzzy logic application to out-of-step protection of generators," *Proceedings of 2001 IEEE/PES Summer Meeting*, Vancouver, BC, Canada, 15–19 July 2001, 2001, vol. 2, pp. 927–932.

[22] I. Kamwa, S.R. Samantaray, and G. Joos, "Development of rule-based classifiers for rapid stability assessment of wide-area post-disturbance records," *IEEE Transactions on Power Systems*, 24(1), 2009, pp. 258–270.

[23] S. Guillaume, "Designing fuzzy inference systems from data: An interpretability-oriented review," *IEEE Transactions on Fuzzy Systems*, 9, 2001, pp. 426–443.

[24] N. Amjady, and S. F. Majedi, "Transient stability prediction by a hybrid intelligent system," *IEEE Transactions on Power Systems*, 22(3), 2007, pp. 1275–1283.

[25] L.S. Moulin, A.P. Alves da Silva, M.A. El-Sharkawi, and R.J. Marks II, "Support vector machines for transient stability analysis of large scale power systems," *IEEE Transactions on Power Systems*, 19(2), 2004, pp. 818–825.

[26] F.R. Gomez, A.D. Rajapakse, U.D. Annakkage, and I.T. Fernando, "Support vector machine based algorithm for post-fault transient stability status prediction using synchronized measurements," *IEEE Transactions on Power Systems*, 26(3), 2011, pp. 1474–1483.

[27] Z. Li, and W. Wu, "Phasor measurements-aided decision trees for power system security assessment," *2009 IEEE Computer Society 2nd Int. Conf. on Information and Computing Science*, pp. 358–361.

[28] J.A. Huang, S. Harrison, L. Wehenkel, G. Vanier, F. Lévesque, and A. Valette, "Operation rules determined by risk analysis for special protection systems at Hydro-Quebec," *CIGRÉ Session 2004*, Paper B5-205, Paris, pp. 1–8.

[29] M. Sadeghi, M.A. Sadeghi, S. Nourizadeh, A.M. Ranjbar, and S. Azizi, "Power system security assessment using AdaBoost algorithm," *Proceedings of the North American Power Symposium (NAPS 2009)*, Starkville, Mississippi, October 2009.

[30] T. Hastie, R. Tibshirani, and J. Friedman, *The Elements of Statistical Learning*, 2nd ed., New York: Springer-Verlag, 2009, 745 p.

[31] J. Abonyi, and B. Feil, *Cluster Analysis for Data Mining and System Identification*, Basel-London-Berlin: Birkhauser, 2007, 319 p.

[32] G. Trudel, J.-P. Gingras, and J.-R. Pierre, "Designing a reliable power system: The Hydro-Québec's integrated approach," *IEEE Proceedings of the Special Issue on Energy Infrastructure Defense Systems*, 93(5), 2005, pp. 907–917.

[33] I. Kamwa, A.K. Pradhan, G. Joos, and S.R. Samantaray, "Fuzzy partitioning of a real power system for dynamic vulnerability assessment," *IEEE Transactions on Power Systems*, 24(3), 2009, pp. 1–10.

[34] V. Kolluri, S. Mandal, M.Y. Vaiman, M.M. Vaiman, S. Lee, and P. Hirsch, "Fast fault screening approach to assessing transient stability in Entergy's power system," *2007 IEEE/PES General Meeting*, Tampa, FL, USA, 24–27 June 2007.

[35] C.W. Liu, M.C. Su, S.S. Tsay, and Y.J. Wang, "Application of a novel fuzzy neural network to real-time transient stability swings prediction based on synchronized phasor measurements," *IEEE Transactions on Power Systems*, PWRS-14(2), 1999, pp. 685–692.
[36] I. Kamwa, R. Grondin, E.J. Dickinson, and S. Fortin, "A minimal realization approach to reduced-order modelling and modal analysis for power system response signals," *IEEE Transactions on Power Systems*, PWRS-8(3), 1993, pp. 1020–1029.
[37] J.F. Hauer, "Application of Prony analysis to the determination of modal content and equivalent models for measured power system response," *IEEE Transactions on Power Systems*, 6(3), 1991, pp. 1062–1068.
[38] L. Breiman, J.H. Friedman, R.A Olshen, and C.J. Stone, *Classification and Regression Trees*, New York: Chapman & Hall, 1984.
[39] L. Rokach, and O. Maimon, "Top-down induction of decision trees classifiers—A survey," *IEEE Transactions on Systems, Man, and Cybernetics—Part C*, 35(4), 2005, pp. 476–487.
[40] S.R. Safavian, and D. Landgrebe, "A survey of decision tree classifier methodology," *IEEE Transactions on Systems, Man and Cybernetics*, 21, 1991, pp. 660–674.
[41] Y. Yuan, and M.J. Shaw, "Induction of fuzzy decision trees," *Fuzzy Sets and Systems*, 69, 1995, pp. 125–139.
[42] J. Zeidler, and M. Schlosser, "Continuous-valued attributes in fuzzy decision trees," *Proceedings of Information Processing and Management of Uncertainty in Knowledge-Based Systems*, Granada, 1996, pp. 395–400.
[43] E.C.C. Tsang, X.Z. Wang, and Y.S. Yeung, "Improving learning accuracy of fuzzy decision trees by hybrid neural networks," *IEEE Transactions on Fuzzy Systems*, 8(5), 2000, pp. 601–614.
[44] X.Z. Wang, B. Chen, G. Qian, and F. Ye, "On the optimization of fuzzy decision trees," *Fuzzy Sets and Systems*, 112, 2000, pp. 117–125.
[45] K. Crockett, Z. Bandar, D. Mclean, and J. O'Shea, "On constructing a fuzzy inference framework using crisp decision trees," *Fuzzy Sets and Systems*, 157 (1), 2006, pp. 2809–2832.
[46] D.S. Siroky, "Navigating random forests and related advances in algorithmic modeling," *Statistical Reviews*, 3, 2009, pp. 147–163. [online]: http://projecteuclid.org/DPubS?service=UI&version=1.0&verb=Display&handle=euclid.ssu.
[47] Rattle (the R Analytical Tool to Learn Easily), by D. Williams, ver. 2.3, 2008: http://rattle.togaware.com/.
[48] K. Alcheikh-Hamoud, N. Hadjsaid, Y. Bésanger, and J.P. Rognon, "Decision tree based filter for control area external contingencies screening," *2009 Bucharest PowerTech*, Bucharest, Romania, 2009.
[49] S. Kar, S.R. Samantaray, and M.D. Zadeh, "Data-mining model based intelligent differential micro-grid protection scheme," *IEEE Systems Journal*, 11(2), 2017, pp. 1161–1169.
[50] D. Achlerkar, S.R. Samantaray, and M. Sabarimalai Manikandan, "Mode decomposition and decision tree based detection and classification of power

quality disturbances in grid-connected distributed generation system," *IEEE Transactions on Smart Grid*, 9(4), 2018, pp. 3122–3132.

[51] S. Kar, and S.R. Samantaray, "Data-mining based intelligent anti-islanding protection relay for distributed generations," *IET Generation, Transmission & Distribution*, 8(4), 2014, pp. 629–639.

[52] S.R. Samantaray, "Ensemble decision trees for high impedance fault detection in power distribution network," *International Journal of Electrical Power and Energy Systems, Elsevier Science*, 43(1), 2012, pp 1048–1055.

[53] S.R. Samantaray, "A data mining model for protection of FACTs based transmission line," *IEEE Transactions on Power Delivery*, 28(2), 2013, pp. 612–618.

[54] S.R. Samantaray, I. Kamwa, and G. Joos, "Ensemble decision trees for PMU based wide-area security assessment in the operations time-frame," *IET Generation, Transmission & Distribution*, 4(12), 2010, pp. 1334–1348.

[55] Smart grid: Enabler of the new energy economy. *U.S. Department of Energy*, December 2008 [Online]. Available: https://www.energy.gov/oe/articles/smart-grid-enabler-new-energy-economy, Electric Advisory Committee.

[56] R.E. Banfield, L.O. Hall, K.W. Bowyer, and W.P. Kegelmeyer, "A comparison of decision tree ensemble creation techniques," *IEEE Transactions on Pattern Analysis and Machine Intelligence*, 29(1), 2007, pp. 173–180.

[57] S. Kar, and S.R. Samantaray, "A fuzzy rule-based approach for intelligent protection of micro-grid," *Electric Power Components and Systems*, 43(18), 2015, pp. 2082–2093.

[58] S.R. Samantaray, K. El-Arroudi, G. Joos, and I. Kamwa, "Fuzzy rule-based approach for islanding detection in distributed generation," *IEEE Transactions on Power Delivery*, 25(3), 2010, pp. 1427–1433.

[59] I. Kamwa, S.R. Samantaray, and G. Joos, "On the accuracy vs transparency trade-off of data-mining models for fast-response PMU-based catastrophe predictors," *IEEE Transactions on Smart Grid*, 3(1), 2012, pp. 152–161.

[60] J. Abonyi, J.A. Roubos, and F. Szeifert, "Data driven generation of compact. accurate, and linguistically sound fuzzy classifiers based on a decision tree initialization," *International Journal of Approximate reasoning*, 32, 2003, pp. 1–21.

Chapter 12

Cyber-physical attack detection approaches in the wide-area measurement system

Zakaria El Mrabet[1] *and Prakash Ranganathan*[2]

Phasor measurement units (PMUs) as part of wide-area measurement systems (WAMS) can improve situational awareness of the electric grid. Typical measurements include time-stamped parameters such as frequency, magnitude and phase angle of voltage, and current parameters. Over 2,500 PMUs are already deployed in the North American grid, each generating 30–120 samples per second—around 1.5 TB of streaming data monthly. Processing such huge volume of PMU data necessitates substantial computational resources and fast algorithms. This chapter reviews state-of-the-art cyber and physical anomalies that can impact WAMS integrity and availability. The chapter also examines current PMU anomaly detection approaches for both historical and streaming environment, including conventional methods, machine learning and hybrid approaches, weighing their advantages and limitations. The chapter highlights the importance of missing data and concept drifts.

12.1 Introduction

In 2015, the Ukraine power system has experienced an outage for several hours impacting 225,000 utility customers. One year later, another cyber-attack was reported on a power control system in another city in Ukraine shutting down substations that controlled 200 MW of capacity [1]. These two succeeding events illustrate the threats posed by cyber-attacks and their malicious ability in disturbing the power system and making a negative social-economic impact. One of the reasons behind these cyber-attacks and several major blackouts, especially in North America, is the lack of system awareness and reliable measurements [2].

Smart Grid harnesses the power of information technology (IT) to deliver intelligently the power to the end-customer and meet wisely the environmental requirements. It allows two-way communications between the end-user and the utility and facilitate the integration of renewable technology. Smart Grid is a

[1]NRG Energy, Inc., Houston, TX, USA
[2]School of Electrical Engineering and Computer Science (SEECS), University of North Dakota, USA

distributed, interconnected, and interdependent system. Thus, an event in one part of the grid could impact operations elsewhere. It could also go beyond the grid to affect other system that relies on stable power. Thus, time-synchronized measurements is needed to obtain an accurate real-time estimation of the state of the grid and react quickly to the suspicious events [3].

12.2 Wide-area measurement system

The WAMS-based PMUs provide a situational awareness, operation, and reliability of the power system network [4]. It plays a major role in the Smart Grid network by providing a near time-synchronized measurement, which enables the grid operator to analyze the operating status of the power grid. Such a feature is achieved through the utilization of the synchrophasor that synchronizes the critical measurements received from different grid components widely distributed in the power grid with a precise clock through the global positioning system (GPS). The PMU is regarded as the nervous system of the power grid as it is capable of providing time-synchronized measurements of the grid status that enable the grid operators to react quickly to accidental and unexpected events.

Since the time-synchronized measurements produced by the PMUs are sent to phasor data concentrator (PDC) or a control center, it is important to protect all the components in the synchrophasor communication system. This system is called the PMU-based WAMS. A typical WAMS is composed of PMUs, and local and regional PDC, as illustrated in Figure 12.1.

The PMU receives the three-phase voltages and currents coming from the power system network in the form of an analog signal (Figure 12.2). This signal is filtered through an anti-aliasing module to limit the bandwidth of the incoming signal and then converted into a digital signal via the (A/D) converter module. The digital signal is fed to the central processing unit (CPU) to compute the phase and magnitude of the signal using the discrete Fourier transformer in conjunction with the timed GPS signal. The output signal is time-synchronized using a sampling clock that is

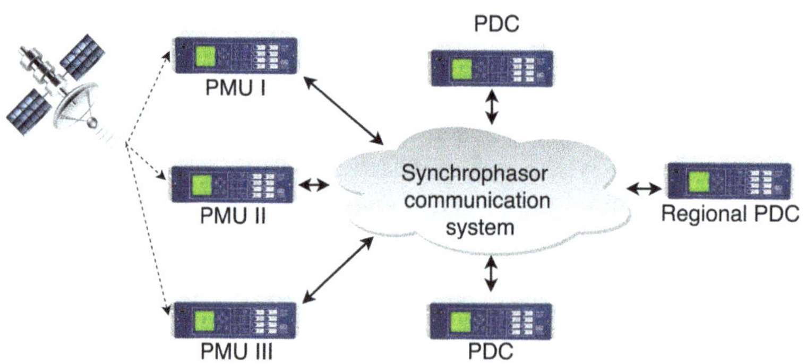

Figure 12.1 A typical PMU-based WAMS architecture

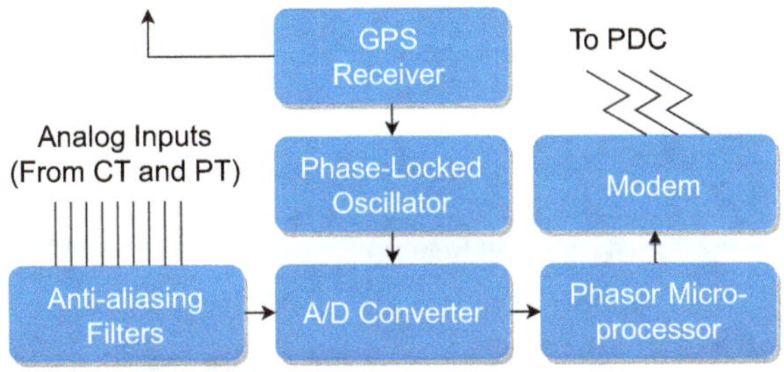

Figure 12.2 Phasor measurement unit components

phase-locked to the one-pulse-per-second provided by the GPS receiver. The pulse signals from the satellite are phase-locked with the sampling clock through the phase locked oscillator (PLO) module. PLO divides the one pulse per second signal from GPS into the required number of pulses per second for sampling [5,6]. The time-stamped signal is transmitted to the local PDC. Then, the measurements from several local PDCs are reported to the regional PDC as shown in Figures 12.1 and 12.2.

The communication between the PMUs and PDC is defined through various communications technology, including power line communication (PLC), optical fiber communication, microwave communication, cellular communication, and satellite communication. The end-to-end communication between the PDC and the PMU is also based on the IEEE C37.118.2-2011 standard which defines the format of the message for exchanging data between the PMU and the PDC or between PDCs. There are four types of message types: data, configuration, header, and command frames. The header, configuration, and data frames are sent from the PMU to the PDC, while the command frames are sent from the PDC to the PMU. The header frame provides information about the PMU and helps the PDC identify the PMU. In the configuration frame (CFG), there are three types: CFG-1 which provides information about the capacity of the PMU, such as reporting rates and noise suppression, and CFG-2 and CFG-3 which provides information about the currently reported measurements. The data frame includes the real-time synchrophasor data measured by the PMU including the current, frequency and voltage, and amplitudes.

Figure 12.3 shows a typical real-time communication between a PMU and a PDC through the IEEE C37.118.2-2011 standard. The PDC initiates the communication by sending a command frame to turn on the transmission and sends a request for a CFG. The PMU responds to this request, and then the PDC makes another request for synchrophasor data. The PMU sends back the requested data which is included in a data frame format. In the data frame parquets, there are various fields that serve in initiating and terminating the communications. For instance, the SYNC field marks the beginning of the frame, a Frame Size field provides the length of the entire frame, an IDCODE field is used as an ID to identify the frame, a SOC field provides information about the time stamp, a FRACSEC field which presents the time quality or time of measurements for data frames, and then a cyclic redundancy check which marks the end of the data frame [7,8].

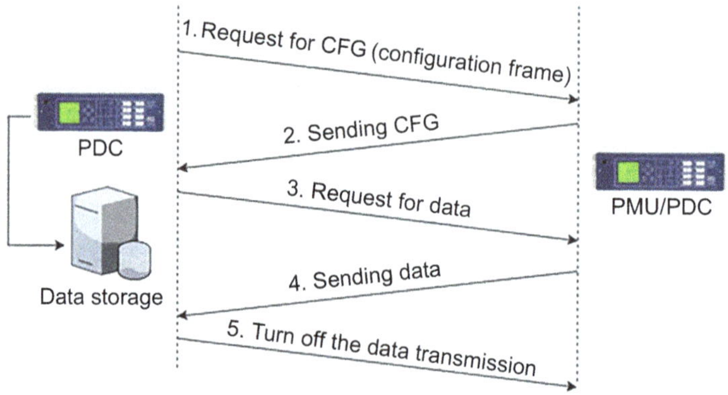

Figure 12.3 PMU-PDC communication via the IEEE C37.118.2 protocol

12.3 Cyber-physical anomalies on PMU

The AWAMS system incorporates computation and networking with physical proves and relies on PMUs to monitor the power system, creating complex feedback loops between the physical and cyber worlds [9]. The cyber and physical systems are interconnected via IT and operational technology (OT). IT refers to the application of networks for storing and transmitting data, while OT refers to monitoring and controlling physical devices deployed on the field [10].

Broadly speaking, anomalies correspond to patterns and abnormal behavior in data. These nonconforming patterns are also referred to as outliers, discordant observations, and exceptions [11]. In the WAMS context, anomalies refer to all malfunctions either intentionally or unintentionally caused by physical or cyber events. Here, we classified anomalies into two classes: cyber and physical; all anomalous events that are directed at compromising the IT network are considered cyber-anomalous events, while those that target the OT network are termed physical malicious events. The physical and cyber anomalies target the availability and integrity of the WAMS.

12.4 Physical anomalies

12.4.1 WAMS availability anomalies

Anomalies that target the WAMS availability include physical faults in power systems. These faults are classified into balanced and unbalanced faults [5,6]. An unbalanced fault, also known as an asymmetrical fault, is a commonly occurring fault in the power system and can be of a series or shunt type. In the series fault type, the voltage and frequency values increase, while the current level decreases at the faulty phases. In shunt faults, the current level rises, while the frequency and voltage levels decrease at the faulty phases. There are several shunt fault types: single line to ground (SLG), line to line (LL), double line to ground (DLG), and three-phase to the ground (LLL). An SLG occurs when a transmission line phase touches a neutral wire or the ground; the DLG or LL fault occurs when two or more phases make a connection with the ground; the cause of this fault is mostly due to natural weather events to storms or high winds or fallen trees. Sometimes, the cause can be due to an equipment failure, the line connecting the remained phases, or a failing tower. According to [6,7] the likelihood of occurrence of each fault type is 70% for SLG, 15% for LL, 10% for DL, and 5% for LLL. Although there is a low likelihood of LL fault occurrence, it is considered a severe fault that can cause a rise in fault current magnitude, thereby resulting in outages or large damage to grid assets.

12.4.2 WAMS integrity anomalies

This category includes the anomalies that compromise the WAMS integrity by modifying or injecting falsified data, i.e. false data injection (FDI). FDI is designed to alter data stored in the control center or transmitted via communications infrastructure, which ultimately compromises the integrity of the power systems or even their availability. An attack of this type was first introduced by authors in [12] in connection with DC system models, in which the authors assumed that the adversary had knowledge of the power system topology and parameters and could alter meter measurements while remaining consistent with physical laws such as Kirchhoff's circuit law to bypass the bad data detection process [10]. Such an attack could result in inaccurate estimates of the power system state, which is essential for a variety of power system applications, including economic dispatch and contingency analysis, which in turn could result in the state estimator sending incorrect values to the operator and leading to unstable system states, voltage collapses, and eventually physical and economic losses [13]. The authors of [14] examined the implications of a hidden FDI attack at the RTU level on AC state estimation and concluded that it had two negative consequences.

If the data is altered in a way that the state estimation schemes cannot detect as false, there will be an incorrect perception of the observable state of the system, which may lead the grid operator to take actions that endanger the system's security. Despite the detection of an attack, parts of the system may become unobservable, meaning the state estimator is unable to estimate changes in the state values (e.g., voltage magnitudes and voltage angles), placing the transmission grid at risk from a local physical attack. It may already be too late to prevent an outage of a greater part of the system by the time the effects of the physical attack have propagated into the rest of the system in which the state is observable. It has also been concluded that the FDI attacker using a DC model has a greater chance of introducing errors in the measurements and ultimately triggering bad data detection. Thus, the nonlinearity of the power flow equations provides advantages to the system operator to detect this type of FDI attack. However, if the attacker knows the estimated state, they can launch an AC FDI attack without being detected by the AC state estimation. But in reality, it is difficult for an attacker to obtain the same estimated state as the operators. Thus, authors in [15] proposed a sufficient condition for an imperfect and undetectable FDI attack. Authors in [16] proposed the design of blind FDI attacks based on little to no knowledge of the Smart Grid topology which would significantly reduce the attack cost.

Authors in [17] have suggested another FDI attack on the electricity market, using small changes to the price signals to increase the difference between the generated and consumed power. This attack is extended here [18], in which the attacker can inject false pricing data at any time and repeatedly over some time, resulting in over-generation, economic losses, and poor power quality.

12.5 Cyber anomalies

12.5.1 WAMS availability anomalies

Jamming attack targets mainly the availability of the WAMS where an adversary exploits the shared nature of the wireless network and sends a random or continuous flow of packets to keep the channel busy and then prevents legitimate devices from communicating and exchanging data [19]. Due to its time-critical nature, Smart Grid requires a highly available network to meet the quality of service requirements and such an attack can severely degrade its performance [19]. Authors in [20] proposed a jamming attack named maximum attacking strategy using spoofing and jamming (MAS-SJ) that targets mainly the wireless Smart Grid network (WSGN). Because WSGN is important for monitoring the Smart Grid along with the PMUs, which play a key component by providing time-synchronized data of power system operating states [20], attacks like MAS-SJ can disturb the operation of the system or even make it unavailable [21].

Denial of service is another attack aimed to adversely affect the availability of a system. In the context of a Smart Grid, DoS attacks are typically used to prevent the control center from receiving sensor measurements or actuators from receiving control commands using a variety of methods, including jamming the communication channel by transmitting a signal with high transmission power to flood the targeted channel and eventually block it or violating network protocols to increase packet collisions. While the system operator can detect the attack due to the loss of measurement data, they are unable to stop it due to the inability to send control signals to the actuators [53,54]. Authors in [8] proposed a DoS-based attack known as the puppet attack that targets the AMI network by exploiting a vulnerability in the dynamic source routing protocol, then exhausting the communication network bandwidth, causing a drop in packet delivery of 10–20%. The time-delay-switch attack [55] is another DoS attack that introduces a delay in the control system leading to instability in the system.

12.5.2 WAMS integrity anomalies

A study conducted by Daniel *et al.* [3] showed how a GPS spoofer was able to manipulate the PMU readings and cause a plant to trip. Through their experiments, they demonstrate that an attacker can manipulate the PMU time stamp by injecting a falsified set of the GPS signal into the antenna of the PMU's time reference receiver. Injecting timing error for a few microseconds was enough for the PMU to violate the maximum phase error allowed by the applicable standards, Such errors can provide a false perception of the status of the grid, leading to unnecessary control actions [3]. Therefore, it is important to protect the PMU from such manipulations.

Viruses are programs that are used to infect a specific device or system and compromise its integrity. Worms are self-replicating programs that take advantage of a network to spread, replicate, and infect other devices and systems [21,22]. Trojan horses are programs that appear to carry out legitimate functions on the target system, but run malicious code in the background [21,22]. In June 2010, Roel Schouwenberg, an analyst at Kaspersky Lab, detected Stuxnet, the first worm that attacks SCADA systems [23]. Stuxnet, a worm of 500 KB, exploited many zero-day vulnerabilities, which were not yet disclosed by the software owner. It infected at least 14 industrial sites based in Iran, including a uranium-enrichment plant. More than 1 year later, two more worms that targeted industrial control systems were discovered, Duqu and Flame. Unlike Stuxnet, Duqu was designed to gather and steal information about industrial control systems. Flame, on the other hand, was created to be used in cyber espionage in industrial networks. It has been found in Iran and other Middle East countries [22,24].

Man-in-the-middle (MITM) are other data integrity attacks where an attacker inserts themselves between two legitimate devices and listens for, performs an injection, or intercepts the traffic between them. Upon connecting to these devices, the attacker retransmits the traffic between them; the two legitimate devices appear to communicate directly, however, they are communicating through a third device [21,22]. In the advanced metering infrastructure (AMI), an MITM attacker falsifies the actual energy consumption by intercepting the data exchanged between the customer and the utility via the ZigBee protocol. This protocol is widely used for short-rand wireless data transfer, particularly for wireless low-power devices, but it suffers several vulnerabilities [25]. Thus, the attacker can easily compromise this weakest point of the system and then tap on it to launch more advanced attacks from within the network.

12.6 Anomaly detection approaches

Several approaches for detecting cyber and physical anomalies have been reported in the literature [26,27]. Some authors have proposed conventional approaches such as impedance-based and traveling wave techniques, while others have proposed machine learning-based approaches, which can be categorized into supervised, unsupervised, semi-supervised, and reinforcement learning (RL). Other authors have proposed hybrid approaches comprising conventional and machine learning approaches. In this section, we will examine all these various approaches and analyze their advantages and limitations.

12.7 Conventional approaches

Conventional approaches rely on traveling wave [57] and impedance-based methods [35,59] for detecting an anomaly in the grid. The traveling wave approach needs high-speed data acquisition equipment, a GPS, sensors, and a transient fault recorder to detect the transient waveform for fault location. The location of the fault is computed by tracking "time-tagging the arrival of the traveling wave at each end of the line and comparing against the time difference to the total propagation time of the line with the help of GPS" [57]. This approach has several advantages, as the approach is not impacted by excessive resistance, load variance, reflection, grounding resistance, refraction of the traveling waves, or series capacitor bank [57]. However, the accuracy of the approach relies on capacitance and line inductance. Unlike the time wave method, the impedance-based approaches [35,59] are simple and easy to implement, as they require only measurement data that include fault voltages and fault currents collected from the digital fault recorder or relays to compute the impedance. The accuracy of this approach can be affected in the case of a grounded fault, where the fault resistances can reach higher values.

12.8 Machine learning-based approaches

The machine learning-based techniques used to detect anomalies in the Smart Grid are categorized as supervised, unsupervised, semi-supervised, and RL [28,29]. Supervised learning algorithms build regression or classification models from a set of input features and their corresponding outputs. Although there is a wide range of supervised models, including logistic regression and neural networks as well, they usually fall into one of the following categories: decision tree (DT) models, function models, Bayesian models, and ensemble models [30]. Unsupervised learning is the process of discovering hidden patterns and associations within unlabeled data. The most common unsupervised learning algorithm is clustering [31]. Semi-supervised learning refers to a combination of supervised learning and unsupervised learning, and it is typically used in cases where training data is insufficient [32]. RL refers to the process of learning directed toward a specific objective in which an agent learns by interacting with an unknown environment, typically in a try-and-error way. In this process, the agent receives feedback from the environment in the form of a reward (or punishment); it then uses this feedback to train itself, as well as to acquire experience and knowledge about the environment [33]. Figure 12.4 illustrates the various anomaly detection approaches in PMU data. Before describing these various ML categories, we will provide preprocessing ML approaches for preparing unbiased PMU training data and assist the models in converging quickly. The preprocessing includes dimensionality reduction and imputing missing data.

12.8.1 PMU data preprocessing

12.8.1.1 Dimensionally reduction for historical PMU data

PMUs generate a substantial amount of data exhibiting multi-dimensionality (e.g., time, location, voltage, current, frequency, power flow, and phase angle). It is necessary to reduce data dimensionality to make use of it in real-time decision-making [34]. Authors in [35] proposed a PCA-based dimensionality reduction

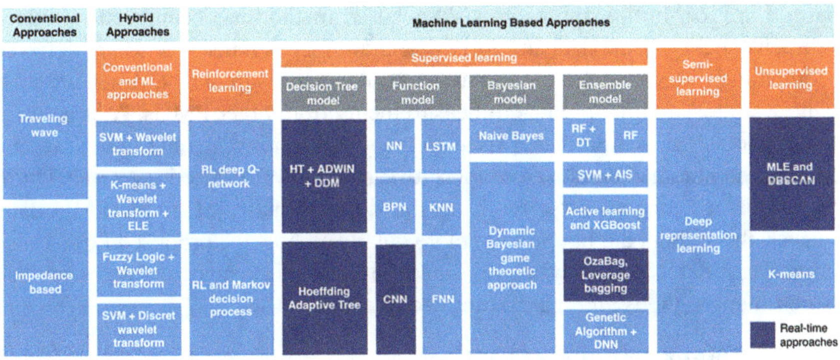

Figure 12.4 Taxonomy of anomaly detection approaches in the WAMS

method to keep only the relevant features and detect power system events at an early stage. In [36], PMU data are represented in spatio-temporal streams. It is highlighted that each PMU measurement contains many categories of variables such as voltage magnitude, power flow, and frequency. However, it is assumed that each type of measurement is independent. This clearly shows the need for higher-order generalizations of vectors and matrices to represent multivariable Spatio-temporal PMU data. Tensors (e.g., multi-way arrays) provide a natural representation for such data [36,37]. Tensors have found applications in many areas such as traffic predictions [37], social network predictions [38], wireless network predictions [39], and visual data completion [40]. Using tensors, the assumption of measurement independence can be relaxed. Analyzing all modes of data and capturing correlations among multiple modes will lead to more accurate event detection and operation. The patterns of events that occur in the power system are revealed by the PMU data. For instance, the event involving loss of generation/load in one of the regions can be seen as transients in the frequency graph, whereas the ap-change position of transformers can be manifested as transients in voltage magnitude rather than frequency or voltage angle [41]. These events may be originating from the power system itself or due to cyber-attacks. In either case, voltage and current transients may have adverse effects on the health and reliability of power equipment. Thus, the timely detection of such events would help in implementing appropriate control actions in order to maintain the resiliency, reliability, and security of the power system.

12.8.1.2 Dimensionality reduction for streaming PMU data

Authors in [15] have proposed an online reduction method for stream synchrophasor data based on the principal component analysis (PCA). The synchrophasor data dimension is reduced by finding a correlation between features such as voltage, frequency, and current, then representing it with their principal components. The proposed approach has been tested on real PMU data generated by SEL412 and GE n60 PMUs deployed on a Real-Time Digital Power System Simulator of a 4-bus and three-generator system. Another PCA-based linear approach has been proposed in [16]. In this approach, a PCA-based method has been used to reduce the dimensionality of the data and early detect anomaly events. Events are detected by computing the error between a small set of PMU measurements, which are used as a base vector, and the predicted measurements from other PMUs. A large error between the actual and predicted measurements indicates a change in the system. Based on the obtained results from these two studies, PCA-based approaches provide promising techniques for reducing the dimensionality of online PMU data. However, the authors did not provide more details on how the proposed PCA-based approach can detect anomaly events since no analysis comparison was provided between their methods and the existing techniques.

12.8.1.3 Imputing missing PMU data

PMU data suffer from data quality issues such as data redundancy, bad, flawed, or missing data. Therefore, it is crucial to implement effective processes and tools to detect and deal with these data quality issues [42]. Thus, data quality is required for many power system applications to maintain the resilience of the power system. However, the quality of PMU data can be affected by errors that can occur at any stage of processing, from sensing the physical quantity to be measured to the display/analysis [43].

The PMU data frame exchanged between the PMU and the PDC can reveal some significant information about the quality of data. Based on the IEEE C37.118 standards, the data frame contains the real-time data measured or calculated by the PMU/PDC. The message includes an identification header, message length, the source ID of the message, a time stamp, and detailed status information regarding the data and its source and quality. Thus, if there are two data frames with the same identification header and the same timestamp, then it can be regarded as duplicated data.

While aggregating the data from several PMUs, the PDC may attempt to fill the unavailable data by providing a filler and setting the STAT flag in the data frame either to 10 or 11. This may indicate that the reported data is invalid. The missing data can be due to network dropout, database storage errors, or missing data from the PMU itself. It can also be detected from the PMU data frame when the STAT flag reports a value of 01. Getting these data quality flags is necessary, but not sufficient for identifying all the bad data.

Missing data is typically handled in two ways: deletion processing and imputation processing [44]. Deletion processing is indeed effective in eliminating missing observations in many cases as well as being easy to implement. However, some studies have found that it can severely lower the performance of subsequent models and create bias in experiments [45]. Imputation, in contrast to deletion, maintains the length of a time series by estimating missing values. Statistical interpolation is one of the most straightforward imputation methods. Various imputing PMU data are provided in Figure 12.5. The data gaps are filled using either mean, median, or mode of existing observations [46]. These techniques consist of recovering the missing values by computing the mean and median of the existing variables along with conserving the distribution of the data. The main

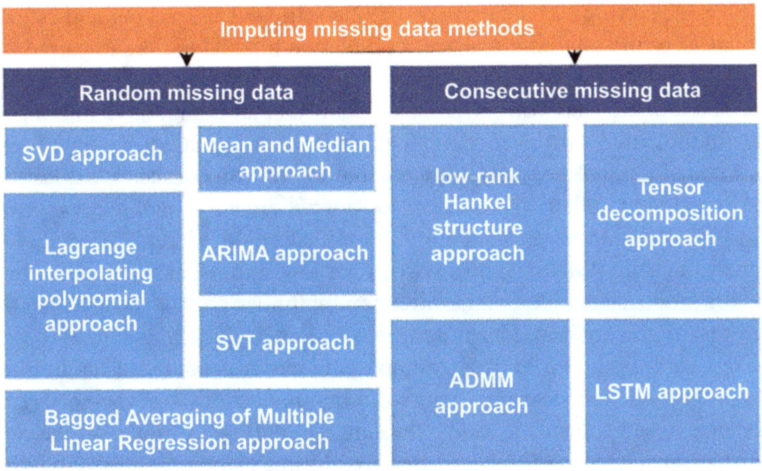

Figure 12.5 Taxonomy of imputing PMU data approaches

drawback of these techniques is that they reduce the variance in the data and they may also delete some important frequency events such as transient or impulse. In [47], the authors proposed a dynamic regression model comprising a regression and an autoregressive integrated moving average model (ARIMA). Using the linear relationship between neighbor and correlated stations, this model adjusts residuals by fitting an ARIMA structure. However, ARIMA is not sufficient because it needs to be combined with additional approaches such as the Kalman filter to recover the missing data. Additionally, authors in [48] have proposed an approach called bagged averaging of multiple linear regression based on a bootstrapping aggregation technique in machine learning. It consists of combining multiple linear regression models into one framework to complete the missing data. It resolves the issues of the mean and ARIMA approaches; however, it is inefficient in dealing with PMU data which includes nonlinear relation between the frequency measurement data. In [22], the authors proposed a recovery approach designed specifically for state estimate application. This approach produces satisfactory results with a low rate of reported data, roughly five reports per second. However, a typical PMU reports the phasor information to the control center with a frequency of 60 samples per second. Thus, this approach is inefficient in dealing with and fulfilling the missing data with a high rate of reported data. Authors of [50] used Lagrange interpolating polynomial method to recover the lost PMU data. The authors assume that the PMU network uses an optical network with a loss of maximum of three continuous packets. Authors in [51] propose a singular value decomposition method that is based on projecting and lifting a subspace to impute missing data. In [52], the authors applied the singular value thresholding (SVT) method to recover missing data, and it proved to be effective for up to 30% of the missing PMU. Nonetheless, these approaches have certain limitations. SVT-based methods, for instance, have high computational complexity, and are only suitable for recovering PMU data in an offline setting. The tracking and estimation performances of the low-rank matrices completion method will be impaired during transients, which occur during disturbances. The online low-rank matrix completion methods should start with PMU measurements that do not have missing index values. Additionally, and even though these approaches have shown prominent results with random missing data, they are inefficient with consecutive missing data [45]. Alternatively, other approaches considered imputing consecutive missing data. In [53], the authors presented a method that could address consecutive data loss across multiple PMU channels based on a low-rank Hankel structure. In [54], the authors proposed the alternating direction multiplier method, which also results in good results for missing data in multiple PMUs. In [55], the authors describes how to recover missing PMU data by employing tensor decomposition. In [45], the authors proposed a deep learning theory through long short-term memory (LSTM) and instance-based transfer learning through TrAdaBoost. Similarly, the authors proposed a transferred LSTM-based iterative estimation model [56] for imputing consecutive missing data. Based on the analysis of the existing approaches, it turns out that ensemble learning produces satisfactory results in terms of fulfilling the missing PMU data without losing any valuable features or changing the data distribution.

12.8.2 ML categories

12.8.2.1 Supervised learning

Several authors have proposed supervised function-based models for detecting anomalies using PMU data [28–57]. Most of these approaches were trained on data including input features such as voltage, current, phase angle, and fault location as output. These supervised-based learning models can be classified into classification and regression models, which in turn can be further categorized into function-based, DT-based, Bayesian, and ensemble models. It is important to note that some of these models are only suitable for the offline application, while others are considered online models as they are capable of handling streaming data and providing responses in near real time.

Function-based models

Function-based models include models that rely on activation functions, such as neural networks, or kernels, such as support vector machine (SVM), to build a mapping function between the inputs and the output. Authors in [28] proposed a back propagation-based neural network (BPN) to estimate fault location in distribution networks. Here, fault current was selected as a key feature to train the NN model. A Levenberg–Marquardt algorithm (also known as damped lease square) is applied to BPN for faster convergence. Then, the BNN model was deployed to run on the DIgSILENT Power Factory 13.2. Similarly, a feed-forward NN (FNN)-based approach is proposed in [58]. Here, fault voltages and fault currents are selected as two features to train the model. A sigmoid activation function was used to normalize the data. Their results showed a detection error of less than 3%. Another NN-based approach was proposed in [59] to estimate fault distances from substation(s). The selected input features include three-phase voltage, current, fault conditions, and active power gathered from substation(s). This approach was trained on different fault locations, resistances, and loads. The approach was tested on an IEEE 34-bus system and yielded promising results, even under dynamic changes in network topology. Additionally, this approach showed more tolerance to noise. Although high accuracy was reported for NN-based approaches in the above-mentioned studies, the training time required for NN is longer and does not suit dynamic or real-time environments. On the contrary, the SVM-based approach is faster and relatively accurate, even for larger-size data. However, it requires careful selection of appropriate kernel types and hyper-parameters.

In [60], the authors proposed a convolutional neural network (CNN)-based approach using bus voltages. This method has been trained and tested on IEEE 39-bus and IEEE 68-bus systems under uncertain conditions for system observability and measurement quality. Their results show that CNN can localize the faulted line even in low visibility (7% of buses) conditions. Another recurrent neural network (RNN) has been proposed in [61] to deal with the electricity stealth issue. Specifically, a deep autoencoder has been coupled with a LSTM-based sequence-to-sequence (seq2seq) structure to capture false data injected in the AMI network.

A KNN-based approach for detecting faults in a photovoltaic (PV) system is proposed in [62]. This approach has been trained and tested on data generated from a developed PV model. The reported results show a classification accuracy of 98.70% with an error value ranging between 0.61% and 6.5%. Another KNN-based approach algorithm was applied to classify three-phase faults (3LG), voltage oscillation, and voltage sag scenarios [63]. However, the model accuracy was not provided. In [11], the authors proposed a distributed intrusion detection system (IDS) in a multi-layer RNN architecture of smart grids. In their approach, the IDSs were deployed at three levels of the smart grid network, home area network, neighborhood area network, and wide area network. IDSs were based on the SVM and artificial immune system (AIS) where each classifier was trained and tested on the NSL_KDD dataset [12]. The obtained results illustrated that SVM outperformed AIS algorithms in terms of false positive rate (FPR) and false negative rate (FNR). However, the proposed classifiers are not suitable for streaming data since they require loading the entire dataset into memory to process it. In addition, the used dataset does not include any data related to the power system faults.

Decision tree-based models

Several online DT-based approaches have been proposed to deal with streaming PMU data and detect the eventual anomalies. For instance, authors [70] proposed a Hoeffding tree (HT) combined with two concept drift detectors: drift detection method (DDM) and adaptive sliding windows (ADWIN) for building a fast DT that is adaptable to changes. This model has been trained on a synthetic dataset where different attack scenarios were simulated. The dataset contains normal and anomaly events, and physical and cyber events. The physical event includes power system faults, while the cyber event includes trip command injection, 1LG fault replay, and replay disabled attacks. The model HT+ADWIN+DDM has been trained and tested on this heterogeneous dataset and reported a classification accuracy greater than 98% for binary classification.

In [19], the authors developed a transfer HT with ADWIN (THTA) for detecting anomaly events in the streaming PMU data. HT has been trained on normal and anomaly events, while ADWIN has been included at the leaf of the HT to make the model adaptable to the concept drift, whether gradual or abrupt. Similarly, authors in [65] have proposed a Hoeffding adaptive tree (HAT)-based approach for detecting events from continuous streams of PMU data within computational boundaries of memory and processing time. The authors have conducted three experiments to generate a training dataset of binary classes. In the first one, synchrophasor data with 3LG has been generated and load fluctuations have been injected by changing the true power (P) and reactive power (Q) at a regular interval to evaluate the ability of the HAT model in adapting to the concept drift. In the second one, two classes of events have been generated: fault and normal classes. The fault class includes the SLG faults, while the normal class includes normal power system operations without changing the load conditions. In the third experiment, non-adaptive HT with a fixed size has been evaluated on the previously generated dataset to illustrate the impact of the tree size on classification accuracy. Based on the reported results, HT showed the concept drift compared to traditional DTs J48 and REPTree. However, in these two studies, the duration of the physical fault events was not considered while generating the training synchrophasor dataset. Some faults have to be monitored for a certain time to identify their natures. For instance, a frequency deviation of 0.15–1.0 Hz is considered an inter-area oscillation only if the fluctuation lasts for 60 s [66]. Otherwise, it is considered a normal power system operation. In addition, some physical events are time-sensitive and if they are not early detected then they may lead to cascaded outages. Thus, it is necessary to include the event duration as a feature in the training dataset.

Bayesian-based models

Authors in [67] proposed a Bayesian-based approach for detecting false injected data. Here, the authors proposed a dynamic Bayesian game-theoretic approach for detecting FDI attacks with incomplete information with a 98% detection rate. Other authors in [68] proposed a Naïve Bayed model along with RF, DT, and logistic regression for detecting various cyber anomalies. The reported results showed satisfactory results in terms of detecting attacks in an offline and online network.

Ensemble models

Authors in [69] proposed an ensemble approach based on the active learning-based extreme gradient boosting (AL-XGBoost). This approach was developed to detect FDI attacks in a cyber-physical energy system. The average detection accuracy obtained with the AL-XGBoost method is 0.9784 and 0.9845 for IEEE 57- and 118-bus test systems, which is higher than SVM and KNN. Authors in [70] proposed an AdaBoost combined with a genetic algorithm (GA) and deep neural network (DNN) to detect electricity theft attacks. The model has shown superiority over SVM, ANN, and RF based on accuracy, and true positive rates and FPR. In [86], authors discuss a fault line identification and localization approach using RF and DT classifiers. Here, the models were trained on an IEEE 68 bus system. The generated data consists of several fault types including a three-phase short circuit (TP) and line-to-ground (LG). The reported experiment results show that classification accuracy of 91%. Another RF-based approach was proposed in [71]. Here, the model was trained on three-phase current and voltage data, validated using an IEEE-34 system, and achieved an accuracy above 90%.

Another online ensemble approach-based IDS has been proposed in [13]. The proposed IDS has been deployed at three levels of the AMI: smart meter, AMI headend, and data concentrator. In their study, seven data stream mining algorithms have been selected: AccuracyUpdatedEnsemble, ActiveClassifier, LeveragingBag, LimAttClassifier, OzaBagAdwin, OzaBagASHT, and SingleClassifierDrift. These algorithms have been trained and tested on the KDD Cup 1999 dataset [14] and a performance analysis comparison has been conducted based on several metrics, including accuracy, the Kappa, the FPR, and the FNR. Based on the obtained results, ActiveClassifier and SingleClassifierDrift reported satisfactory results in terms of memory and time so they are suitable candidates for smart meters that have limited computational resources. LeveraginBag provided higher accuracy and lower FPR but it is demanding in terms of memory. Thus, it is an appropriate candidate for the data concentrator which possesses a moderate amount of memory. For the data concentrator, which requires a response time of a fraction of a second, ActiveClassifier can be used since it processes the data quickly with modest accuracy, FPR, and FNR. However, the dataset used for training these models is not based on power system communication scenarios. Instead, it is based on the communication scenarios in computer networks. Thus, the reported results do not reflect the actual output of the studied data stream classifiers in the power system context.

12.8.2.2 Unsupervised learning

Authors in [72] proposed an unsupervised learning model for real-time event classification and fault localization in synchrophasor data. Their methodology relies on three processes. The first process focuses on removing bad data from collected PMU measurements using the maximum likelihood estimation (MLE) approach. In the second process, the events are classified using a combination of density-based spatial clustering of applications with noise (DBSCAN), and logic rules are generated using a physics-based decision tree (PDT) method. This PDT method uses parameters such as active power, reactive power, and fault event types. The third process reports localizing events in real time using a graph theory. Finally, a score metric is computed using Shannon entropy and descriptive statistical parameters (e.g., standard deviation, range, mean difference, and crest factor). Three case studies have been considered using metrics such as precision and recall, and their reported result shows that their proposed data cleansing approach outperforms Chebyshev and K-means methods, with a 95% precision. Additionally, the average run-time taken for their classification algorithm is around 0.09 s for a typical window size of 30 samples involving five PMU sensors.

Authors in [29] proposed another unsupervised K-mean-based model for detecting DoS attacks, including UDP and ICMP flooding attacks. The proposed model was able to cluster the normal and abnormal behaviors using unlabeled data, but the model's performances were not provided.

12.8.2.3 Semi-supervised and reinforcement learning

Authors in [73] proposed a semi-supervised and deep representation learning for detecting anomalies in WAMS including short-circuit fault, remote tripping command injection attacks, and data injection attacks. Authors in [74] proposed a multi-agent RL-based approach for detecting the simple and coordinated FDI attacks on the distributed control layer in a DC microgrid. Similarly, authors in [75] modeled the anomaly detection problem as a partially observable Markov decision process (POMDP) problem and developed a universal online detection algorithm based on RL for detecting FDI, DoS, and jamming attacks. The model was evaluated in an IEEE 14-bus system and showed a precision above 99%.

12.8.2.4 Hybrid approaches

Several hybrid methods combining conventional and ML approaches were proposed to detect anomalies in PMU data. For instance, authors in [76] proposed a wavelet transform and SVM to locate faults in transmission lines which can be described in three stages. In the first stage, voltage and current values emitted by a transmitter were used to locate the fault; the second phase fed a multi-class SVM model to the training based on selected influential features, and classification of fault location was done using a regression approach. Here, the fault classification error is below 1% for all fault types and specifically 0.26% for SLG, 0.74% for LLG, 0.20% for LL, and 0.39% for LLLG. Authors in [77] have proposed an event location estimation (ELE) algorithm for the wide-area monitoring system for PMU data. Their approach relies on clustering and wavelet analysis to detect and localize events in real time. In this work, the network is initially divided into several clusters, where each cluster is defined as an electrical zone using K-means. Next, a wavelet-based event detection approach is used to detect and localize event occurrences by tracking any large (e.g. event magnitudes) disturbance levels. Once the event is detected, its magnitude is defined using a modified wavelet energy value, and its location is estimated at each electrical zone. The authors implemented the ELE approach in a real-world setting containing 32 dynamic events with an excellent localizing accuracy value. It is important to note that the authors did not consider data quality issues in the PMU measurements. Some probable causes for data quality issues could be irregular sampling or data rate, bandwidth challenges, and time synchronization errors.

In [78], the authors discuss a wavelet decomposition technique combined with fuzzy logic to identify both faulty line(s) and their locations in a multi-terminal high-voltage direct current (HVDC) network. In their paper, wavelet coefficients of both positive and negative currents were initially computed and then fed to a fuzzy logic-based voting system to identify the faulty line(s). Once the line is identified, a traveling wave-based algorithm is used to determine the exact fault location using the Daubechies wavelets. However, wavelet transform gives us an advantage called multi-resolution analysis which helps in analyzing signals at various scales. In HVDC systems, this is essential as faults may develop and result in fast alterations of current and voltage levels. By analyzing the wavelet coefficients, it is feasible to detect patterns that resemble known faulty conditions such as short circuits or line outages. A discrete wavelet transform (DWT) combined with SVM for fault detection in distribution networks has been proposed in [79]. Here, features are extracted using SVM and DT and then optimized using a GA. Their model performance was evaluated on two active distribution networks (e.g., IEEE 13-bus and IEEE 34-bus systems), and the authors claim that their model outperforms the probabilistic neural network.

Systems based on fuzzy logic provide a flexible alternative to dealing with uncertainty in the PMU data. Rather than the traditional binary classification approach, or even a nominal fit algorithm with graph indicators, fuzzy logic uses memberships to describe system likelihoods based on real-time data which are very useful in situations related to anomaly detection where precise models are hard to model. In addition, fuzzy systems can be integrated with other methods as neural networks or wavelets to create a hybrid model causing anomaly detection in PMU networks more flexible and context-aware.

After analyzing the various techniques, it can be observed one way or another that each technique seems to have its inherent characteristics and weaknesses for being used on the PMU data process. The wavelet-SVM approach is a reliable algorithm to detect high-frequency faults; in other words, this method can be employed when real-time detection of class I or IV fault types is vital. The ELE algorithm is particularly good at pinpointing events across a network of PMUs. In contrast, in the presence of uncertain or imprecise data which is a commonly encountered challenge within PMU environments, fuzzy-based approaches help perform significantly better. Wavelet-SVM and ELE methods are soft computing techniques, fast with high accuracy; hence, they have their own significance but the ability of fuzzy logic to deal with non-binary information makes it useful for imprecise datasets (incomplete datasets) or noise in datasets. The key to these various strategies is that there is no one-size-fits-all, and a mixed model that combines the strengths of many approaches often gives us more robust results.

A summary of all these anomaly detection approaches along with their advantages and limitations is provided in Table 12.1.

12.9 Evaluation performance metrics

Several computational bottlenecks will be encountered when processing large volumes of PMU streaming data. Process limitation arises as PMU data streams are continuous and we cannot afford to store all the live feeds in memory for training or even evaluation. Further, the real-time nature of PMU data needs rapid processing, which leads to CPU/GPU resources overloading particularly for involved algorithm participants such as neural networks or ensemble models. Latency is also a major

Table 12.1 Anomaly detection approaches for PMU data

Category	Approach	Fault types	Anomalous event type	Advantages	Limitations
Conventional	Traveling wave [57]	Physical	Single-phase grounding fault	High tolerance to the excessive resistance, load variance, reflection, grounding resistance, refraction of the traveling wave or series capacitor bank	The accuracy of the model relies heavily on capacitance and line inductance
	Impedance-based		Unbalanced faults	Simple and easy to implement	The accuracy can be affected in the case of a grounded fault, where the fault resistances can reach higher values.
Machine learning	Back propagation-based neural network [60]	Physical	1 phase, 2 phase, and 3 phases faults	The model is fast to converge and requires few features	The selected features are insufficient to capture various anomaly signatures.
	KNN [86]	Physical	Open circuit faults, line-line (LL) faults, partial shading with and without bypass diode faults and partial shading with inverted bypass diode faults in real time	Fault location accuracy reaches 98.70% with an error between 0.61% and 6.5%.	The study considered the PV data only
	Feed-forward NN (FNN) [67]	Physical	Shunt faults	The detection error is less than 3%. High tolerance to the fault resistance, fault type, and fault location.	The fault location was not considered and they are unsuitable for real-time applications
	NN [68]	Physical	Ground fault, short circuit	High tolerance to noise	
	CNN [69]	Physical	TP, LG, double line to ground (DLG) and line to line (LL) faults	Optimal localization estimation even under low visibility (7% of buses)	

(Continues)

MLE + DBSCAN [80]	Physical	Three-phase fault (3LG), P load increased, Q load increased	The proposed data cleansing approach outperforms Chebyshev and K-means and achieve a precision of 95%. Less than 0.9 s to classify events for a typical window size of 30 sample data.	
RF+ DT [78]	Physical	3LG, line to ground (LG), line-to-line to ground, line-to-line (LL), loss of line	Fault location detection accuracy is 91%	
RF [79]	Physical	Single-phase to ground faults	Fault location detection accuracy is 90.96% in distribution systems	
HT+ADWIN +DDM [71]	Cyber and physical	Symmetric, unsymmetrical faults, and trip command injection attack	Classification accuracy is greater than 94% for multiclass and greater than 98% for binary class.	Fault location and duration were not considered. This model is not suitable for real-time network.
HAT [72]	Cyber and physical	3LG, SLG	Optimal adaptability to the concept drift events	
OzaBag and Active classifier [25]	Cyber	DoS, unauthorized access from a remote machine, to local super user (root) privileges, surveillance, and probing	ActiveClassifier and SingleClassDrift algorithms reported satisfactory results in terms of time processing and memory consumption	
RL deep Q-network [82]	Cyber and physical	FDI attack	Effective in detecting simple and coordinated FDI attacks	
SVM+AIS [58]	Cyber	DoS attack and unauthorized access from a remote machine	The detection accuracy is high.	Processing time is high and it is not suitable for streaming data. The fault duration was not considered.
POMPD [83]	Cyber and physical	FDI, DoS, and jamming attacks	Model efficient in detecting various attacks with a 99% precision	Model in unsuitable for real-time deployment

(Continues)

Table 12.1 (Continued)

Category	Approach	Fault types	Anomalous event type	Advantages	Limitations
	LSTM [70]	Cyber and physical	FDI attack and electricity theft	The model outperforms the Naïve Bayes, SVM, and ARIMA in terms of detection rate and false alarm with an improvement of 4–21% and 4–13%, respectively	
	NB, RF, DT, LR [75]	Cyber	Scanning attack	Results showed that these models are efficient in detecting SCADA cyberattacks in real time.	Neither attack location nor duration were considered.
	Dynamic Bayesian game-theoretic approach [74]	Physical	Load measurement	The model is effective in detecting anomalies with incomplete information and with a detection rate of 98%.	
	AL-XGBoost [76]	Physical	FDI attack and cascading failure	The model reaches a detection accuracy of 99% and outperforms SVM and KNN.	
	Semi-supervised deep representation learning [81]	Cyber and physical	Short-circuit fault, remote tripping command injection attack, and FDI attack	The model outperforms the supervised model in terms of the true positive rate.	Model suffers from a high false positive rate compared to supervised ML models.
	AdaBoost with GA and DNN [77]	Physical	FDI attack	Model is superior to SVM, RF, and NN in terms of detecting electrical theft with 94.8% of true positive rate.	The attack model was not provided.
	K-means [61]	Cyber	DoS attack	Model is efficient in handling unlabeled data.	Model's performances were not provided.

(Continues)

Hybrid	SVM+ wavelet transform [84]	Physical	SLG, LLG, LL	The fault classification error is below 1% for all fault types.	The fault duration was not considered and the model is not suitable for streaming. power system data.
	K-mean + wavelet transform + ELE [85]	Physical	32 dynamic events including generator and line trips	Fault location accuracy attain 100%	The fault duration was not considered and the processing time is high, which makes them inappropriate for streaming data.
	Fuzzy logic + wavelet transform [86]	Physical	Negative to ground and the positive to ground faults	The error between the actual fault location and the predicted one is low then 0.002%	
	SVM+ discrete wavelet transform [87]	Physical	High impedance fault	Fault location accuracy is 98.27% for IEEE 13-bus and 98.29% for the IEEE 34-bus test systems.	

issue as any delays in the data processing may cause severe lag and anomaly detection failures leaving our grid unstable. Dealing with these bottlenecks has yielded some solutions that can occur from parallel processing, distributed computing, or hardware acceleration for GPUs. We can also make use of lightweight machine learning models and approximate solutions (such as sketching algorithms) to process data in real time with reasonable tradeoffs on accuracy versus speed.

12.9.1 Supervised offline ML approaches

For evaluating the supervised offline ML approach's performances in the matter of detecting the cyber and physical anomalies, several metrics can be considered including the accuracy, Kappa, probability of detection, miss detection, and fault alarms. The accuracy evaluates the models in terms of distinguishing accurately the anomalous and normal events. It is expressed as:

$$Accuracy = \frac{TP + TN}{TP + TN + FP + FN} \tag{12.1}$$

where TP is the true positive and corresponds to the number of cyber/physical anomalies correctly classified as anomalies. TN is the true negative and corresponds to the number of normal events classified as normal. FP is the false positive and corresponds to the number of normal events misclassified as anomalous events. FN is the false negative corresponding to the number of anomalous events that are incorrectly classified as normal events. The TP, TN, FP, FN can be derived from the confusion matrix. The accuracy may provide overly prediction results with imbalanced classes. The Kappa metrics can be used to address this issue and control the instances that may have been correctly classified by chance. It is expressed as

$$Kappa = \frac{(Accuracy - random\ accuracy)}{(1 - random\ accuracy)} \tag{12.2}$$

where $Accuracy$ is given by (12.1), and $random\ accuracy$ is defined as

$$random\ accuracy = \frac{(TN + FP) * (TN + FN) + (FN + TP) * (FP + TP)}{Total^2} \tag{12.3}$$

Here, $Total$ is $TP + TN + FN + FP$. The probability of detection (Pd) is the number of anomalous events divided by the total number of anomalous events in the dataset. It is given by

$$Pd = \frac{TP}{TP + FN} \tag{12.4}$$

The probability of false alarm (Pfa) is the number of events that are incorrectly classified as anomalies, while in fact, they are normal. It is expressed as

$$Pfa = \frac{FP}{FP + FN} \tag{12.5}$$

The probability of miss detection (*Pmd*) is the number of events that are incorrectly classified as normal, while in fact, they are anomalies. It is given by

$$Pmd = \frac{FN}{TP + FN} \tag{12.6}$$

12.9.2 Unsupervised ML approaches

For evaluating the unsupervised and regression ML models, the mean average error (MAE) and mean squared error (MSE) can be selected. The MAE is the average of the absolute differences between the actual and predicted value, and it is given by

$$MAE = \frac{1}{n}\sum_{i=1}^{n}|\hat{y} - y| \tag{12.7}$$

where $\hat{y}$ is the predicted value, y is the actual value, and n is the number of instances or cases. Unlike MAE, MSE has the benefit of penalizing for significant errors because it averages the squared differences between the actual value and the predicted one, expressed as

$$MAE = \frac{1}{n}\sum_{i=1}^{n}(\hat{y} - y)^2 \tag{12.8}$$

12.9.3 Supervised online ML approaches

With streaming PMU data, it is challenging to store the entire dataset and split it into training, validation, and testing data to evaluate the real-time ML models. Therefore, other evaluation techniques are considered for evaluating these models' performances, particularly, Holdout, interleaved test-then-train, and prequential. In Holdout, a section of the incoming data is used to train the model, and small test cases are used to compute the performance. In interleaved test-and-train, each instance of the data stream is used for testing and training the model. Prequential is similar to interleaved test-and-train and uses a sliding window or a decaying factor.

12.10 Research directions

Loading and processing an open-ended source of fast and huge volumes of PMU data is challenging in terms of affording the required computational resources. Additionally, the conventional ML approaches, such as DTs, used for extracting insights are no longer relevant since they require loading and scanning the entire dataset [80]. Thus, two important criteria have to be satisfied by any ML approach when it comes to dealing with streaming PMU data: building the model with a limited dataset as it is challenging to store the entire streaming PMU measurements in memory, and the model must adapt itself when the data distribution changes gradually or abruptly over time, which could be due to load changes [81].

Otherwise, the model built in the past will no longer be consistent with the data received in the present and its performance will decrease. Developing a classifier that meets these requirements is challenging, so approximate solutions can be considered with an associated error to be minimized [82,83]. The development of lightweight middleware, ensemble methods, and adaptive learning algorithms represents a promising direction for enhancing the capability of ML models in this critical area.

Due to the high potential of unexpected events that could occur in the distribution of the streaming PMU data, it is important to deal with these unexpected events and help the ML model in adapting and classifying these unknown events; this phenomenon is known as concept drift. A concept drifts when the law underlying the data changes gradually or abruptly. Thus, the model built in the past is no longer consistent with the data received in the present. This abrupt phenomenon could be due to different reasons including noise, the influence of the environment, variation in characteristics not considered in the model, and seasonal alterations. Several approaches have been proposed for tackling such events. These approaches include the DDM, early drift detection methods (EDDM), linear four rate (LFR), just in time (JIT), and ensemble methods [84–86]. DDM and EDDM don't require to store the data but they are susceptible to false alarms. DDM evaluates the proportion of errors and constructs a binomial distribution over these errors, while EDDM expands on this idea by specifically considering the margin of misclassification [94,95]. These methods may have difficulty in environments with quickly changing, non-stationary data seen on PMU data streams. The use of error rates could create a high false positive, which may prompt unnecessary model updates that do not reflect real drifts [96,97]. In addition, the statistical assumptions of these methods do not necessarily hold in the dynamic time-varying nature of PMU data [98], making them late or misfiring to efficiently detect a significant change.

LFR [84] is constant in space complexity and can deal with imbalance classes as it uses different error types separately, but it still suffers from the labeling cost. The gradual shifts could likely be detected by the LFR method but would not detect PMU data streams more suitable to abrupt changes. The LFR may not react quickly enough to capture rapid fluctuations that can occur in electrical grid data, thus causing its performance deterioration [98]. Similarly, JIT techniques have been proposed to practically update models with real-time data; however, due to the thousands of PMUs producing substantial volume velocity, relying solely on PMUs for updates can be prohibitively slow [99]. JIT [86] does not require labeled data and can detect abrupt change but it is inefficient with the gradual drift. Ensemble methods [25] are effective in detecting the recurring concepts, but they use a large batch, which makes them ineffective in identifying precisely the change location. Ensemble methods, although generally robust, may suffer from some weakness when applied to PMU data streams. These often leverage the diversity of classifiers to increase detection accuracy. If the ensemble members are trained on similar data, they might converge and not generalize well to new patterns in the stream of ordered points [100]. However, it decreases the diversity of classifiers, which can negatively affect their ability to detect concept drift. This is especially problematic in environments like PMU applications [101] where data distribution may rapidly and unpredictably change. The above problems showcase the necessity for adaptive techniques such as ADWIN, which can adapt to non-stationary data streams.

ADWIN turns out to be an optimal concept drift detector since it is better for change localization and uses a dynamic window that adapts its size according to the change rate observed in the data within the window. When the data is stationary, the window size dynamically increases, and it decreases when a change is detected. Thus, it is important to include the ADWIN model in online ML approaches to deal with the potential concept drifts.

12.11 Conclusion

In this chapter, a comprehensive overview of the cyber and physical anomalies was presented, along with an analysis of their impact on the availability and integrity of the WAMS. A critical review of current anomaly detection approaches was conducted. These approaches were divided into conventional, machine learning, and hybrid approaches, and then their strengths and limitations were assessed. Lastly, it provided insight into a variety of statistical and machine learning-based approaches to deal with missing PMU data and concept drifts.

Acknowledgments

Dr. Prakash Ranganathan thanks his graduate students Zakaria El Mrabet and Sriram Prabhakara Rao for their contributions.

References

[1] R. J. Campbell, "Electric grid cybersecurity," Congressional Research Service, 2018. Available from: https://crsreports.congress.gov/product/pdf/R/R45312.

[2] D. Lee and D. Kundur, "Cyber attack detection in PMU measurements via the expectation-maximization algorithm," in *2014 IEEE Global Conference on Signal and Information Processing (GlobalSIP)*, 2014, pp. 223–227. doi:10.1109/GlobalSIP.2014.7032111.

[3] D. P. Shepard, T. E. Humphreys, and A. A. Fansler, "Evaluation of the vulnerability of phasor measurement units to GPS spoofing attacks," *Int. J. Crit. Infrastruct. Prot.*, vol. 5, no. 3–4, pp. 146–153, 2012, doi:10.1016/j.ijcip.2012.09.003.

[4] C. Chapman, P. Guan, T. LeRoy, E. Miller, and D. Park, "Wide area measurement system utilizing open source tools," *Proc. Natl. Conf. Undergrad. Res. NCUR*, 2013, p. 9.

[5] E. Jamil, M. Rihan, and M. A. Anees, "Towards optimal placement of phasor measurement units for smart distribution systems," in *2014 6th IEEE Power India International Conference (PIICON)*, 2014, pp. 1–6. doi:10.1109/POWERI.2014.7117773.

[6] B. Appasani and D. K. Mohanta, "A review on synchrophasor communication system: Communication technologies, standards and applications," *Prot. Control Mod. Power Syst.*, vol. 3, no. 4, pp. 1–17, 2018, doi:10.1186/s41601-018-0110-4.

[7] H. Gharavi and B. Hu, "Synchrophasor sensor networks for grid communication and protection," *Proc. IEEE Inst. Electr. Electron. Eng.*, vol. 105, no. 7, pp. 1408–1428, 2017, doi:10.1109/JPROC.2017.2696881.

[8] "IEEE standard for synchrophasors for power systems," *IEEE Std C37118-2005 Revis. IEEE Std 1344-1995*, pp. 1–65, 2006, doi:10.1109/IEEESTD.2006.99376.

[9] B. Tekinerdogan, D. Blouin, H. Vangheluwe, M. Goulão, P. Carreira, and V. Amaral, *Multi-Paradigm Modelling Approaches for Cyber-Physical Systems*. Cambridge, MA: Academic Press, 2020.

[10] H. Zhang, B. Liu, and H. Wu, "Smart grid cyber-physical attack and defense: A review," *IEEE Access*, vol. 9, pp. 29641–29659, 2021, doi:10.1109/ACCESS.2021.3058628.

[11] H.-H. Hsu, C.-Y. Chang, and C.-H. Hsu, *Big Data Analytics for Sensor-Network Collected Intelligence*. Cambridgc, MA: Morgan Kaufmann, 2017.

[12] Y. Liu, P. Ning, and M. K. Reiter, "False data injection attacks against state estimation in electric power grids," *ACM Trans. Inf. Syst. Secur. TISSEC*, vol. 14, no. 1, p. 13, 2011.

[13] G. Liang, J. Zhao, F. Luo, S. R. Weller, and Z. Y. Dong, "A review of false data injection attacks against modern power systems," *IEEE Trans. Smart Grid*, vol. 8, no. 4, pp. 1630–1638, 2017, doi:10.1109/TSG.2015.2495133.

[14] G. Hug and J. A. Giampapa, "Vulnerability assessment of AC state estimation with respect to false data injection cyber-attacks," *IEEE Trans. Smart Grid*, vol. 3, no. 3, pp. 1362–1370, 2012, doi:10.1109/TSG.2012.2195338.

[15] J. Zhao, L. Mili, and M. Wang, "A generalized false data injection attacks against power system nonlinear state estimator and countermeasures," *IEEE Trans. Power Syst.*, vol. 33, no. 5, pp. 4868–4877, 2018, doi:10.1109/TPWRS.2018.2794468.

[16] J. Kim, L. Tong, and R. J. Thomas, "Subspace methods for data attack on state estimation: A data driven approach," *IEEE Trans. Signal Process.*, vol. 63, no. 5, pp. 1102–1114, 2015, doi:10.1109/TSP.2014.2385670.

[17] R. Tan, V. Badrinath Krishna, D. K. Yau, and Z. Kalbarczyk, "Impact of integrity attacks on real-time pricing in smart grids," in *Proceedings of the 2013 ACM SIGSAC Conference on Computer & Communications Security*, 2013, pp. 439–450.

[18] S. Maharjan, Q. Zhu, Y. Zhang, S. Gjessing, and T. Basar, "Dependable demand response management in the smart grid: A Stackelberg game approach," *IEEE Trans. Smart Grid*, vol. 4, no. 1, pp. 120–132, 2013.

[19] Z. Lu, W. Wang, and C. Wang, "From jammer to gambler: Modeling and detection of jamming attacks against time-critical traffic," in *INFOCOM, 2011 Proceedings IEEE*, 2011, pp. 1871–1879.

[20] K. Gai, M. Qiu, Z. Ming, H. Zhao, and L. Qiu, "Spoofing-jamming attack strategy using optimal power distributions in wireless smart grid networks," *IEEE Trans. Smart Grid*, vol. 8, no. 5, pp. 2431–2439, 2017, doi:10.1109/TSG.2017.2664043.

[21] Z. El Mrabet, N. Kaabouch, H. El Ghazi, and H. El Ghazi, "Cyber-security in smart grid: Survey and challenges," *Comput. Electr. Eng.*, vol. 67, pp. 469–482, 2018.

[22] A. A. Cárdenas, S. Amin, and S. Sastry, "Research challenges for the security of control systems," in *Proceedings of the 3rd Conference on Hot Topics in Security*, USA, 2008, pp. 1–6.

[23] J. Qin, M. Li, L. Shi, and X. Yu, "Optimal denial-of-service attack scheduling with energy constraint over packet-dropping networks," *IEEE Trans. Autom. Control*, vol. 63, no. 6, pp. 1648–1663, 2018, doi:10.1109/TAC.2017.2756259.

[24] P. Yi, T. Zhu, Q. Zhang, Y. Wu, and J. Li, "A denial of service attack in advanced metering infrastructure network," in *2014 IEEE International Conference on Communications (ICC)*, 2014, pp. 1029–1034.

[25] A. Sargolzaei, K. Yen, and M. Abdelghani, "Delayed inputs attack on load frequency control in smart grid," in *ISGT 2014*, 2014, pp. 1–5. doi:10.1109/ISGT.2014.6816508.

[26] E. D. Knapp and R. Samani, *Applied Cyber Security and the Smart Grid: Implementing Security Controls into the Modern Power Infrastructure*. Amsterdam: Elsevier, Syngress, 2013.

[27] D. Kushner, "The real story of stuxnet," *IEEE Spectr.*, vol. 50, no. 3, pp. 48–53, 2013, doi:10.1109/MSPEC.2013.6471059.

[28] F. M. Cleveland, "Cyber security issues for advanced metering infrastructure (AMI)," in *2008 IEEE Power and Energy Society General Meeting – Conversion and Delivery of Electrical Energy in the 21st Century*, 2008, pp. 1–5. doi:10.1109/PES.2008.4596535.

[29] O. Olawumi, K. Haataja, M. Asikainen, N. Vidgren, and P. Toivanen, "Three practical attacks against ZigBee security: Attack scenario definitions, practical experiments, countermeasures, and lessons learned," in *2014 14th International Conference on Hybrid Intelligent Systems*, 2014, pp. 199–206. doi:10.1109/HIS.2014.7086198.

[30] G. Ma, L. Jiang, K. Zhou, and G. Xu, "A method of line fault location based on traveling wave theory," *Int. J. Control Autom.*, vol. 9, no. 2, pp. 261–270, 2016.

[31] Y. Zhang, L. Wang, W. Sun, R. C. Green II, and M. Alam, "Distributed intrusion detection system in a multi-layer network architecture of smart grids," *IEEE Trans. Smart Grid*, vol. 2, no. 4, pp. 796–808, 2011, doi:10.1109/TSG.2011.2159818.

[32] G. A. Ajenikoko and S. O. Sangotola, "An overview of impedance-based fault location techniques in electrical power-transmission network," *Int. J. Adv. Eng. Res. Appl. IJA-ERA*, vol. 2, no. 3, pp. 2454–2377, 2016.

[33] S. F. Alwash, V. K. Ramachandaramurthy, and N. Mithulananthan, "Fault-location scheme for power distribution system with distributed generation," *IEEE Trans. Power Deliv.*, vol. 30, no. 3, pp. 1187–1195, 2014.

[34] S. A. M. Javadian, A. M. Nasrabadi, M.-R. Haghifam, and J. Rezvantalab, "Determining fault's type and accurate location in distribution systems with DG using MLP neural networks," in *2009 International Conference on Clean Electrical Power*, 2009, pp. 284–289.

[35] D. M. Menon and N. Radhika, "Anomaly detection in smart grid traffic data for home area network," in *2016 International Conference on Circuit, Power and Computing Technologies (ICCPCT)*, 2016, pp. 1–4. doi:10.1109/ICCPCT.2016.7530186.

[36] C. R. Rao and V. N. Gudivada, *Computational Analysis and Understanding of Natural Languages: Principles, Methods and Applications*. Amsterdam: Elsevier, 2018.

[37] P. Skoda and F. Adam, *Knowledge Discovery in Big Data from Astronomy and Earth Observation: Astrogeoinformatics*. Amsterdam: Elsevier, 2020.

[38] S. Armoogum and X. Li, "Big data analytics and deep learning in bioinformatics with Hadoop," in A. K. Sangaiah, A. Annamalai, and P. Arunkumar (eds), *Deep Learning and Parallel Computing Environment for Bioengineering Systems* (pp. 17–36). Cambridge, MA: Elsevier, 2019.

[39] M. Naeem, S. T. H. Rizvi, and A. Coronato, "A gentle introduction to reinforcement learning and its application in different fields," *IEEE Access*, vol. 8, pp. 209320–209344, 2020, doi:10.1109/ACCESS.2020.3038605.

[40] R. J. Hyndman, X. A. Liu, and P. Pinson, "Visualizing big energy data: Solutions for this crucial component of data analysis," *IEEE Power Energy Mag.*, vol. 16, no. 3, pp. 18–25, 2018, doi:10.1109/MPE.2018.2801441.

[41] L. Xie, Y. Chen, and P. R. Kumar, "Dimensionality reduction of synchrophasor data for early event detection: Linearized analysis," *IEEE Trans. Power Syst.*, vol. 29, no. 6, pp. 2784–2794, 2014, doi:10.1109/TPWRS.2014.2316476.

[42] L. Chu, R. Qiu, X. He, Z. Ling, and Y. Liu, "Massive streaming PMU data modelling and analytics in smart grid state evaluation based on multiple high-dimensional covariance test," *IEEE Trans. Big Data*, vol. 4, no. 1, pp. 55–64, 2017, doi:10.1109/TBDATA.2017.2715165.

[43] H. Tan, Y. Wu, B. Shen, P. J. Jin, and B. Ran, "Short-term traffic prediction based on dynamic tensor completion," *IEEE Trans. Intell. Transp. Syst.*, vol. 17, no. 8, pp. 2123–2133, 2016, doi:10.1109/TITS.2015.2513411.

[44] M. Nakatsuji, Q. Zhang, X. Lu, B. Makni, and J. A. Hendler, "Semantic social network analysis by cross-domain tensor factorization," *IEEE Trans. Comput. Soc. Syst.*, vol. 4, no. 4, pp. 207–217, 2017, doi:10.1109/TCSS.2017.2732685.

[45] M. H. Zayani, V. Gauthier, and D. Zeghlache, "Improving link prediction in intermittently connected wireless networks by considering link and proximity stabilities," in *2012 IEEE International Symposium on a World of Wireless, Mobile and Multimedia Networks, WoWMoM 2012*, 2012, pp. 1–10. doi:10.1109/WoWMoM.2012.6263701.

[46] J. Liu, P. Musialski, P. Wonka, and J. Ye, "Tensor completion for estimating missing values in visual data," *IEEE Trans. Pattern Anal. Mach. Intell.*, vol. 35, no. 1, pp. 208–220, 2013, doi:10.1109/TPAMI.2012.39.

[47] S. S. Negi, N. Kishor, K. Uhlen, and R. Negi, "Event detection and its signal characterization in PMU data stream," *IEEE Trans. Ind. Inform.*, vol. 13, no. 6, pp. 3108–3118, 2017, doi:10.1109/TII.2017.2731366.

[48] NASPI, "PMU data quality: A framework for the attributes of PMU data quality and quality impacts to synchrophasor applications," 2017. Available from: https://www.naspi.org/sites/default/files/reference_documents/PARTF_WhitePaper_20170314_Final_PNNL_26313.pdf.

[49] IEEE Power & Energy Society, "IEEE Standard for Synchrophasor Measurements for Power Systems," IEEE Std C37.118.1-2011 (Revision of IEEE Std C37.118-2005), pp. 1–61, 2011, doi:10.1109/ieeestd.2011.6111219.

[50] S. Armijo-Olivo, S. Warren, and D. Magee, "Intention to treat analysis, compliance, drop-outs and how to deal with missing data in clinical research: A review," *Phys. Ther. Rev.*, vol. 14, no. 1, pp. 36–49, 2009.

[51] Z. Chen, H. Xu, P. Jiang, *et al.*, "A transfer learning-based LSTM strategy for imputing large-scale consecutive missing data and its application in a water quality prediction system," *J. Hydrol.*, vol. 602, p. 126573, 2021.

[52] J. Ma, J. C. Cheng, F. Jiang, W. Chen, M. Wang, and C. Zhai, "A bi-directional missing data imputation scheme based on LSTM and transfer learning for building energy data," *Energy Build.*, vol. 216, p. 109941, 2020.

[53] P. Tencaliec, A.-C. Favre, C. Prieur, and T. Mathevet, "Reconstruction of missing daily streamflow data using dynamic regression models," *Water Resour. Res.*, vol. 51, no. 12, pp. 9447–9463, 2015, doi:10.1002/2015WR017399.

[54] N. T. Le and W. Benjapolakul, "A data imputation model in phasor measurement units based on bagged averaging of multiple linear regression," *IEEE Access*, vol. 6, pp. 39324–39333, 2018.

[55] D. Duan, L. Yang, and L. L. Scharf, "Phasor state estimation from PMU measurements with bad data," in *2011 4th IEEE International Workshop on Computational Advances in Multi-Sensor Adaptive Processing (CAMSAP)*. IEEE, 2011, pp. 121–124. doi:10.1109/CAMSAP.2011.6135902.

[56] C. Huang, F. Li, L. Zhan, *et al.*, "Data quality issues for synchrophasor applications Part II: Problem formulation and potential solutions," *J. Mod. Power Syst. Clean Energy*, vol. 4, no. 3, pp. 353–361, 2016, doi:10.1007/s40565-016-0213-8.

[57] J. M. Lim and C. L. DeMarco, "Bad data detection and estimation in high dimensional measurement data," in *2017 IEEE Power & Energy Society General Meeting*, 2017, pp. 1–5.

[58] J.-F. Cai, E. J. Candès, and Z. Shen, "A singular value thresholding algorithm for matrix completion," *SIAM J. Optim.*, vol. 20, no. 4, pp. 1956–1982, 2010.

[59] Y. Hao, M. Wang, J. H. Chow, E. Farantatos, and M. Patel, "Modelless data quality improvement of streaming synchrophasor measurements by exploiting the low-rank Hankel structure," *IEEE Trans. Power Syst.*, vol. 33, no. 6, pp. 6966–6977, 2018.

[60] M. Liao, D. Shi, Z. Yu, Z. Yi, Z. Wang, and Y. Xiang, "An alternating direction method of multipliers based approach for PMU data recovery," *IEEE Trans. Smart Grid*, vol. 10, no. 4, pp. 4554–4565, 2018.

[61] D. Osipov and J. H. Chow, "PMU missing data recovery using tensor decomposition," *IEEE Trans. Power Syst.*, vol. 35, no. 6, pp. 4554–4563, 2020.

[62] J. Ma, J. C. P. Cheng, Y. Ding, *et al.*, "Transfer learning for long-interval consecutive missing values imputation without external features in air pollution time series," *Adv. Eng. Inform.*, vol. 44, p. 101092, 2020, doi:10.1016/j.aei.2020.101092.

[63] Z. Ni, S. Paul, X. Zhong, and Q. Wei, "A reinforcement learning approach for sequential decision-making process of attacks in smart grid," in *2017 IEEE Symposium Series on Computational Intelligence (SSCI)*, 2017, pp. 1–8.

[64] Y. Aslan, "An alternative approach to fault location on power distribution feeders with embedded remote-end power generation using artificial neural networks," *Electr. Eng.*, vol. 94, no. 3, pp. 125–134, 2012.

[65] F. Dehghani and H. Nezami, "A new fault location technique on radial distribution systems using artificial neural network," in *22nd International Conference and Exhibition on Electricity Distribution (CIRED 2013)*, 2013, pp. 1–4.

[66] W. Li, D. Deka, M. Chertkov, and M. Wang, "Real-time faulted line localization and PMU placement in power systems through convolutional neural networks," *IEEE Trans. Power Syst.*, vol. 34, no. 6, pp. 4640–4651, 2019.

[67] A. Takiddin, M. Ismail, U. Zafar, and E. Serpedin, "Deep autoencoder-based anomaly detection of electricity theft cyberattacks in smart grids," *IEEE Syst. J.*, vol. 16, no. 3, pp. 4106–4117, 2022, doi:10.1109/JSYST.2021.3136683.

[68] S. R. Madeti and S. N. Singh, "Modeling of PV system based on experimental data for fault detection using kNN method," *Sol. Energy*, vol. 173, pp. 139–151, 2018.

[69] M. Al Karim, M. Chenine, K. Zhu, L. Nordstrom, and L. Nordström, "Synchrophasor-based data mining for power system fault analysis," in *2012 3rd IEEE PES Innovative Smart Grid Technologies Europe (ISGT Europe)*, 2012, pp. 1–8. doi:10.1109/ISGTEurope.2012.6465843.

[70] U. Adhikari, T. H. Morris, and S. Pan, "Applying Hoeffding adaptive trees for real-time cyber-power event and intrusion classification," *IEEE Trans. Smart Grid*, vol. 9, no. 5, pp. 4049–4060, 2018, doi:10.1109/TSG.2017.2647778.

[71] NASPI "Using Synchrophasor Data for Oscillation Detection," 2017. Available from: https://www.naspi.org/sites/default/files/reference_documents/crstt_oscillation_detection_20180129_final.pdf.

[72] M. Tian, Z. Dong, and X. Wang, "Analysis of false data injection attacks in power systems: A dynamic Bayesian game-theoretic approach," *ISA Trans.*, vol. 115, pp. 108–123, 2021, doi:10.1016/j.isatra.2021.01.011.

[73] M. A. Teixeira, T. Salman, M. Zolanvari, R. Jain, N. Meskin, and M. Samaka, "SCADA system testbed for cybersecurity research using machine learning approach," *Future Internet*, vol. 10, no. 8, p. 76, 2018, doi:10.3390/fi10080076.

[74] W. Xue and T. Wu, "Active learning-based XGBoost for cyber physical system against generic AC false data injection attacks," *IEEE Access*, vol. 8, pp. 144575–144584, 2020, doi:10.1109/ACCESS.2020.3014644.

[75] Z. Qu, H. Liu, Z. Wang, J. Xu, P. Zhang, and H. Zeng, "A combined genetic optimization with AdaBoost ensemble model for anomaly detection in buildings electricity consumption," *Energy Build.*, vol. 248, p. 111193, 2021, doi:10.1016/j.enbuild.2021.111193.

[76] A. Zainab, S. S. Refaat, D. Syed, A. Ghrayeb, and H. Abu-Rub, "Faulted line identification and localization in power system using machine learning techniques," in *2019 IEEE International Conference on Big Data (Big Data)*, 2019, pp. 2975–2981.

[77] H. Okumus and F. M. Nuroglu, "A random forest-based approach for fault location detection in distribution systems," *Electr. Eng.*, vol. 103, no. 1, pp. 257–264, 2021, doi:10.1007/s00202-020-01074-8.

[78] S. Pandey, A. Srivastava, and B. Amidan, "A real time event detection, classification and localization using synchrophasor data," *IEEE Trans. Power Syst.*, vol. 35, no. 6, pp. 4421–4431, 2020, doi:10.1109/TPWRS.2020.2986019.

[79] R. Qi, C. Rasband, J. Zheng, and R. Longoria, "Detecting cyber attacks in smart grids using semi-supervised anomaly detection and deep representation learning," *Information*, vol. 12, no. 8, p. 328, 2021, doi:10.3390/info12080328.

[80] A. J. Abianeh, Y. Wan, F. Ferdowsi, N. Mijatovic, and T. Dragičević, "Vulnerability identification and remediation of FDI attacks in islanded DC microgrids using multiagent reinforcement learning," *IEEE Trans. Power Electron.*, vol. 37, no. 6, pp. 6359–6370, 2022, doi:10.1109/TPEL.2021.3132028.

[81] M. N. Kurt, O. Ogundijo, C. Li, and X. Wang, "Online cyber-attack detection in smart grid: A reinforcement learning approach," *IEEE Trans. Smart Grid*, vol. 10, no. 5, pp. 5174–5185, 2019, doi:10.1109/TSG.2018.2878570.

[82] S. Ekici, "Support vector machines for classification and locating faults on transmission lines," *Appl. Soft Comput.*, vol. 12, no. 6, pp. 1650–1658, 2012.

[83] D.-I. Kim, A. White, and Y.-J. Shin, "PMU-based event localization technique for wide-area power system," *IEEE Trans. Power Syst.*, vol. 33, no. 6, pp. 5875–5883, 2018, doi:10.1109/TPWRS.2018.2824851.

[84] A. Hossam-Eldin, A. Lotfy, M. Elgamal, and M. Ebeed, "Combined traveling wave and fuzzy logic based fault location in multi-terminal HVDC systems," in *2016 IEEE 16th International Conference on Environment and Electrical Engineering (EEEIC)*, 2016, pp. 1–6.

[85] Y. Mohammadnian, T. Amraee, and A. Soroudi, "Fault detection in distribution networks in presence of distributed generations using a data mining-driven wavelet transform," *IET Smart Grid*, vol. 2, no. 2, pp. 163–171, 2019.

[86] N. Dahal, O. Abuomar, R. King, and V. Madani, "Event stream processing for improved situational awareness in the smart grid," *Expert Syst. Appl.*, vol. 42, no. 20, pp. 6853–6863, 2015, doi:10.1016/j.eswa.2015.05.003.

[87] A. Bifet, R. Gavaldà, G. Holmes, and B. Pfahringer, *Machine Learning for Data Streams: With Practical Examples in MOA*.MIT Press, 2018.

[88] A. Bifet and R. Kirkby, *Data Stream Mining: A Practical Approach*. Waikato: Center for Open Source Innovation 2009.

[89] A. Muallem, S. Shetty, J. W. Pan, J. Zhao, and B. Biswal, "Hoeffding Tree algorithms for anomaly detection in streaming datasets: A survey," *J. Inf. Secur.*, vol. 8, no. 4, pp. 339–361, 2017.

[90] H. Wang and Z. Abraham, "Concept drift detection for streaming data," in *2015 International Joint Conference on Neural Networks (IJCNN)*, IEEE, 2015, pp. 1–9.

[91] M. Baena-García, J. del Campo-Ávila, R. Fidalgo, A. Bifet, R. Gavaldá, and R. Morales-Bueno, "Early drift detection method," 2016.

[92] G. Ditzler and R. Polikar, "Incremental learning of concept drift from streaming imbalanced data," *IEEE Trans. Knowl. Data Eng.*, vol. 25, no. 10, pp. 2283–2301, 2013, doi:10.1109/TKDE.2012.136.

[93] B. Mirza, Z. Lin, and N. Liu, "Ensemble of subset online sequential extreme learning machine for class imbalance and concept drift," *Neurocomputing*, vol. 149, pp. 316–329, 2015, doi:10.1016/j.neucom.2014.03.075.

[94] C. Cámara, P. Peris-López, L. González-Manzano, and J. Tapiador, "Real-time electrocardiogram streams for continuous authentication," *Appl. Soft Comput.*, vol. 68, pp. 784–794, 2018, doi:10.1016/j.asoc.2017.07.032.

[95] P. Sidhu, A. Ravi, D. Malik, and M. Bhatia, "An approach to handle novel classes: Weighted Novel Class Detection algorithm," in *2015 Annual IEEE India Conference (INDICON)*, 2015, pp. 1–6, 2015. https://doi.org/10.1109/indicon.2015.7443720.

[96] I. Rabiu, N. Salim, M. Nasser, A. Da'u, T. Eisa, and M. Dalam, "Drift detection method using distance measures and windowing schemes for sentiment classification," *Comput. Mater. Con.*, vol. 74, no. 3, pp. 6001–6017, 2023, doi:10.32604/cmc.2023.035221.

[97] R. Gemaque, A. Costa, R. Giusti, and E. Santos, An overview of unsupervised drift detection methods. *Wiley Interdiscip. Rev. Data Min. Knowl. Discov.*, vol. 10, no. 6, p. e1381, 2020, doi:10.1002/widm.1381

[98] S. Yu, Z. Abraham, H. Wang, M. Shah, Y. Wei, and J. Príncipe, Concept drift detection and adaptation with hierarchical hypothesis testing. *J. Frank. Inst.*, vol. 356, no. 5, pp. 3187–3215, 2019, doi:10.1016/j.jfranklin.2019.01.043.

[99] V. Mittal and I. Kashyap, Online methods of learning in occurrence of concept drift. *Int. J. Comput. Appl.*, vol. 117, no. 13, pp. 18–22, 2015, doi:10.5120/20614-3280.

[100] K. Goel and S. Batra, Dynamically updated diversified ensemble-based approach for handling concept drift. *Turk. J. Elec. Eng. Comp. Sci.*, vol. 28, no. 1, pp. 556–574, 2020, doi:10.3906/elk-1810-122.

[101] P. Gonçalvès, S. Santos, R. Barros, and D. Vieira, A comparative study on concept drift detectors. *Expert Syst. Appl.*, vol. 41, no. 18, pp. 8144–8156, 2014, doi:10.1016/j.eswa.2014.07.019.

Index

www.ingramcontent.com/pod-product-compliance
Lightning Source LLC
LaVergne TN
LVHW011837200926
848484LV00002B/367